New Developments in Marine Biotechnology

New Developments in Marine Biotechnology

Edited by

Y. Le Gal

National Museum of Natural History and
College of France
Concarneau, France

and

H. O. Halvorson

University of Massachusetts
Boston, Massachusetts

With the editorial assistance of

Anne-Marie Lambert

Springer Science+Business Media, LLC

Library of Congress Cataloging-in-Publication Data

New developments in marine biotechnology / edited by Y. Le Gal and H.O. Halvorson.
p. cm.
"Proceedings of the 4th International Marine Biotechnology Conference, held September 22-29, 1997, in Sorrento, Paestum, Oranto, and Pugnochiuso, Italy"--T.p. verso.
Includes bibliographical references and index.
ISBN 978-1-4419-3300-3 ISBN 978-1-4757-5983-9 (eBook)
DOI 10.1007/978-1-4757-5983-9
1. Marine fishes--Molecular aspects--Congresses. 2. Marine biotechnology--Congresses. 3. Fishery resources--Management--Congresses. I. Le Gal, Yves. II. Halvorson, Harlyn O. III. International Marine Biotechnology Conference (4th : 1997 : Sorrento, Italy, etc.)
QL620.N49 1998
572.8'1177--dc21 98-24800
CIP

Proceedings of the 4th International Marine Biotechnology Conference, held September 22 – 29, 1997, in Sorrento, Paestum, Otranto, and Pugnochiuso, Italy

ISBN 978-1-4419-3300-3

Originally published by Plenum Press, New York in 1998
Softcover reprint of the hardcover 1st edition 1998

http://www.plenum.com

10 9 8 7 6 5 4 3 2 1

PREFACE

Past efforts to colonize the environment and domesticate living species, coupled with scientific research, have resulted in the possession (but not always the real control) by humans of any available terrestrial space. However, oceans, which represent up to two-thirds of the surface of the planet, had not been really approached until the middle of this century. As oceanographic science develops, the picture of a rich, diverse, complex and also, in many respects, specific marine life, is coming into view.

In a broad sense, marine biotechnologies can be understood as the various means or techniques of managing marine living systems for the benefit of mankind. The first goal we have is for marine life to provide biomass for food. However, today it is not certain that a significant increase of total world fisheries' catches will be possible in the future. There are several ways to address this. First, we need to generate better, more complete, or different uses of the biomass actually fished. This is mainly a matter of upgrading fish and fish wastes. Second, we need to artificially grow the living species. This falls within the scope of cell cultivation and of aquaculture. Both approaches have to be appreciated simultaneously in terms of biology, ecology, and economy. In both approaches, profit improvements are linked to the introduction of biotechnological methods and to the use of biotechnological processes.

The main characteristics of fished biomasses is that they still exist and are readily available. They can be considered a huge reservoir of molecules: polysaccharides, enzymes, fats, etc., exhibiting physical, chemical, or biological activities of interest for various purposes. The main problem (and it is not a minor one), in terms of techniques and cost, is to isolate and purify these molecules. The second issue in biomass treatment is mass cultivation of marine organisms. It is now clear that trying to reproduce biomass intensively and artificially cannot easily yield profits, unless we use a series of biotechnological tricks that will permit a drastic lowering of the costs. During the last 10 years, another important problem has emerged. This is the spreading of pathogenic organisms in overcrowded sea farms. Within a short period of time, sea farms could be almost completely destroyed by marine viruses, microorganisms, or parasites about which we have little information.

Solutions to these problems represent real strategic tasks for the marine biotechnologists requiring basic research in developmental biology, genetics, gene enginering, endocrinology, pathology, and immunology of species as different as flatfish, salmon, shrimps, abalone, among others.

Biodiversity is largely a reflection of the very specific aspects of marine life. An early trend consisted of limiting the scope of marine biotechnologies to the production of biological models that facilitate the study of general mechanisms. These studies feed our knowledge and understanding of life that is built on an unique pattern. In contrast, they also favor the exploitation of structural, developmental, and biochemical specificities. Marine biotechnologies reveal their genuine potential in offering the investigation and exploitation of molecules and mechanisms for which we do not know of any terrestrial counterparts. Marine biotechnology is by nature multidisciplinary and clearly incorporates new technologies from molecular biology and chemical analysis to bioreactor technology.

Marine biotechnologies also deal with environmental management. The first step in any kind of management involves a diagnosis of the condition of a systems. The past decade has been marked by considerable progress in using rapid and sensitive methods for estimating biological responses to human-induced changes in the environment. Many of these methods now use molecular probes, nucleic acids, immunoreagents, or enzymatic biosensors that allow us to record efficiently a considerable number of data. A main problem is how to handle this huge quantity of information, to use it, and to forecast the evolutionary trends of an estuary, a bay, a sea, or an ocean.

Finally, one of the most promising goals for marine biotechnologies will be the possibility of using sophisticated biological tools for managing marine ecosystems. Controlling natural production of useful species will be less costly than trying to rear completely demanding species. Understanding the tenuousness of the relationship between planktonic species and their environment will perhaps give us an insight on climatic changes and on the biological future of the planet.

The domains covered by marine biotechnologies are vast and range over various overlapping disciplines, from the molecular approaches of developmental biology and biodiversity to the chemistry of natural substances. New fields are rapidly evolving and are helping to successively emphasize specific areas of biological sciences.

With its biphasic unfolding, the format of the fourth edition of International Marine Biotechnology Conference (IMBC'97) was original and successful, as it enabled the presentation of straightforward reports and constructive discussions.

With more than sixty selected papers organized in eight sections, this book covers the present state of the art in marine biotechnologies.

HH and YLG

TRIBUTE TO NINO SALVATORE

The International Marine Biotechnology Conferences represent an assembly of interdisciplinary scientists and technologists with a common interest in Marine Science. Nino Salvatore was one of these. He joined the IOC to plan for IMBC'94 in Tromsoe, Norway and quickly demonstrated that he was one of those rare individuals in the scientific community who made an almost instantaneous impression on any person fortunate enough to be acquainted with him. His high standards and enthusiasm were widely felt—from the revitalizing of the Stazione Zoologica in Naples, to science policy in the EU, to support for biotechnology, developmental biology, and molecular biology. Prof. Salvatore was a strong enthusiast for basic research and its application to solving problems of the day.

During the IMBC'94 meeting, the lack of an European organization to deal with international and European collaboration became evident. Characteristically, Nino Salvatore saw the need to establish such an organization. He organized an *ad hoc* meeting and a decision was made to go ahead. The European Society for Marine Biotechnology was formed, and its first President, Dr. Jan Olafsen, is a member of IOC and was our host in Tromsoe for IMBC'94.

When the decision was made to hold IMBC'97 in Italy, Dr. Salvatore applied his energy and enthusiam to its organization, financial support, and his wish to do something different. An international program committee, chaired by Dr. Frank Gannon, developed a program based on peer review of submitted abstracts. The mobility of the meeting is an expression of Nino Salvatore's desire to have as many people and scenarios involved as possible because of the diverse subject areas that need to be covered in biotechnology. If people cannot come to the conference, the conference will visit them. He also had in mind to permit as many of his countrymen to participate as possible while at the same time broadening the picture of the scope of this interdisciplinary subject area in Italy in the minds of foreign conference participants.

Science has lost a visionary person with a remarkable character. Individuals do make a difference and Prof. Salvatore. He will be missed. The IMBC'97 is dedicated to him. We seek your help in making this meeting a success and thereby honoring Gaetano Salvatore.

Harlyn O. Halvorson

ACKNOWLEDGMENTS

High Patronage of the President of the Italian Republic

Under the aegis of the European Union

Under the auspices of:

Presidenza del Consiglio dei Ministri
Ministero dell'Università e Della Ricerca Scientifica e Tecnologica
Ministero dei Beni Culturali e Ambientali
Consiglio nazionale delle Ricerche
Regione Campania
Regione Puglia
Amministrazione Provinciale di Napoli
Amministrazione Provinciale di Salerno
Amministrazione Provinciale di Foggia
Comune di Sorrento
Comune di Capaccio/Paestum
Comune di Otranto
Università degli Studi di Napoli Federico II
Seconda Università di Napoli
Università degli Studi di Lecce
Unione degli Industriali della Provincia di Napoli

With the support of:

American Society for Microbiology
Biotechnology Center of Excellence Corp, USA
Department of Energy, USA
Massachusetts Foundation for Excellence in Marine and Polymer Science
National Science Foundation, USA
National Institutes of Health, USA
Office of Ocean and Atmospheric Research
Policy Center for Marine Biosciences and Technology, USA
Society for Industrial Microbiology, USA
United States Department of Agriculture

With the contribution of:

Camera di Commercio Industria Artigianato E Agricoltura, Lecce
Camera di Commercio Industria Artigianato E Agricoltura, Foggia
Ente Provinciale per il Turismo, Lecce

CONTENTS

Section 3: Aquaculture

Section 7: Biodiversity, Environmental Adaptation, and Evolution

Section 8: Biomarkers, Symbiosis, and Viruses

Section 9: Workshops

1

BIOTECHNOLOGY: BIOLOGY OR TECHNOLOGY?

Keynote Lecture

Arthur Kornberg

INTRODUCTION TO DR. ARTHUR KORNBERG

We are privileged to open this conference by one of our most distinguished contributors to science and a spokesman for science, Dr. Arthur Kornberg. Following a brilliant career in Biotechnology at NIH and Washington University in St. Louis, he undertook a study of the mechanism of DNA replication which attracted international attention and led to his Nobel Prize. Moving to Stanford University, he established one of the most outstanding Departments in Molecular Biochemistry. Their graduates and Post Doctoral Fellows are found worldwide in major research universities. After retiring as chair, Dr. Kornberg continued to provide leadership to the scientific community, guidance to government and interpretations about science to the public. A strong supporter of the need for basic research, his vision on how this is accomplished and how this is translated to solve practical problems has been widely received. Recognizing this, and the interdisciplinary nature of Marine Biotechnology, it was Nino Salvatore's wish to have Prof. Kornberg open IMBCC'97 with a keynote address "Biotechnology: Biology or Technology?"

H. O. H.

KEYNOTE LECTURE

During this 20th century with its succession of microbe hunters, vitamin, enzyme and gene hunters, and golden ages in medical science, the current age of gene hunting is undeniably the most golden. We have an inexhaustible supply of genes and simple and efficient techniques to track and capture them. We are participating in the most revolutionary advance in the history of biological and medical science. The term revolutionary is generally overused, but not here. The effects of this advance on medicine, agriculture, industry, and basic science have not been exaggerated.

New Developments in Marine Biotechnology, edited by Le Gal and Halvorson.
Plenum Press, New York, 1998.

Yet even more revolutionary but generally unnoticed, is a development which, lacks a name, has no obvious applications, but will surely lead to even more remarkable and unanticipated practical applications. I refer to the coalescence, the confluence and the merging of the numerous basic biologic and medical sciences into a single, unified discipline which has emerged because it is expressed in a single universal language, the language of chemistry.

The breakthrough of recombinant DNA and genetic engineering was based on the discoveries of enzymes that make, break and seal DNA. All these basic advances were made in academic laboratories built and supported almost entirely by funds from the NIH. For thirty years, my research on the biosynthesis of the building blocks of nucleic acids, their assembly in DNA replication and the training of well over a hundred young scientists, was funded with many millions of dollars without any promise or expectation that this research would lead to marketable products or procedures. No industrial organization had, or would ever have, the resources or disposition to invest in such long-range, apparently impractical programs. We carried out these studies to satisfy a need to understand basic processes in cellular function. Yet to my great pleasure, such studies of the replication, repair and rearrangements of DNA have had many practical benefits.

The pathways of assembling DNA from its building blocks have been the basis for the design of most drugs used today in the chemotherapy of cancer, AIDS, Herpes and autoimmune diseases. These studies are also crucial to understanding the repair of DNA, so important in the aging process, for understanding mutations and the origin of some cancers. It may seem unreasonable and impractical, call it counterintuitive, even to scientists to solve an urgent problem, such as a disease, by pursuing apparently unrelated questions in basic biology or chemistry. Yet, the pursuit of understanding the basic facts of nature has proven throughout the history of medical science to be the most practical, the most cost-effective route to successful drugs and devices. Investigations that seemed totally irrelevant to any practical objective have yielded most of the major discoveries of medicine—X-rays, penicillin, polio vaccine, recombinant DNA and genetic engineering. All these discoveries have come from the pursuit of questions in physics, chemistry and biology, unrelated at the outset to a specific medical or practical problem.

With regard to industrial inventions, there is a common saying: "Necessity is the mother of invention." Not true! Rather, the reverse has proven to be true: invention is the mother of our necessities. Inventions only later become necessities!

Quite clearly, even industrial inventions emerge from a creative process. As such they are haphazard rather than goal-oriented. The lessons to be learned from this history should be crystal clear. It is crucial for a society, a culture, a company, a university, to understand the nature of the creative process and to provide for its support. No matter how counter-intuitive it may seem, basic research is the lifeline of practical advances in medicine; pioneering inventions are the source of industrial strength. The future is not predicted it is invented.

Of course it is important that basic discoveries be promptly and wisely applied to solve practical problems. The recent applications of biotechnology to medicine have given us major insights into diabetes, cancer and other metabolic diseases. Will these approaches and techniques be equally effective when applied to the human brain and behavior? I am sure they will. That human behavior is a matter of chemistry and neurons is hard for some to accept, including physicians. We must remind them that the same chemical language that describes the functions of the heart and liver surely applies to the basic operations of the brain.

The overriding issue in biomedical science, as I see it, is how to give our abundant scientific talent the resources to exploit the extraordinary new technologies to advance knowledge. Currently, a pervasive mood among productive biomedical scientists makes them fear for continued grant support, persuades them to choose safe and practical projects over the untried and adventurous, and tempts their interest in commercial ventures. This is clearly a state that discourages young people from entering science and drives others to abandon science for business, law and other pursuits.

The independence of an American scientist to initiate and pursue his own research program in the biomedical sciences has been achieved because the NIH awarded research grants to the individual scientist who is not indebted to a department head, a dean or to university politics. The university has no choice but to give the scientist independence in order to compete for them, for their teaching contributions and the very considerable income from indirect costs attached to their grants. Yet, it should be said, that the very competition for grantees is an essential ingredient of the success of the NIH grants program. It depends on the fact that the private and public universities are free from centralized government controls, something virtually unique in the United States.

The expansion of research grants in many countries is highly laudable, but unfortunately the old mechanisms often prevail. In Japan, a very large sum is awarded to "centers of excellence" in which the director can exert authority over what is done and who does it. In Europe, research programs, especially in the smaller countries, rely on grants from the European Union. The EU requires that investigators from three or more countries find a consensus project that can be parcelled up among them. This leaves no room for a scientist to do something utterly original and unpopular, and much time is wasted on bureaucratic maneuverings. Recent reports indicate that in the United Kingdom the Medical Research Council is planning to consolidate grants along similar lines. Here in Italy, the powerful baronial organization of research granting agencies perpetuates fragmentation and favoritism.

Another problem I want to consider includes conflicts within our science, conflicts that can reduce our effectiveness: these include conflicts between the cultures of chemistry and biology, confusion in biotechnology between biology and technology, and big science versus little science. In each of these conflicts, philosophical differences are overlaid by strong economic, social and political forces.

First is the rift for more than a century between the cultures of chemistry and biology. The emergence of biochemistry early in this century might have bridged the gap between chemistry and biology but it didn't. Nor has the recent popularity of genetic chemistry.

Chemists continue to synthesize and analyze small molecules with ever greater precision, but they neglect the biological macromolecules: the proteins, nucleic acids and polysaccharides; these seem to them too complex or too mundane. Too few chemists exploit the borderland, in which they can find rich harvests in the vast and awesome biological chemistry evolved for over a billion years.

Biologists for their part avoid enzymology. To them, enzymes are faceless components of kits or gene products inferred from sequences recognized by motifs and homologies. Biologists are so enthralled by the mysteries of evolution, development, aging and diseases, that they resist reductionist chemical approaches and focus instead on the vital phenomena they create by altering the genomes of their favorite organisms.

Biologists should realize that the history of science is littered with vitalistic pronouncements that reduction has gone as far as it can go. With the application of chemical techniques of ever-increasing sensitivity and precision, they can gain a deeper and 4-dimensional understanding of biologic events.

Another conflict is found in the increasing influence of biotechnology enterprises. Genetic engineering and associated technologies have been enormously successful. Yet we must be aware that the very term biotechnology, adopted as a euphemism for genetic engineering, may blur the important distinction between biology, a quest for knowledge, as opposed to technology, the application of that knowledge for products and profit.

Scientists and academic institutions involved in biotech enterprises are likely to be distracted from their central mission: the pursuit of the basic understanding of nature. I am especially concerned with another problem. There is an illusion created by the financial and research successes of a few biotech ventures that a significant fraction of basic advances can be supplied by industry. Although such achievements are laudable, they represent only a tiny fraction, perhaps five percent of the basic knowledge needed to combat diseases, advances which can come only from the massive federal support of research and training.

Lastly, I want to mention still another conflict, the conflict between big and little science. Of course, we need both. There are projects that require large and expensive equipment and several disciplines to use it effectively. My concern is that with the worldwide expansion of big science, little science will vanish. As I view the steady growth of collective science and big science, the greatest danger I see is a dampening of individual creativity and reversion to the old politics—the inevitable local politics that infects every group and institution.

I want to recognize what deserves the most emphasis and what unites us all. It is our unconflicted and overriding devotion to the culture of science. We must make it clear to the public that science is great, although scientists are still people. As people, they are no different from others: dentists, lawyers, artists, businessmen. Scientists are just as prey to the human failings of arrogance, greed, dishonesty and psychopathy. What does set them apart from others is the discipline of science, a practice that demands exact and objective descriptions of progress, evidence that can be verified or denied by others.

It is the discipline of science that enables all of us ordinary people, whether we be chemists, biologists or physicians, to go about doing the ordinary things, which, when assembled, reveal the extraordinary intricacies and awesome beauties of nature. Science not only permits us to contribute to grand enterprises, but also offers a changing and endless frontier for exploration.

This frontier for exploration has given me an opportunity to probe an utterly new area after having worked on DNA for 40 years. A few years ago, I described my fascination with another polymer, one which was surely on earth before nucleic acids and proteins. It was likely a precursor and catalyst in the synthesis of RNA, DNA and proteins and is now conserved in every bacterial, fungal, plant and animal cell. It is a chain of hundreds of phosphates linked by the high-energy bonds found in ATP. Because of the antiquity of poly P and its apparent lack of any functions, it was dubbed a "molecular fossil." My mission has been to restore the fossil to life. We have discovered many functions of poly P, the most intriguing is that in E. coli it is essential for the elaborate adaptations that the organism makes for its survival after exponential growth. Mutants lacking poly P die off quickly. Simply put, poly P is essential for graceful aging in E. coli. The enzyme that makes poly P in bacteria is highly conserved. These include M. tuberculosis, Helicobacter pylori, Neisseria meningitidis and other pathogens, and also cyanobacteria and streptomyces. We are working with medical microbiologists to determine the influence of poly P on the virulence of these pathogens and the production of antibiotics in Streptomyces.

People wonder whether the computer revolution and other advanced technologies have altered the way we do bioscience research. Can our research now be engineered and pursued by formula? Not yet. The technical tools are indispensable, but the practice of science remains in essence highly creative and its province is Nature. Sir Karl Popper, an eminent philosopher of science and society, who died three years ago in London, said that "next to music and art, science is the greatest, most beautiful and most enlightening achievement of the human spirit." I would place science first.

We probe the inexhaustible mysteries of Nature from a variety of directions, and with different intensities and styles. These probings are determined by our emotions, our moods and our cultural heritage, much as these influence tne artist. The major discoveries in bioscience are more often intuitive or serendipitous than the result of logical analysis.

The machines we use produce images and compositions of objective precision. But this should not obscure the fact that we use these machines as tools, with tastes as distinctive as those that painters use their palettes, composers their notes, and authors their words in creating their images of Nature. Seneca, the great Roman statesman and philosopher, once said: "All art is but imitation of Nature." What we try to do in science is to get ever closer to Nature. In the art of medicine, we try to find for the individual a harmonious niche in Nature.

REFERENCE

FASEB Journal, 1997, **11**:1209–1214

2

THE PARADOX OF GROWTH ACCELERATION IN FISH

José de la Fuente,* Isabel Guillén, and Mario P. Estrada

Mammalian Cell Genetics Division
Centro de Ingeniería Genética y Biotecnología
P. O. Box 6162, Havana, Cuba

1. INTRODUCTION

Growth is a complex and tightly regulated process in fish. The growth hormone (GH) is a polypeptide playing a key role in the process of growth and is synthesized mainly by somatotrophos in the anterior pituitary gland. Release of GH from the pituitary gland is thought to be controlled primarily by hypothalamic factors. Once in the circulation, a substantial proportion of the GH appears to bind to a specific binding protein, probably responsible for the control of the hormone half life in the circulation. After binding to specific cell receptors, GH stimulates, primarily in the liver, Insulin-like growth factor (IGF-I and IGF-II) synthesis and secretion to elicit the growth promoting action in an autocrine and paracrine fashions. IGF also elicits a negative feedback on the secretion of GH in the pituitary gland in tilapia (Guillén et al., in press).

Growth acceleration has been reported for tilapia and other fish species. However, these results did not address the question regarding the levels of ectopic GH required to achieve maximal growth acceleration without causing detrimental effects to the animals. This is a fundamental question to understand the process of growth in fish and to effectively manipulate this process.

2. EXPERIMENTAL PROCEDURES

The details of the experiments considered here have been published elsewhere (Guillen et al., 1996; Hernandez et al., 1997).

* E-mail: jose.delafuente@cigb.edu.cu

New Developments in Marine Biotechnology, edited by Le Gal and Halvorson.
Plenum Press, New York, 1998.

3. RESULTS AND DISCUSSION

3.1. The Administration of High Doses of Recombinant tiGH Results in Growth Inhibition in Tilapia

Cloned cDNA for tilapia GH (tiGH) was expressed in *E. coli*. After purification, recombinant tiGH was correctly folded and biologically active. The growth of juvenile tilapia (*O. hornorum*) was followed after intraperitoneal injections of recombinant tiGH (0, 0.1, 0.5 and 2.5 μg/g body weight (gbw); 13 tilapia per group) at intervals of 7 days. The control group received injections of vehicle plus 5 μg BSA/gbw. The level of growth acceleration (in %) was determined at the end of the experiment by comparing the mean weight of tilapia in the experimental groups with the control tilapia. A Student t-Test was used to compare the results. In the week 9 of the experiment, the group receiving 0.5 μg tiGH/gbw showed a 6% growth acceleration (p=0.05) whereas the group receiving 0.1 μg/gbw showed a 2% growth acceleration. In the group with the highest dose, a growth retardation of 1% was recorded.

These results evidenced a dose-dependent effect of tiGH administration on the growth performance of juvenile tilapia at the doses of 0.1 and 0.5 μg tiGH/gbw. However, the injection of 2.5 μg tiGH/gbw produced a negative effect on the growth performance.

3.2. Low Level Ectopic Expression of tiGH Results in Growth Acceleration in Transgenic Tilapia

To assay the effect of different expression patterns of ectopic tiGH, 4 lines of transgenic tilapia (*O. hornorum*) were generated with chimeric constructs containing the tiGH cDNA, 5′ regulatory sequences derived from the human cytomegalovirus (CMV) or Rous sarcoma virus (RSV), polyadenylation sites from the SV40 and the first intron from the trout GH gene (INT) (de la Fuente et al., 1995; Hernández et al., 1997; Table 1).

Table 1. Summary of the characterization of transgenic tilapia lines

Tilapia lines[a]		tiGH RNA levels[b]	tiGH protein levels[c]	Growth acceleration[d]	References[e]
CMV>tiGH>CAT>SV40	F1 (P1 × wt)	5	10	82% (p=0.001)	1,2
	F2 (F1 × F1)	+	+	55% (p=0.009)	3
	$F2^{-/+}$	+	+	62% (p=0.005)	3
	$F2^{+/+}$	+	+	31% (p=0.07)	3
RSV>+INT>tiGH>SV40	[F1 (P1 × wt)]	240	78	0%	1,4
CMV>–INT>tiGH>SV40	[F1 (P1 × wt)]	23	723	0%	1,4
CMV>+INT>tiGH>SV40	[F1 (P1 × wt)]	8	60	3.4% (p=0.006)	1,4
Non-transgenic control siblings		0	0	—	—

[a]Hybrids (predominantly *O. hornorum* although the breeding history is not known). Wt, wild type hybrid *O. hornorum* tilapia; –/+, heterozygous; +/+, homozygous.

[b]RNA levels (in arbitrary units) were calculated by summarizing the results of Northern blot analyses in the liver, gonads and muscle. Signals in the X-ray films were scanned and normalyzed against gliceraldehyde 3 phosphate dehydrogenase. +, denotes presence of tiGH RNA in muscle samples analyzed by *in situ* hybridization.

[c]Tilapia ectopic GH protein levels (in arbitrary units) were calculated by summarizing the exposure time required for photography (employing an Olympus exposure control unit) in gonad, heart and muscle tissue sections after immunohistochemical analysis with anti-tiGH-anti rabbit IgG-Rhodamine conjugate. Values were normalized against the control. +, denotes presence of tiGH in non-quantitated tissue sections.

[d]Growth acceleration in transgenic tilapia compared to non-transgenic siblings (Student t-Test).

[e]1, de la Fuente et al., 1995; 2, Martínez et al., 1996; 3, Guillén et al., 1996; 4, Hernández et al., 1997.

Different patterns of ectopic expression of tiGH were detected in gonad, liver, brain, heart and muscle cells of transgenic tilapia lines by RNA and/or protein analysis (Table 1; Guillén et al. 1996; Hernandez et al., 1997). Transgenic lines with lower ectopic tiGH mRNA levels were the only showing growth acceleration. Small variations in the tiGH levels resulted in big changes in the level of growth acceleration, suggesting that the ectopic expression of tiGH promoted growth only at low expression levels (Table 1), a fact that was also noticed in the experiment described before injecting different doses of recombinant tiGH. Furthermore, 4 month old transgenic homozygous ($F2^{+/+}$) and heterozygous ($F2^{-/+}$) tilapia and non-transgenic siblings were studied for 3 months grown communally in the same pond (Guillen et al., 1996). The results suggested a transgene-dosage effect (Table 1).

Groups working with relatively weak promoters have reported growth acceleration in transgenic salmon (Devlin et al., 1994) and carp (Zhu, 1992) while reports employing the strong RSV promoter in transgenic carps showed only modest levels of growth acceleration (Zhang et al., 1990). Furthermore, Zhang et al. (1990) reported that transgenic common carp carrying RSV>rtGH1cDNA and expressing the transgene at low levels grew faster than those containing higher rtGH levels. These studies suggested that high levels of GH may produce inhibitory effect on growth (Zhang et al., 1990; Lu et al., 1992). These results are in accordance with those reported here for transgenic tilapia.

3.3. A Model for Growth Acceleration in Tilapia

The results obtained by us resulted in a paradox: high exogenous GH levels did not promote growth, but rather could produce a growth retardation effect. However, low levels of exogenous GH result in growth acceleration. This "exogenous growth hormone to accelerate growth in fish paradox" remains the "French red wine paradox"; non is as negative as too much, but little could be beneficial.

The effect of ectopic tiGH levels over a certain value resembled in tilapia the physiological situation of low condition factor (because, for example, of low food availability; Sumpter et al., 1992; Guillén et al., in press). It has been reported that fish with a low condition factor have elevated GH plasma concentrations and low levels of IGF-I, factors that

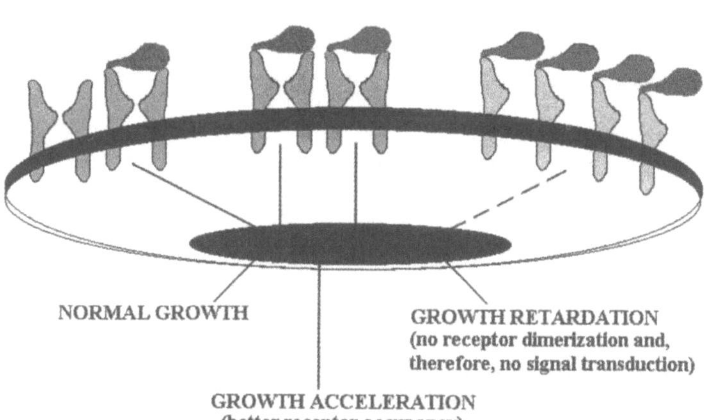

Figure 1. A model for our hypothesis on the effect on growth of ectopic GH in tilapia. Elevated GH levels could down-regulate the signal-transducing capacity of GH and/or IGF receptors, thereby not accelerating growth. Low ectopic GH levels could accelerate growth by permitting a better receptor occupancy, thus optimizing growth-promoting activity.

result in growth retardation (Sumpter, 1992; Guillén et al. in press). In the two transgenic lines with higher ectopic tiGH mRNA levels no growth acceleration was recorded (Table 1).

These results conduced as to the hypothesis that elevated GH levels could down-regulate the level, or more likely the signal-transducing capacity, of GH and/or IGF receptors, thereby not accelerating growth. An excess in GH circulating levels could prevent the formation of the active GH-receptor complex by inhibiting the necessary for biological activity dimerization reported for the human GH receptor (Wells, 1996). Low ectopic GH levels could accelerate growth by permitting a better receptor occupancy, thus optimizing growth-promoting activity. This hypothesis resulted in the model depicted in Figure 1. However, alternative pathways may be also affected with the ectopic expression of GH, affecting the process of growth in fish.

ACKNOWLEDGMENTS

We would like to thank colleagues from our group for fruitful discussions. The work from our group was partially supported by the International Centre for Genetic Engineering and Biotechnology Collaborative Research Program (project CRP/CUB93-05).

REFERENCES

de la Fuente J, Martínez R, Estrada MP, Hernández O, Cabrera E, García del Barco D, Lleonart R, Pimentel R, Morales R, Herrera F, Morales A, Guillén I., Piña JC (1995) Towards growth manipulation in tilapia (*Oreochromis* sp.): generation of transgenic tilapia with chimeric constructs containing the tilapia growth hormone cDNA. J. of Marine Biotechnology 3:216–219.

Devlin RH, Yesaki TY, Biagi CA. Donaldson EM. (1994). Extraordinary salmon growth. Nature 371: 209–210.

Guillén I, Martínez R, Hernández O, Estrada MP, Pimentel R, Herrera F, Morales A, Rodríguez A, Sánchez V, Abad Z, Hidalgo Y, Lleonart R, Cruz A, Vázquez J, Sánchez T, Figueroa J, Krauskopf M and de la Fuente J (1996) Aquaculture Biotechnology Symposium Proceedings.Edited by Edward M. Donaldson and Don D.MacKinlay.International Congress on the Biology of Fishes. San Francisco State University July 14–18, pp. 63–72.

Guillén I, Estrada MP, Morales R, Melamed P and de la Fuente J. (1997). The interrelation of body growth with growth hormone, insulin-like growth factor and prolactin levels in juvenile tilapia (*Oreochromis aureus*). Minerva Biotecnologica (in press).

Hernández O, Guillén I., Estrada MP, Cabrera E, Pimentel R., Piña JC, Abad Z, Sánchez V, Hidalgo Y, Martínez R., Lleonart R. de la Fuente J (1997) Characterization of transgenic tilapia lines with different ectopic expression of tilapia growth hormone. Molecular Marine Biology and Biotechnology (in press).

Lu, JK, Chen T.T, Chrisman C.L., Andrisani OM. Dixon JE. (1992) Integration, expression and germ-line transmission of foreign growth hormone genes in medaka (*Oryzias latipes*). Molecular Marine Biology and Biotechnology 1(4/5): 366–375.

Martínez R, Estrada MP, Berlanga J, Guillén I., Hernández O, Cabrera, E, Pimentel R., Morales R, Herrera F, Morales, A, Piña, JC, Abad Z, Sánchez V, Melamed P, Lleonart R. de la Fuente, J (1996) Growth enhancement in transgenic tilapia by ectopic expression of tilapia growth hormone. Molecular Marine Biology and Biotechnology 5(1): 62–70.

Sumpter J.P. (1992) Control of growth of rainbow trout (*Oncorhynchus mykiss*). Aquaculture 100: 299–320.

Wells JA (1996) Binding in the growth hormone receptor complex. Proc Natl Acad Sci USA 93: 1–6

Zhang P, Hayat M, Joyce C, Gonzalez-Villasenor LJ, Lin CM, Dunham R, Chen TT, Powers, DA (1990) Gene transfer, expression and inheritance of pRSV-rainbow trout GH-cDNA in the common carp, *Cyprinus carpio* (Linnaeus). Mol.Repro.Dev. 25: 3–13.

Zhu Z (1992) Generation of fast growing transgenic fish: Methods and Mechanisms. In: Transgenic Fish. Hew, C.L. and Fletcher, G.L. (eds.). World Scientific Publishing Co., Singapore, pp. 92–119.

3

GENE TRANSFER IN ZEBRAFISH ENHANCED BY NUCLEAR LOCALIZATION SIGNALS

Philippe Collas and Peter Aleström*

Department of Biochemistry
Norwegian College of Veterinary Medicine
P.O. Box 8146 Dep., N-0033 Oslo, Norway

1. INTRODUCTION

Transgenic fish are routinely produced by injection of plasmid DNA into eggs (reviewed by Flechter and Davies, 1991). In zebrafish, the pronuclei of fertilized eggs are not visible, thus the DNA is injected into the cytoplasm or the yolk. Cytoplasmically injected DNA can integrate into the genome and be transmitted through the germline (Stuart et al., 1988; 1990). Frequencies of transgene integration and germline transmission may be as high as 25% (Stuart et al., 1988; Culp et al., 1991) but are often in the order of a few percents.

The low frequency of transmission of a transgene to F1 generation in zebrafish may be explained by factors controlling DNA stability, nuclear import and chromosome integration. Transgenes are often degraded or rearranged (Iyengar and MacLean, 1995), often remain extrachromosomal (Stuart et al., 1990), and when integrated, may be transcriptionally silent (Culp et al., 1991). Since the DNA is injected into the egg cytoplasm, rapid embryonic cell divisions may favor transgene integration at later stages of development, leading to a high degree of mosaicism (Flechter and Davies, 1991). To achieve nuclear uptake and chromosome integration of the DNA, high numbers ($>10^6$) of vectors are generally injected. This results in high concentrations of foreign DNA within the embryo, leading to increased risks of toxicity (Flechter and Davies, 1991).

Several methods have been developed to improve the efficiency of transgene integration into the host genome. They include (1) the use of pseudotyped viruses (zebrafish; Lin et al., 1994), (2) transgene integration mediated by a retroviral integrase protein (zebrafish; Ivics et al., 1993), (3) integration by transposable elements (*Drosophila*; Kaufman and Rio, 1991), and (4) binding of DNA to nuclear proteins (mammalian cells; Wienhues et al., 1987) or nuclear localization signals (zebrafish; Collas et al., 1996a; Collas and

* *Author for correspondence. Phone: +47 22 96 45 71; Fax: +47 22 60 09 85

New Developments in Marine Biotechnology, edited by Le Gal and Halvorson.
Plenum Press, New York, 1998.

Aleström, 1997), as a means of facilitating uptake of transgene DNA by the host nucleus. In this communication, we review data from our laboratory showing that NLS binding to DNA enhances DNA uptake by zebrafish embryo nuclei and increases frequencies of transgene expression, integration and germline transmission .

1.1. NLS Peptides Enhance Nuclear Uptake of Transgene DNA

Several systems have been used to investigate nucleocytoplasmic transport, in which synthetic NLSs direct nuclear import of non-karyophilic proteins to which they are cross-linked.

1.1.1. NLSs Bind to DNA. The specificity of NLS for nuclear import, together with the positive charge of the peptide, make NLSs good candidates for binding plasmid DNA. Synthetic NLSs analogous that of SV40 T antigen were bound by ionic interactions to a 5.5 kb plasmid carrying a luciferase reporter gene. Binding was assessed in a gel retardation assay (Collas et al., 1996a).

1.1.2. DNA–NLS Complexes Are Imported by Nuclei in Embryos. A procedure based on blastomere fractionation was described to separate zebrafish embryo nuclei from cytoplasmic fractions (Collas and Aleström, 1997). With isolated nuclei in hand, it was possible to examine by PCR the presence of injected DNA in the nuclear fraction, using primers that specifically amplify a fragment of the plasmid injected ("l" fragment). One can also include an additional primer pair in the PCR reaction, that amplifies a zebrafish control genomic sequence ("c" fragment). By densitometric measurement of the ratio of PCR signal intensity of "l" to "c" (l/c ratio), one can quantify the relative amounts of DNA present in the nuclear fraction.

Using this approach, nuclear targeting of DNA by NLS peptides was demonstrated in a time course measurement of DNA import into nuclei (l/c ratios) following cytoplasmic injection of 10^4 complexes per egg (Fig. 1). Plasmid DNA injected in the center of the yolk was detected in nuclei 30 min after injection, and nuclear uptake of DNA proceeded over several hours. Naked DNA or DNA bound to nuclear import deficient reverse NLS peptides (revNLS) was not readily imported into nuclei. NLS specifically directed DNA import to nuclei, as shown in a competition experiment using albumin-NLS conjugates. Relative amounts of plasmid DNA imported into zebrafish embryonic nuclei have also

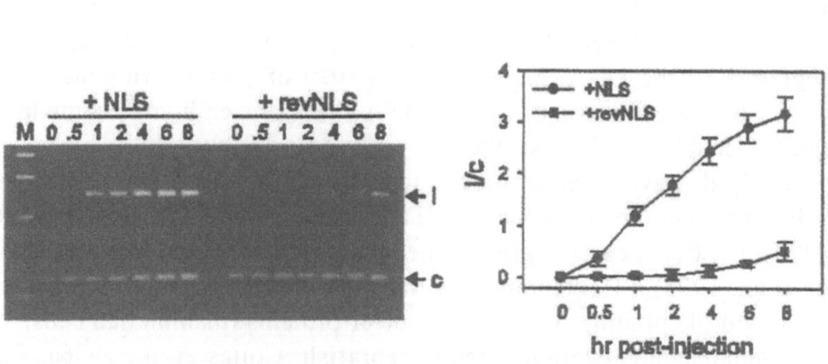

Figure 1. Enhanced kinetics of DNA import into nuclei after binding to NLSs. Eggs were injected in the center of the yolk with 10^4 plasmid copies bound to NLSs or revNLSs at a peptide:DNA molar ratio of 100:1. Nuclear fractions were prepared and the presence of plasmid DNA examined by PCR using control *(c)* and plasmid-specific *(l)* primers. *Left*, representative gel. *Right*, mean ± SD of l/c ratios of three separate experiments.

been estimated by fluorescence *in situ* hybridization (Collas et al., 1996b). These results indicate that NLS binding to DNA promoted rapid transport of DNA to nuclei under conditions where naked DNA was not imported, or imported slowly and in limited amounts. The amount of DNA imported into nuclei was affected by altering the coupling ratio of NLS to DNA. In our hands, an NLS:DNA molar ratio of 100:1 provided both high embryo survival rate and efficient nuclear import of DNA (Collas and Aleström, 1997).

1.1.3. DNA–NLS Complexes Are Targeted to Nuclei in Vitro. We have developed an *in vitro* assay to examine the conditions promoting nuclear import of transgene DNA (Collas and Aleström, 1996). The system consists of sea urchin pronuclei incubated in a fertilized zebrafish egg extract containing fluorescently (EtBr)-labeled plasmid DNA or DNA–NLS complexes. The extract was shown to support (i) binding of DNA–NLS complexes to the nuclear membrane in the absence of ATP, and (ii) translocation of the complexes inside the nucleus when ATP was added. Nuclear import of DNA *in vitro* did not occur without NLS or with the revNLS peptide, and was competed by albumin-NLS conjugates. Nuclear import of DNA–NLS likely occurred via nuclear pores, as blocking pore function inhibited this process.

1.2. NLS Increases Transgene Expression Frequency

A consequence of efficient nuclear import of DNA-NLS complexes is a dramatic increase in transgene expression frequency (Collas et al., 1996a). Our results indicate that transient luciferase expression was affected by the number of DNA-NLS complexes injected, and by the molar ratio NLS to DNA. Below 10^3 complexes, no expression occurred regardless of the NLS:DNA binding ratio. Injection of 10^3 and 10^4 DNA–NLS complexes prepared at an NLS:DNA molar ratio of 100:1 induced expression in 30% and 70% of embryos, respectively. At 10^6 copies, the frequency of luciferase expression was 100% with naked DNA, but decreased with DNA–NLS complexes, probably due to embryo death following transfer of excessive DNA into nuclei. Effects of NLS on luciferase expression were similar with supercoiled and linear DNA. Injection of 10^4 copies of DNA bound to revNLS did not induce transgene expression.

NLS peptides would constitute a valuable tool in transgenesis if not only DNA uptake by nuclei, but also integration into chromosomes and germline transmission of the transgene were improved. Recent data from our laboratory suggest that the use DNA-NLS complexes increases the proportion of transgenic zebrafish founders, increases the frequency of germline transmission of the transgene to F1, and dramatically reduces the incidence of mosaicism in transgenic founder fish (Collas and Aleström, manuscript submitted).

2. CONCLUSION

It is clear from these studies that the main advantage of using NLS peptides in transgenesic work resides in the rapid transfer of transgene DNA from the cytoplasm to the nucleus. Consequently, NLSs are expected to minimize the time between transgene injection and genome integration. This aspect is particularly important when working with embryos dividing rapidly such as zebrafish. Early transgene integration during embryo development should in turn reduce the incidence of mosaicism in transgenic founders.

Can NLS peptides be of interest in other transgenesis applications? In species or cell types where the DNA is delivered directly into the nucleus, NLSs are of less obvious use, but in all species requiring cytoplasmic injection, DNA–NLS complexes may be beneficial. Other applications where NLSs may prove beneficial include gene therapy and DNA vaccination. To this end, NLS peptides may also be included in macromolecular complexes that allow direct transfer of DNA molecules across the plasma membrane. This possibility remains to be investigated.

3. SUMMARY

Nuclear localization signals (NLSs) are short peptides required for nuclear import of karyophilic proteins. We have taken advantage of the nuclear targeting property of NLS to improve the efficiency of nuclear uptake of transgene DNA and of transgenesis in zebrafish. Synthetic NLSs are bound to plamid DNA by ionic interactions. Cytoplasmic injection of DNA–NLS complexes in zebrafish eggs enhances the rate and amount of DNA taken up by nuclei. Nuclear import of DNA–NLS complexes can be duplicated *in vitro*. Use of NLS increases transient expression frequency of the transgene. We suggest that NLSs may be a valuable tool to improve transgenesis efficiency in fish and other marine species.

ACKNOWLEDGMENT

Work of the authors was supported by EU grant Bio2-CT94-2039.

REFERENCES

Collas, P., and Aleström, P. 1996. Nuclear localization signal of SV40 T antigen directs import of plasmid DNA into sea urchin male pronuclei in vitro. Mol. Reprod. Devel. **45:**431–438.

Collas, P., Husebye, H., and Aleström. 1996a. The nuclear localization sequence of the SV40 T antigen promotes transgene uptake and expression in zebrafish embryo nuclei. Transgen. Res. **5:**451–458.

Collas, P., Sørensen, A., and Aleström, P. 1996b. Fluorescence in situ hybridization analysis of NLS-mediated transgene uptake by zebrafish embryo nuclei. Mol. Biol. Cell **7(suppl):**334a.

Collas, P., and Aleström, P. 1997. Rapid targeting of plasmid DNA to zebrafish embryo nuclei by the nuclear localization signal of SV40 T antigen. Mol. Mar. Biol. Biotech. **6:**48–58.

Culp, P., Nüsslein-Volhard, C., and Hopkins, N. 1991. High-frequency germ-line transmission of plasmid DNA sequences injected into fertilized zebrafish eggs. Proc. Natl. Acad. Sci. USA. **88:**7953–7957.

Flechter, L.G., and Davies, P.L. 1991. Transgenic fish for aquaculture. In J.K. Setlow, (ed.), *Transgenic Fish for Aquaculture in Genetic Engeneering*, Plenum Press, New York, pp. 331–370.

Ivics, Z., Iszvack, Z., and Hacket, P. 1993. Enhanced incorporation of transgenic DNA into zebrafish by a retroviral integration protein. Mol. Mar. Biol. Biotech. **2:**162–173.

Iyengar, A., and Maclean, N. 1995. Transgene concatemerization and expression in rainbow trout *(Oncorhynchus mykiss)*. Mol. Mar. Biol. Biotech. **4:**248–254.

Kaufman, P.D., and Rio, D.C. 1991. Germline transformation of *Drosophila melanogaster* by purified P element transposase. Nuc. Acid Res. **19:**6336.

Lin, S., Gaiano, N., Culp, P., Burns, J.C., Friedman, T., Yee, J.-K., and Hopkins, N. 1994. Integration and germline transmission of a pseudotyped retroviral vector in zebrafish. Science 265:666–669.

Stuart, G.W., McMurray, J.V., and Westerfield, M. 1988. Replication, integration and stable germ-line transmission of foreign sequences injected into early zebrafish embryos. Development **103:**403–412.

Stuart, G.W., Vielkind, J.R., McMurray, J.V., and Westerfield, M. 1990. Stable lines of transgenic zebrafish exhibit reproducible patterns of transgene expression. Development **109:**577–584.

Wienhues, U., Hosokawa, K., Hoveler, A., Siegmann, B., and Doerfler, W. 1987. A novel methods for transfection and expression of reconstituted DNA-protein complexes in eukaryotic cells. DNA **6:**81–89.

4

GENE TRANSFER IN RED SEA BREAM (*PAGROSOMUS MAJOR*)

Peijun Zhang,[1] Yongli Xu,[1] Zongzhu Liu,[1] Yuan Xiang,[1] Shaojun Du,[2] and Choy L. Hew[2]

[1]Institute of Oceanology
Academia Sinica
Qingdao 266071, China
[2]Department of Clinical Biochemistry and Biochemistry
University of Toronto
Toronto, Canada

1. INTRODUCTION

Since the first paper reporting gene transfer in goldfish was published 12 years ago (Zhu et al., 1985), significant progress has been made in studies of transgenic fish during the past decade. Several methods of gene transfer applied to more than ten fish species were developed and quite a few kinds of chimeric gene including growth hormone (GH), antifreeze protein (AFP) and various report genes were introduced into fish eggs to generate transgenic fish. The GH gene transgenic salmon with 13 times increase in its body weight was produced by using microinjection (Du et al., 1992) and the AFP gene transgenic goldfish with cold tolerance was generated by oocyte microinjection (Wang et al., 1995). These dramatic successes in transgenic fish study make investigators to be more and more interested in producing superior fish stocks with improved characteristics such as faster growth, freeze and cold tolerance as well as disease resistance.

With its attraction to consumers and high commercial value, red sea bream has become an important economical fish species in mariculture in China. However, it is hard to increase the culture yield of this species owing to its low growth rate. In order to solve this problem,we tried several years to make fast growing transgenic red sea bream by using an "all fish" GH gene construct as the chimeric DNA and electroporation as the transfer method.

The reason to use "all fish" GH gene construct is that our purpose is to make transgenic fish for human consumption. The choice of chimeric DNA should take into consideration of consumer acceptance and avoid potential health hazards. Therefore, the constructs containing human growth hormone, mouse metallothionein-1 (mMt-1) promoter or rous sarcoma virus (lRSV) promoter are not suitable because they may get regulatory obstacles and possible rejection in the market.

New Developments in Marine Biotechnology, edited by Le Gal and Halvorson.
Plenum Press, New York, 1998.

The reason to use electroporation as gene transfer method is also of consideration for aquaculture application. Microinjection of chimeric DNA into the fertilized eggs has been proved to be an efficiant method for producing transgenic fish (Ozato et al., 1986; Maclean et al., 1987;Zhang et al., 1990). However, it is hard to operate with large quantity of fish eggs owing to its need of strict training, time-consuming and labor-intensitivity. By using electroporation, we can treat more than ten thousand fish eggs in ten minutes. Here we report the successful trial to produce transgenic red sea bream with large scale.

2. MATERIALS AND METHODS

2.1. Plasmid Construction

The ocean pout antifreeze promoter-salmon growth hormone fusion gene was constructed by ligating the 2.2 kb BamHI-BglII fragment from lOP5 containing the opAFP promoter (Hew et al., 1988) with a 73 bp BglII-PstI synthetic linker. This DNA fragment was then ligated with a 0.7 kb PstI-StuI fragment containing chinook salmon GH coding sequence, and linked to a 1 kb HpaI-HindIII fragment from lOP5,which included the opAFP gene polyadenylation and the transcription termination signals at the stuI site. The final construct was designated opAFP-GHc.

2.2. Collection of Eggs and Sperm

Sexually mature red sea bream (*Pagrosomus major*) females were injected with carp pituitary extract (5 mg/kg body weight) dissolved in 0.7% NaCl solution 12 h prior to stripping. Eggs and sperm were stripped and mixed with a feather to fertilize. The fertilized eggs were collected and electroporated before first cleavage.

2.3. Electroporation

The electrod chamber with a size of 5 cm long, 1 cm wide and 1 cm deep was filled with 800 eggs and 1 ml of DNA solution (50 μg/min 1.5 NaCl solution). The pulse length was 50 μs, and the pulse interval is 1 ms. The voltage of the power supply varying between 100 v and 500 v. The pulse numbers were 6, 8, 10 and 12 respectively. After electroporation ,the treated eggs were immediatly transfered to the normal sea water at 20°C.

2.4. PCR Analysis

Three primers were used in PCR analysis to screen the genomic DNA of presumptive transgenic fish. The sequences of these primers are: Primer A, +275′-GTCAGAAGTCTC-AGCrACAGC-3′+47 sense strand; Primer B,+861 5′-ATCTCAACAGTCTCCACAGGT-3′+881 antisense strand; Primer D,+339 5′-ACAGAAGTCCAGCAGGAATAT-3′+359 antisense strand. The position refers to the distance to the TATA box of the transgene. To isolate genomic DNA from fin (2 month old) or whole body (one month old) of the electroporated or control fish, the fish body or fin was homogenized in 200 μl of 1x PCR buffer (50 mM KCI, 10 mM Tris, pH 8.8, 15 mM MgCI, 0.1% Triton X-l00) with 0.1 mg/ml proteinase K incubated at 55°C for 2 h, followed by adding 200 μl of 10 mM NaOH and heating in a boiling water bath for 2 min. The sample was centrifuged at top speed in a rnicrofuge for 5 min and the supernatant was ready for PCR. PCR was carried out in a 50 μl reactin solution containing lx PCR buffer, 1 μM of each primer, four deoxyribonucleotide triphosephates at 0.2 mM each, and 1.5 units of Tag DNA polymerase for 30 cycles. Each cycle includes 1 min at 92°C, 1 min at 60°C,and 2 min at 72°C. Ten μl of the PCR amplified DNA sample was loaded on a 1.0% agarose gel to analyze by electrophoresis.

2.5. Western-Immunoblot Analysis

Serum samples collected from 31 three-month-old PCR positive fish were analyzed by Western-Immunoblot. Ten μl of serum was subjected to the SDS-polyacrylamide gel electrophoresis in 10–20% acrylamide gradient. After blot transfer, the salmon GH was reacted with monoclonic chinook salmon GH antibody and horseradish oxidase labeled goat-anti-rabbit antibody successively and detected by reaction with substrate.

3. RESULTS

3.1. Survival and Hatching Rate

The survival rate of red sea bream eggs after electroporation with a conditon of voltage, 350 V; pulse number, 10; pulse length, 50 μs and pulse interval, lms is 79.2%. The hatching rate of the surviving treated eggs is 92.1% which is similar to the untreated control eggs (92.8%).

3.2. Integration of the Foreign Gene

PCR analysis of one-month-old presumptive transgenic fish indicated that out of 126 randomly selected individuals, 37 were found to carry the foreign sequence, showing the integration rate of 29.3%, while analysis of two-month old fish showed the integration rate was 38%.

3.3. Expression of the Integrated Foreign Gene and Growth Detection

Western-Immunoblot analysis indicated that out of 31 three-month-old PCR-positive transgenic fish, seven individuals contained the chinook salmon GH in their blood, showing the expression rate of 22.6%.

From the age of 2-month, we detected growth of the treated and untreated fish by measuring their body length and body weight monthly (Table 1). At the age of 7-month,

Table 1. Growth rates of the transgenic and control red sea bream

	Months after hatching				
	3	4	5	6	7
Treated fish					
Weight (g)	8.8 ± 0.9	21.7 ± 3.2	38.5 ± 5.7	62.8 ± 10.2	82.7 ± 11.8
Length (cm)	5.7 ± 0.8	8.2 ± 1.0	10.1 ± 1.9	12.7 ± 2.4	15.2 + 2.8
Control fish					
Weight (g)	8.1 ± 1.0	19.9 ± 1.7	31.1 ± 2.9	50.9 ± 5.8	68.1 ± 6.3
Length (cm)	8.4 ± 0.1	7.3 ± 1.6	9.5 ± 1.3	11.4 ± 1.8	13.9 ± 1.6
Transgenic fish					
Weight (g)	8.9 ± 1.3	24.7 ± 3.8	42.4 ± 6.6	69.2 ± 13.6	93.1 ± 18.5
Length (cm)	5.9 ± 0.5	8.8 ± 1.8	11.7 ± 2.2	14.1 ± 3.1	16.8 ± 3.6

the everage amount of body length and body weight of the GH-electroporated group increased 9.3% and 21% respectively comparing with the untreated control group. The PCR positive transgenic fish grow even faster than the tested 2 groups showing 21% and 36% increase in body length and body weight respectively comparing with the untreated control group. The largest Western-immunoblot analysis positive transgenic individual is 2.21 times larger than the everage of control in body weight.

4. ABSTRACT

In order to cultivate fast-growing economic fishes, an "all fish" growth hormone (GH) gene construct containing ocean pout (*Macrozoares americanus*) antifreeze protein (AFP) gene promoter and chinook salmon GH cDNA was introduced into the fertilized eggs of red sea bream (Pagrosomus major) by electroporation. The optimal electric pulse for red sea bream eggs was determined as 350 V/cm, 50 μs and 10 times by single-factor test. Out of 48.0 thousand eggs treated, 38.0 thousand (79.2%) eggs survived and 35.0 thousand (92.1%) eggs hatched. By using specific oligonucleotide primers, polymerase chain reaction (PCR) analysis of genomic DNA isolated from presumptive transgenic fish showed that 37 of 126 randomly selected individuals (29.3%) were found to carry the foreign sequence integrated when tested at age of 30 days, while 38 of 100 (38%) were found when tested at age of 2 months. Western-immuno blot analysis of the serum samples from 31 of 3-month old PCR-positive fish indicated that there were 7 individuals (22.6%) containing the chinook salmon GH in there blood. The everage amount of body lenth and body weight of the thransgenic fish increased 9.3% and 21% respectively comparing with the control group when detected at age of 7 months.

REFERENCES

Du, S.J., Gong, Z., Fletcher, G.L., Shears, M.A., King, M.J., Idler, D.R. and Hew, C.L., 1992, *Biotechnology* **10**: 176–181

Hew, C.L., Wang, N.-C., Joshi, S., Fletcher, G.L., Scott, G.K. 1988, *J. Biol. Chem.* **263**: 12049–12055.

Maclean, N., Penman, D. and Zhu, Z. 1987, *Biotechnology*, **5**: 257–261.

Ozato, K. , Kondoh, H., Inogara, H., Iwamahu, T., Wakamatsu, Y. and Okado, T.S. 1986, *Cell.Diff.* **19**:237–244.

Wang, R., Zhang, P., Gong, Z., Hew, Choy L. 1995, *Mol. Mar. Biol. Biotechnol.*, **4**(1): 2–26.

Zhang, P., Hayat, M., Joyce, C., Gonzalez-Villasenor, L.I., Lin, C.M., Dunham, R.A., Chen, T.T. and Powers, D.A. 1990, *Mol.Reprod.Dev.* **25**: 3–13.

Zhu, Z., Li G., He, L. and Chen, S. 1985, *Z. Angew. Ichthyol.* **1**: 32–34

5

PRODUCTION OF LINES OF GROWTH ENHANCED TRANSGENIC TILAPIA (*OREOCHROMIS NILOTICUS*) EXPRESSING A NOVEL PISCINE GROWTH HORMONE GENE

Azirur Rahman and Norman Maclean*

Division of Cell Science
School of Biological Sciences
Bassett Crescent East
University of Southampton
Southampton SO16 7PX, United Kingdom

1. INTRODUCTION

Since the first introduction of novel genes in fish (Maclean and Talwar, 1984; Zhu *et al.*, 1985), a wide range of fish species have been used in transgenic research either for commercial or academic purposes. For commercial exploitation this technology has been applied in several commercially important fish species to improve desirable genetic traits such as growth (Rokkones *et al.*, 1989; Zhang *et al.*, 1990; Brem *et al.*, 1988; Penman *et al.*, 1990; Gross *et al.*, 1992), cold tolerance (Fletcher *et al.*, 1988), or disease resistance (Anderson *et al.*, 1996) for aquaculture development. Most of these gene transfer experiment have concentrated on growth enhancement using regulatory sequences and coding sequences of distantly related species (Rokkones *et al.*, 1989; Penman *et al.*, 1990; Brem *et al.*, 1988; Zhang *et al.*, 1990). Although successful gene transfer in these fish was observed in only a few cases has germline transmission and expression of transgenes in progeny been convincingly demonstrated (see reviews by Maclean and Rahman, 1994; Gong and Hew, 1995). In studies where the regulatory sequences used were of mammalian origin, low or nil expression was observed (Guyomard *et al.*, 1989; Alam *et al.*, 1996). In other studies, enhanced growth was observed in transgenic carp (Zhang *et al.*, 1990) and in tilapia (Martinez *et al.*, 1996) when growth hormone sequences were driven by a viral promoter. However, the main objective of the production of transgenic fish with GH gene

* To whom correspondence should be made. Tel: + 44 (0) 1703 594403, Fax: + 44 (0) 1703 594269

New Developments in Marine Biotechnology, edited by Le Gal and Halvorson.
Plenum Press, New York, 1998.

is to generate novel strain of growth enhanced fish which could subsequently be used in aquaculture. It is, therefore, important that the gene constructs used for producing transgenic fish be of fish origin, and not from mammalian or viral origin. Moreover, it was also observed that regulatory sequences from fish are found to be more effective than those of mammalian origin in respect to expression of transgenes (Alam *et al.,* 1996) in fish.

Thus Du *et al.,* (1992) and Devlin *et al.,* (1994) have used "all fish" gene construct in producing transgenic salmon. In the former study, transgenic atlantic salmon (*Salmo salar*) were produced using a gene construct (OPAFPcsGH) containing a growth hormone gene from chinook salmon (*Oncorhynchus tschawytcha*) driven by regulatory sequence from an antifreeze gene of the ocean pout (*Macrozoarrces americanus*) and the resulting transgenic fish grew 2–6 times more than the non-transgenic siblings. In latter study, dramatic growth (up to 11 fold over the non-transgenic sibling) in coho salmon (*Oncorhynchus kisutch*) was observed when a gene construct (OnMTGH1) consisting of a sockeye salmon (*Oncorhynchus nerka*) GH gene spliced to metallothionein promoter of the same origin was used.

We here report the successful integration, expression and germline transmission of a "all fish" gene construct (OPAFPcsGH) in tilapia. Due to expression of the salmon growth in transgenic tilapia we have observed enhanced growth in these fish.

We have choosen to work on the tilapia (*O. niloticus*) which is being farmed in more than 60 countries around the world and genetic improvement of fish species by transgenic technique is of significant impact on world fish production. Moreover, tilapia is omnivorous, has short generation time and spawns regularly at 3 weeks intervals, unlike other commercial species which have particular spawning season.

2. MATERIALS AND METHODS

2.1. DNA Construct and Microinjection

The "all fish" gene construct kindly provided by C.L. Hew (University of Toronto, Canada), consists of a chinook salmon growth hormone cDNA spliced to an oceanpout antifreeze gene regulatory sequence (Du *et al.,* 1992). The approximately 3.9 Kb linear insert used for microinjection was separated from vector sequences by restriction with BamHI (Fig. 1). About 250 pl, containing approximately 10^5 copies of the transgene, were microinjected into the cytoplasm of 1–2 cell fertilised eggs using the technique described by Rahman and Maclean (1992).

2.2. Extraction of Genomic DNA

Genomic DNA was extracted from caudal fin clips of adults and also from 3-day-old G1 fry. Prior to fin clipping the fish were anaesthetised using 400 PPM 2-phenoxy-ethanol, and standard procedures for isolation and purification of DNA were followed as previously described in Rahman and Maclean (1992). The concentration of DNA was determined using a fluorometer as per manufacturer's instruction (Hoefer Scientific).

2.3. Detection of Transgene

Detection of transgenes in host genomic DNA was carried out by Southern blotting. Genomic DNA (3 µg) was digested with HindIII at 37°C for overnight. The restricted DNA was subjected to electrophoresis on a 0.8% agarose gel for approximately 14 hours

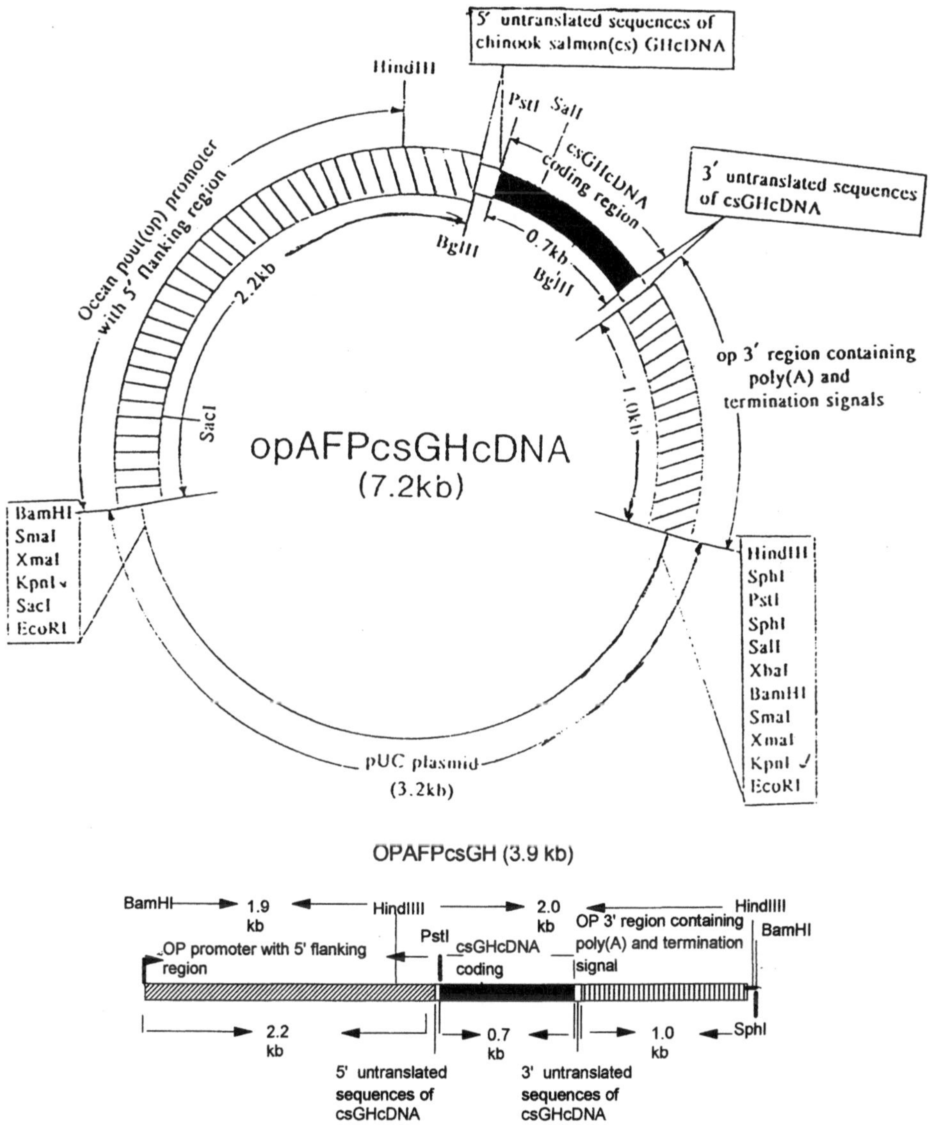

Figure 1. The OPAFPcsGH construct isolated from the plasmid after digestion with BamHI and used for microinjection, showing the restriction sites for HindIII.

after ensuring completion of restriction digestion. The gel was then blotted onto a nylon membrane (Hybond N+, Amersham) using a vacublot apparatus (Vaccugene XL, Pharmacia) according to manufacturer's instructions. The standard procedure of Southern blotting and hybridisation of membranes were carried out as described in Rahmån et al. (1997). The probes used for Southern blot analysis was the entire 3.9 Kb OPAFPcsGH.

2.4. Analysis of Growth Performance

Comparison of growth performance was made between transgenic and non-transgenics from the same batch which were developed from a cross between hemizygous transgenics and wild type controls and reared in the same conditions. After hatching, the fry were reared in a glass tank (0.02 cu.m) and reared for 1–2 months. They were then transferred to a large tank (0.13 cu.m) and reared for a further 4–5 months. All glass tanks were linked in a recirculatory system. The fry were fed with flake food and juvenile and adult fish with commercial trout pellets. Fish were fed 3 times a day to satiation. At the appropriate stage the weight of individual fish was taken to evaluate growth along with fin clip prior to determine transgenic status. Before taking fin clip fish were tagged by using a transponder (Trovan Identification Systems, The Netherlands) which was introduced into the body cavity of fish under anaesthesia. To avoid intestinal injury during insertion of the transponder, these fish were starved for at least 24 hours prior to transponder insertion. A hand held detector was used to identify tagged fish. Condition factor was also determined in some of the G1 and G2 batches of 2 transgenic lines. Condition factor (K) was calculated using following using following formula:

$$K = W/L^3 \times 10^3$$

where W is the weight in grams and L is the total length.

2.5. Radioimmunoassay

Radioimmunoassay of blood samples of transgenic and non-transgenic fish were carried in Pharos Ltdin Belgium. Blood samples were collected from the caudal veins of anaesthetised fish with hypodermic needles containing 50 μl of Cortland's solution with 10 mM EDTA and placed on ice. The blood plasma samples (without label) were sent to Pharos on dry ice to assay for the presence of exogenous salmon GH using salmon specific growth hormone RIA (Poncin and Rousseau, 1994).

3. RESULTS

3.1. Inheritance of OPAFPcsGH Gene Construct

In order to produce lines of transgenic tilapia containing the exogenous growth hormone gene construct we have introduced the "all fish" (OPAFPcsGH) construct into tilapia. We have microinjected this construct into 800 eggs of tilapia from which 118 fish reached to adult stage. Screening for the presence of the transgene was carried out and 7 fish were found to be positive for the transgene in Southern blots of either fin, blood or sperm DNA. These transgenic fish were analysed for the inheritance of the transgene and 3 fish were found to be transmitting the GH genes to their progeny. The transmission rate from founder G0 fish to G1 of these 3 lines are <1%, 8% and 6% in C58, C86 and C118 lines respectively. Southern blot analysis of DNA revealed that integration of either one copy or multiple concatamerized copies of the OPAFPcsGH had occurred at a single site in the host chromosome of 3 lines of transgenic tilapia. The DNA band pattern in these 3 lines were different from each other but identical with those of other individuals of the same line (Fig. 2). Restriction digestion of the OPAFPcsGH construct with HindIII generates approximately a 2.0 Kb internal fragment and a 1.9 Kb flanking sequence (Fig. 1). After digestion of genomic DNA and hybridisation with 3.9 kb OPAFPcsGH probe the internal band (2.0 Kb) should be present in all lanes containing the transgene and this was observed as shown in Fig. 2. The presence of approximately 5 Kb junction fragment

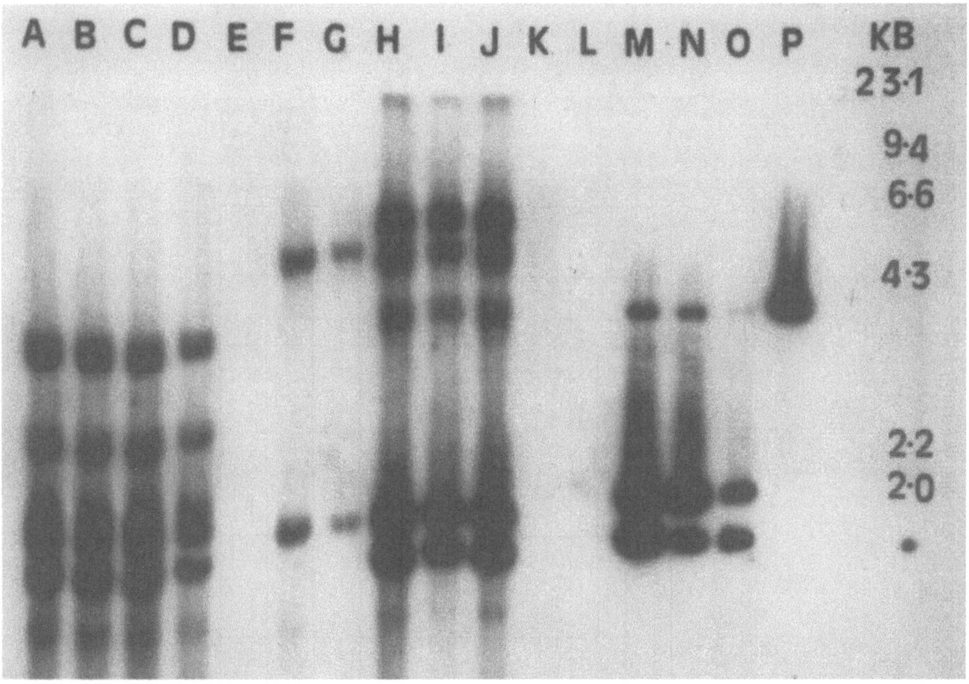

Figure 2. Southern blot analysis of fin DNA from G2 and G1 adult tilapia from the C118 and C86 lines respectively and DNA from whole fry of the C58 line and fin DNA from wild type tilapia. 3 μg of genomic DNA was digested with HindIII and hybridised to entire 3.9 Kb OPAFPcsGH. Lanes A-D Fin DNA from adult transgenic individuals from the C118 line; Lane E: Fin DNA of adult non-transgenic G1 of C86 line; lanes F–G: Fin DNA from adult G1 transgenic individuals from the C86 line; lanes H–J: Whole fry DNA from 3 days old transgenic G1 progeny from the C58 line; lane K: Whole fry DNA from 3 days old non-transgenic G1 progeny from C58 line; lane L: Fin DNA from adult wild type tilapia; Lanes M, N and O: HindIII digested OPAFPcsGH fragments (10 copies, 5 copies and 1 copy/cell respectively): Lane P: 3.9 kb undigested fragment of OPAFPcsGH.

in Fig. 2 lanes F and G instead of 1.9 Kb fragment seen in lanes M–O, suggests that a single copy integration of OPAFPcsGH into the host chromosome of C86 line has occurred. In another case the presence of a 1.9 Kb band (Fig. 2 lanes A–D and H–J) was observed which has originated from the HindIII digestion of head to tail concatamers of OPAFPcsGH construct in C118 and C58 lines indicating the integration of multiple copies of the transgene. G2 generations were produced by crossing transgenic G1 with wild type tilapia and approximately 50% of the progeny were found to be transgenic (Table 1) which suggests that the transgenes were integrated at a single locus in the host chromosome.

3.2. Transgene Expression

The oceanpout antifreeze protein is normally expressed at a low level in most tissues of the ocean pout but is strongly expressed in the liver (Gong *et al.,* 1991). Since the regulatory sequences used were derived form this gene, expression of GH gene should be strongest in the liver. Blood plasma of G0, G1 and G2 of transgenic lines containing chinook salmon GH gene were examined to detect circulating chinook salmon growth hormone in the transgenic tilapia. Low levels of salmon GH were detected in G0 transgenic tilapia of C58 (0.26 ng/ml) and C86 (0.15 ng/ml) lines due to the mosaic distribution of

Table 1. Rate of transgene transmission, level of exogenous salmon GH expression and growth performance of transgenic tilapia

Transgenic lines	GH constructs used	Transmission rate to G1	Transmission rate to G2	Expression of Salmon GH (ng/ml)			Growth performance in average weight (gm)				Condition factor	
				G0	G1	G2	T		NT		T	NT
							G1	G2	G1	G2		
C 118	OPAFPcsGH	39/600 (6%)	304/589 (52%)	0.1 n=1	1.3 n=1	1.02 ± 0.14 n=6		$53.3±4.8^{**}$ n=6		28.7±4.7 n=9		
"								$52±14.1^{*}$ n=5		15.6±2.4 n=8		
"								$79±6.49^{***}$ n=18		18± 1.78 n=32	17.46 ±0.32	17.58 ±0.31
C 86	"	73/916 (8%)	92/188 (49%)	0.15 n=1	1.2 n=1	NA	$81.5±9.5^{*}$ n=2		22.6±1.82 n=49		17.46 ±0.2	18.9 ±0.38
"							$19.5±0.57^{***}$ n=41		4.9±0.13 n=446			
"								$21.4±0.68^{***}$ n=110		6.5±0.2 n=102	15.33 ±0.17	18.06 ±0.3
C58	"	4/494 (0.8%)	--	0.25 n=1	NA	NA	---	----	-----	-----	---	----

t -test : * $P < 0.05$, ** $P < 0.01$, ***$P < 0.001$ T= Transgenic, NT = Non-transgenic, NA = Not analyzed, n= Number of individuals analyzed.

transgene in G0 fish. The levels of chinook salmon GH expression in the G0 fish of C118 line were the same as background levels observed in control fish (<0.1 ng/ml), perhaps due to the absence of the transgene in the liver. However, in G1 and G2 generations of the C118 line the level of salmon GH was 10–13 times higher than the background level (Table 1). Similarly, salmon growth hormone levels (1.2 ng/ml) in the G1 generation of the C86 line were found to be 12 times higher than the background level (<0.1 ng/ml).

3.3. Growth Performance of Transgenic Tilapia

Growth performance was compared between transgenic and non-transgenic fish of full sibling groups reared together in a single tank. Growth assessment was performed on G1 and G2 generations of C86 and C118 lines containing the OPAFPcsGH construct. Growth performance was observed in 3 batches of G2 generations of the C118 line produced by crossing transgenic G1 with a wild type controls. Southern blot analysis was carried out to identify transgenic individuals and 28 out of 84 individuals analyzed were found to be transgenic. The average weight of these transgenics in 3 batches were 2 fold ($P \leq 0.01$), 3 fold ($P \leq 0.05$) and 4 fold ($P \leq 0.001$) greater than their non-transgenic full siblings (Table 1). Similarly, growth performance analysis was carried out in 2 batches of G1 and 1 batch of G2 of C86 line. The data indicates that the average weight of transgenics are 4 times higher ($P \leq 0.001$) than in their non-transgenic siblings (Table 1 and Fig. 3). All the transgenic fish are above the average weight of their non-transgenic siblings.

Figure 3. Photograph showing three 9 month old transgenic (right) and three non-transgenic full siblings (left) from the C86 transgenic line. The fish are representative samples of the mean size of the two groups.

4. DISCUSSION

The present investigation strongly suggests that a chinook salmon GH gene construct (OPAFPcsGH) has been stably integrated, expressed and germline transmitted in transgenic tilapia. We have also demonstrated that expression of chinook salmon GH in transgenic tilapia enhanced the growth of fish. This construct has also been found to express in transgenic salmon (Du *et al.*, 1992) and in transgenic trout (Devlin *et al.*, 1995) where enhanced growth was observed in transgenic fish.

In this study, 3 founder G0 fish were found to express the transgene and transmitted the transgenes to their progeny. Growth performance analysis was carried out on G1 and G2 generations of 2 lines (C86 and C118). In another line (C58) transgenic adult G1 could not be identified due to very low transmission rate (<1%) from G0 to G1, and therefore growth performance of the transgenic could not tested in this line. Since the G0 transgenic fish are mosaic and subsequent generations from G1 transgenic fish are non-mosaic for the transgene, it is possible to assess growth performance of transgenics from G1 and following generations. Growth performance was compared between transgenic and non-transgenic full siblings reared in the same tank to reduce the interaction of other factors on growth (Speare *et al.*, 1995). Enhanced growth in transgenic tilapia was observed in both the C86 and C118 line. Transgenics were found to grow on average 3 times faster than their full non-transgenic siblings in C118 line and on average 4 times faster in C86 line. The condition factor of transgenic and non-transgenic fish of C118 and C86 lines was measured and it was observed that in the C118 line condition factor of transgenics and non-transgenics did not differ significantly (Table 1) which reveals that allometric growth resulted in normal body proportions in the transgenic fish. Growth enhanced transgenic salmon showed similar allometric growth. However, the condition factor of the transgenic fish was found to be lower than the non-transgenic full siblings in C86 line (Table 1) which reveals that transgenic fish grew faster in length than the weight. Due to transgene mosaicism in founder fish, low levels of expression of salmon growth hormone in G0 transgenic tilapia was observed (<0.01–0.25 ng/ml) compare to non-mosaic G1 and G2 generation transgenics (0.6–1.5 ng/ml). Similar levels of exogenous growth hormone have also been reported in transgenic carp, for instance, 2.28–4 ng/ml (Rosochacki *et al.*, 1993) and 1–4 ng/ml (Chen *et al.*, 1990). It has been suggested that GH can improve food and protein conversion by mobilization of lipid so that more of the ingested amino acids are available for protein growth and lipid is used preferentially as an energy source (Weatherly and Gill, 1987). In a study where both transgenic and non-transgenic tilapia of G2 from the C86 line were fed 3% body weight of food the preliminary data suggests that the food conversion ratio is 30% higher in transgenic individuals as compared to their non-transgenic full siblings.

In several studies it has been reported that elevated levels of growth hormone in transgenic animals can cause physiological abnormalities such as diabetes in transgenic sheep (Rexroad *et al.*, 1991)and gastric ulceration and osteochondritis dessicans in pigs (Pursel *et al.*, 1990; Cameron *et al.*, 1994). No major physiological abnormalities have been reported in transgenic fish containing elevated exogenous growth hormone (Zhang *et al.*, 1990; Du *et al.*, 1992). However, it has reported that some large transgenic salmon showed excess growth of opercular bone (Devlin *et al.*, 1995). A slight modification to the shape of opercular and jaw bone was also observed in some of the oldest transgenic tilapia in our study. The lack of physiological abnormalities in transgenic fish containing high levels of exogenous growth hormone may be due to the plastic nature of growth in fish as compared to mammals (Weatherly and Gill, 1987). It has also been observed that low levels of sperm production occurs in male transgenic tilapia and in some cases this may lead to complete cessation of sperm production. It is reported that transgenic mice containing exogenous GH gene have low sperm count (Bertke *et al.,* 1992). Transgenic females of the G2 generations of C118 line were found to be fertile while a G2 female of C86 was sterile. Although the amount of sperm produced by transgenic tilapia is sufficient to fertilise hundreds of eggs *in vitro* it may not be sufficient for efficient fertilisation in the wild. Therefore, genetic contamination by the transgenic males in wild would be probably very low if they escaped accidentally.

5. ABSTRACT

We have generated novel strains of transgenic tilapia containing a growth hormone gene construct: this fish exhibit extraordinary growth enhancement. To produce these lines we have microinjected a chinook salmon growth hormone (GH) gene driven by an ocean-pout antifreeze promoter (OPAFPcsGH) into 800 fertilised eggs of tilapia. Among the founder (G0) transgenic tilapia, 3 fish contained the transgene either in sperm or eggs. Germline transmission revealed that the rate of transgene expression to G1 progeny was <1%, 6% and 8% in these 3 transgenic tilapia lines. Expression of salmon growth hormone was also detected in these founder fish at a low level (due to transgene mosacism/integration site/copy number). However, the level of expression of exogenous salmon growth hormone in nonmosaic transgenic progeny tilapia is found to be 10–13 times higher than the level observed in the founder generation. Growth performance of G1 and G2 generations of 2 transgenic lines suggest that transgenic tilapia have approximately 4 times higher growth ($P < 0.001$) than their non-transgenic full siblings. Very low levels of abnormalities such as slight excess growth of bone in the lower jaw were observed in a few individuals of G1 and G2 generations. Apart from poor sperm production by transgenic males there are no other physiological abnormalities evident in these transgenic tilapia. Transgenic males were found to be produce very low amounts of sperm compared to wild type males and in some cases showed complete sterility.

ACKNOWLEDGMENTS

We are extremely grateful to C. L. Hew (University of Toronto, Canada) and R. H. Devlin (Vancouver, Canada) for providing us with the GH gene constructs. We thank D. Caucheteux and J. Smal (PHAROS, Belgium) for the radioimmunoassay. This work was supported by the Department for International Development (DFID) (formerly Overseas Development Administration (ODA)) UK. Fish Genetics Programme.

REFERENCES

Alam, M.S., Laveder, F.L., Iyengar, I; Rahman, M.A.., Ayad, H.H., Lathe, R., Morley, S.D. and Maclean, N. 1996, *Mol. Reprod. Dev.* **45**:117–122.

Anderson, E.D., Mourich, D.V., Fahrenkrug, S.C., LaPatra, S., Shepherd, J. and Leong, J.-A.C. 1996, *Mol. Mar. Biol. Biotechnol.* **5**(2):114–122.

Bertke, A., Naar, E.M., Johnson, L., May, M.R., Cecim, M., Yun, J.S. and Wagner, T.E. 1992, *J. Reprod. Fertil.* **95**:109–118

Brem, G., Brenig, B., Horstgen- Schwark, G. and Winnacker, E. L. 1988, *Aquaculture* **68**:209–219.

Cameron, E.R., Harvey, M.J.A. and Onions, D.E. 1994, *Transgenic Science*. Br. Vet. J. **150**:9–24
Chen, T.T., Lin, C.M., Zhu, Z., Gonzalez-Villasenor, L.I., Dunham, R.A. and Powers, D.A. 1990, In Church, R.B ed., *Transgenic Models in Medicine and Agriculture*, pp. 127–39. New York: Wiley-Liss
Devlin, R.H., Yesaki, T.Y., Biagi, C.A., Donaldson, E.M., Swanson, P. and Chan, W. 1994, *Nature*. **371**:209–10.
Devlin, R.H., Yesaki, T.Y., Donaldson, E.M., Du, S.J. and Hew, C.L.1995, *Can. J. Fish Aquat. Sci.* **52**:1376–1384.
Du, S.J., Gong, Z., Fletcher, G.L., Shears, M.A., King, M.J., Idler, D.R. and Hew, C.L. 1992, *Bio/Technology*. **10**:176–81.
Fletcher, G.L., Shears, M.A., King, M.J., Davies, P.L. and Hew C.L. 1988, *Can.J.Fish. Aquat. Sci.* **45**:352–357.
Gong, Z; Hew, C.L. and Vielkind, J.R., 1991, *Mol. Mar. Biol. Biotechnol.* , **1**(1): 64–72.
Gong, Z. and Hew, C.L. 1995, *Current Topics Dev. Biol.* **30**:177–214.
Gross, M.L., Schneider, J.F., Moav, N., Moav, B., Alvarez, C., Myster, S.H., Liu, Z., Hallerman, E.M., Hackett, P.B., Guise, K.S., Faras, A.J. and Kapuscinski, A.R. 1992, *Aquaculture* **103**: 253–73.
Guyomard, R., Chourrout, D., Lerox, C., Houdebine, L. and Pourrain, F. 1989, *Biochimie* **71**: 857- 63
Maclean, N. and Talwar, S. 1987, *J. Embryol. Expt. Morphol.* **82** (suppl): 187
Maclean, N. and Rahman, A. 1994, Transgenic fish. In Maclean N. ed., Animals with novel genes, pp 63- 105 .Cambridge University Press, USA.
Martinez, R., Estrada, M.P., Berlanga, J., Guillén, I., Hernández, O., Cabrera, E., Pimentel, R., Morales, R., Herrera, F., Morales, A., Piña, J.C., Abad, J., Sánchez, V., Melamed, P., Lleonart, R. and de la Fuente, J. 1996, *Mol. Mar. Biol. Biotechnol.* **5**(1):62–70.
Penman, D.J., Beeching, A.J., Penn, S. and Maclean, N. 1990, *Aquaculture* **85**:35 -50.
Poncin, A. and Rousseau, P. 1994, Societe Europeenne De Biotechnologie Sa, Scientific Rep. no. Rt-001-94-01: pp 1–3.
Pursel, V.G., Bolt, D.J., Miller, K.F., Pinkert, C.A., Hammer, R.A., Palmiter, R.D. and Brinster, R.L. 1990, *J. Reprod. Fertil.* (suppl) **40**:235–245.
Rahman, M.A. and Maclean, N. 1992, *Aquaculture*. **105**:219–32.
Rahman, M.A., Iyengar, I. and Maclean, N. 1997, *Transgenic Res.* **6**: 1–10
Rexroad, C.E., Maho, K., Bolt, D.J. et al., 1991, *J. Anim. Sci.* **69**:2995–3004.
Rokkones, E., Alestrom, P., Skjervold, H. and Gautvik, K.M. 1989, *J. Comp.Physiol. B* **158**:751–758
Rosochacki, S.J., Bialowas, H., Czlonkowska, M., Guszkiewicz, A., Kossakowski, M., Sadowska, J., Siadkowska, E., Zebrowska, T., Szumiec, J. and Zwierzchowski, L. 1993, *Anim. Sci.* papers Reports **11**(1): 47–58.
Speare, D.J., Macnair, N. and Hammell, K.L. 1995, . *Am. J. Vet. Res.* **56**(10): 1372–1379.
Weatherly, A.H. and Gill, H.S. 1987, *Academic Press, London.* pp 1–429.
Zhang, P., Hayat, M., Joyce, C., Gonzalez-Villasenor, L.I., Lin, C.M., Dunham, R.A., Chen, T.T. and Powers, D.A. 1990, *Mol. Reprod. Dev.* **25**: 3–13.
Zhu, Z.; Li, G; He, L. and Chen, S. 1985, Z, *angew.Ichthyol.* **1**: 31–34.

6

RETENTION OF A FOREIGN GENE TRANSFERRED AS A PROTAMINE–DNA COMPLEX BY ELECTROPORATED SALMON SPERM

F. Y. T. Sin,* J. G. I. Khoo, U. K. Mukherjee, and I. L. Sin

Department of Zoology
University of Canterbury
Private Bag 4800
Christchurch, New Zealand

It has been demonstrated that electroporated salmon sperm can be used as a vector to transfer foreign genes into salmon ova. When 1- to 11-day old embryos were analyzed using the polymerase chain reaction for the presence of the transgene, over 95% of the embryos were found to contain the transgene (Walker *et al.*, 1995). However, when 12 weeks old alevin were analyzed, the percentage of alevins containing the transgene was highly variable between experiments, ranging from zero to over 85% (Symonds *et al.*, 1994). The variability of transgene retention may be caused by the DNase I activity of the host cells. Alternatively, the delivered DNA may fail to enter the nucleus and, thus, be unable to be stably integrated into the fish genome. Weinhues *et al.* (1987) first demonstrated that DNA-protamine complex enhances the translocation of the DNA into the nucleus. Recently, Page *et al.* (1995) co-injected DNA-polylysine complex into the mouse pronuclei. These authors found an increase in integration of the transgene from 0 to 12%. These experiments suggest that the DNA-protein complex may prevent access of the nucleases to the foreign DNA, and hence prevent the foreign DNA from degradation and enhance the chances for the foreign DNA to enter the nucleus.

The aims of this study were to test the effect of protamine on the retention of foreign DNA transferred by chinook salmon (*Oncorhychus tshawytscha*) sperm, and to determine whether the transgene can persist in the older fish and if so, whether the transgene is integrated into the host genome.

A linearized plasmid containing a gene construct consisting of the chinook salmon metallothionein promoter and coho salmon IGF1 cDNA, pchMT-coIGF1, was used in the

* Fax: 03 364 2024. E-mail: fsin@csc.canterbury.ac.nz

New Developments in Marine Biotechnology, edited by Le Gal and Halvorson.
Plenum Press, New York, 1998.

study. Protamine and DNA interaction was tested by gel retardation analysis using 0.7% agarose gel in Tris-borate EDTA buffer, pH 8. DNA to protamine ratios of 1:0, 1:0.25, 1:0.5 and 1:1 were tested. As protamine concentration increased, retardation of the DNA-protamine became evident. At a ratio of 1:1 all the DNA failed to migrate into the gel. This indicates that either all the DNA has formed a protamine-DNA complex which is either too large to migrate into the gel, or had no net negative charge.

The effect of protamine on sperm-DNA association was tested by electroporation of 0.5 ml of fresh milt and 0.5 ml HEPES (pH 6) containing 100 mg of plasmid DNA and 100 mg of salmon protamine. Subsequently, the sperm solution was washed with HEPES twice, and 50 mg of DNase 1 was added to the sperm and incubated for 30 min. The digestion was stopped by heating at 100°C for 10 min. The sperm DNA was extracted by proteinase K and 1% SDS in Tris-HCl, pH 7, overnight, followed by phenol/chloroform extraction (Sambrook *et al.*, 1989). The sperm DNA was quantified at 260 nm and by spotting on agarose gel (Sambrook *et al.*, 1989). The plasmid DNA was detected by PCR using primers 5′-CTGATTAAGTTTTGTATAGT-3′(MT1) and 5′-GTGTTTCCTCGACTTGACTT-3′ (IGF2). A 311 bp fragment of the salmon melanin concentrating hormone was amplified using primers 5′-TGATTCAGTCAGAAGCGGGA-3′(MCH1) and 5′-GACAAGATGCAGCATAG-TAG-3′. This is used as an internal control. PCR was performed in 25 μl reaction volume containing 200 ng DNA, 12.5 mM each dNTP, 0.8 mM each of the primers, 1.8 mM $MgCl_2$, and 1 unit of *Taq* polymerase (Boehringer). Amplification was carried out for 30 cycles, each of 60 sec at 94°C, 60 sec at 56°C, 60 sec at 72°C. When the sperm were soaked in DNA solution, no PCR product was detected after ethidium bromide staining. PCR product was detected in the experiment in which the sperm was electroporated in DNA solution. When PCR was performed on the sperm DNA from sperm which had been electroporated in DNA and protamine the bands were more intense. These differences in band intensity were not due to unequal amount of sperm DNA being used for the PCR reaction because the MCH fragment was similar in staining intensity between the different samples. This would suggest that protamine enhanced sperm-DNA association. The higher concentration of PCR product in the protamine treated sperm group might suggest that protamine protected the plasmid DNA from degradation.

To test the effect of protamine on the retention of the plasmid DNA, 100 μg DNA was mixed with 100 μg of salmon protamine (Sigma), and the DNA-protamine complex was mixed with 0.5 ml of flesh milt and electroporated at 1000 v/cm with two pulses of 27.4 msec each. The milt was used to fertilize 500 ova. A control group was treated similarly, but without protamine. The embryos were allowed to develop and the fish were grown to 15 months before being sacrificed for DNA analysis. PCR analyses of DNA from fin clips, muscle, liver, heart and blood revealed a mosaic distribution of the transgene. Sixteen out of 20 fish from the protamine group were transgenic (80%), while 7 out of 13 fish from the control group were transgenic (54%) (Table 1).

To determine whether the transgene was integrated into the genomic DNA, genomic DNA was isolated from all the tissues by proteinase K/SDS digestion followed by phenol/chloroform extraction. The purified DNA was digested with *Kp*NI. The digested DNA was analysed by Southern blot hybridization using the linearized pMTIFG1 plasmid DNA as a probe. Integration of the transgene was found in the blood of one of the control fish. This was suggested by the sizes of the two fragments produced by *Kp*NI digestion. They were different in size from the expected size of DNA fragments (5831 and 138 bp) obtained by *Kp*N1 digestion of a linearized plasmid DNA. Further, there were more than two bands with varying intensity suggesting that some rearrangements of the transgene had taken place. The major bands detected were 11064, 5785, 4890, 4065, 2506, 2282,

Table 1. Summary showing transgene retention in embryos and 15-month-old fish

Age	DNA (mg/ml)	% Retention	Integration (int/N)	Reference
1–10 days	100	90	ND	Walker et al. (1995)
10 days	100	69	ND	Mukherjee (Unpub.obs.)
6 weeks	100	29	ND	Mukherjee (Unpub.obs.)
12 weeks	20	44–85	none	Symonds et al. (1994)
12 weeks	20	13–33	ND	Walker et al. (1995)
	100	7–15	ND	
	200	7–47	ND	
10 months	200	0	none	Walker (1997)
15 months	100	54	7.7% (1/13)	Present study
	100	80 (+p)	10% (2/20)	Present study

Field strength: 1000 V/cm; pulse length: 27.4 ms; no. of pulse: 2. p, protamine. ND, not determined.

1452, 1157 bp. The intensities of the bands also suggest that multiple copies of the transgene were integrated. In the protamine treated group two samples, blood and muscle, showed integration of the transgene. The bands were 6321 and 8179 bp long, respectively. The shorter band was not detectable, presumably due to the low concentration. The band intensity was similar to the intensity of those bands produced by the endogenous IGF DNA after one week of exposure of the X-ray film.

CONCLUSION

We have demonstrated that a foreign gene transferred by electroporated sperm may integrate into the host genome and the efficiency of integration is similar to those obtained by microinjection. A protamine-DNA complex can be transferred into salmon ova by electroporated sperm and the transgene integrated into the host genome. The efficiency is slightly better than that obtained without protamine.

REFERENCES

Page, R.L., Butler, S.P., Subramanian, A., Gwazdauskas, F.C., Johnson, J.L., and Velander, W.H., 1995, *Transgenic Research* **4**:353–360.

Symonds, J.E., Walker, S.P., Sin, F.Y.T., and Sin, I.L., 1994, *Molecular Marine Biology and Biotechnology* **3**:104–111.

Walker, S.P., Symonds, J.E., Sin, I.L. and Sin, F.Y.T., 1995, *Journal of Marine Biotechnology* **3**:232–234.

Walker, S.P., 1997, Ph.D. Thesis. University of Canterbury.

Weinhues, U., Hosokawa, K., Hoveler, K., Siegmann, B., and Doerfler, W., 1987, *DNA* **6**: 81–89.

7

A NOVEL ANTIOXIDANT DERIVED FROM SEAWEED

W. C. Dunlap,[1] K. Masaki,[2] Y. Yamamoto,[3] R. M. Larsen,[4] and I. Karube[3]

[1]Australian Institute of Marine Science
[2]Toyo Suisan Kaisha Ltd
[3]Research Center for Science and Technology
University of Tokyo
[4]Sir George Fisher Centre
James Cook University

Many organisms inhabiting shallow-water or inter-tidal marine environments are protected from long-term solar damage by producing natural, UV-absorbing sunscreens of the generic class known as mycosporine amino acids (MAAs) (Dunlap *et al.* 1986) These compounds are derived from the shikimic acid pathway (Bently 1990) via 3-dehydroquinic acid and 4-deoxygadusol (4-DG). Mycopsorines were first discovered in terrestrial fungi (Favre-Bonvin *et al.* 1976) and these metabolites are believed to provide UV protection to fungal spores while exposed to solar radiation during atmospheric dispersal (Young and Patterson 1982).

A collaborative study was conducted at RCAST to investigate the possible relationship between UV-absorbing and antioxidant functions of MAAs. We found that imino-mycosporines are oxidatively inert while oxo-carbonyl mycosporines (mycosporine-glycine and mycosporine-taurine) are moderately active. The MAA precursor 4-deoxygadusol, however, was found to be a strong antioxidant with activity comparable to that of ascorbate.

Many species of marine algae are rich in imino-MAAs. In a prior study (Dunlap *et al.*, in prep.) we found that the marine bacterium *Vibrio harveyi* efficiently converts algal MAAs, shinorine and porphyra-334, to mycosporine-glycine in the enteric process of MAA absorption leading to the trophic accumulation of asterina-330 by deposit-feeding holothuroids (Fig. 1) (Shick *et al.* 1992). Further investigation revealed that algal MAAs could be converted in moderate yield to the active 4-DG antioxidant [a metabolite found in the eggs of fish and sea urchins (Chioccara *et al.* 1980, 1986; Planck *et al.* 1981)] on fermentation with either an unidentified *Vibrio* sp. (RML017) or with several species of *Pseudoalteromonas*. Toyo Suisan is developing the commercial production of 4-DG using this "retro-biosynthetic" pathway for intended use as a natural antioxidant in food-processing and cosmetic applications (Masaki *et al.* 1996).

New Developments in Marine Biotechnology, edited by Le Gal and Halvorson.
Plenum Press, New York, 1998.

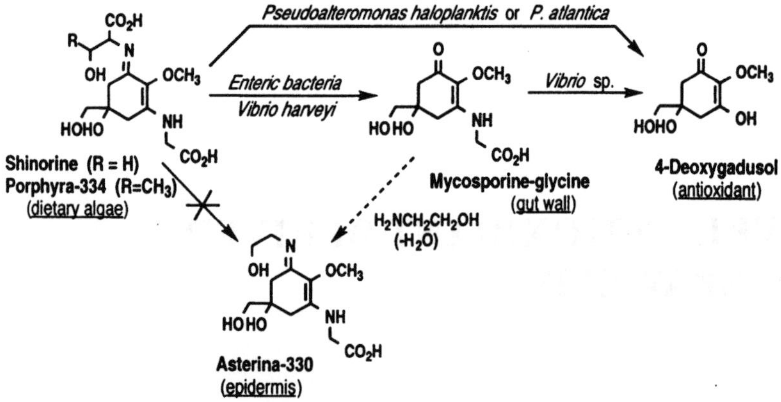

Figure 1. Proposed scheme for the bioconversion of algal MAAs (shinorine and porphyra-334) to asterina-330 in tropical holothuroids, and conversion to the strong antioxidant, 4-deoxygadusol.

Mycosporine-glycine is a predominant MAA component in coral symbiosis (Dunlap and Shick, in press). As well as functioning as a natural sunscreen (λ_{max} = 310 nm), this MAA also has moderate antioxidant activity and may provide some protection against photooxidative stress in the hyperoxic tissues of algal-invertebrate symbiosis (Kuhl *et al.* 1995). The presumed biochemical precursor of MAAs, 4-deoxygadusol (4-DG), is often

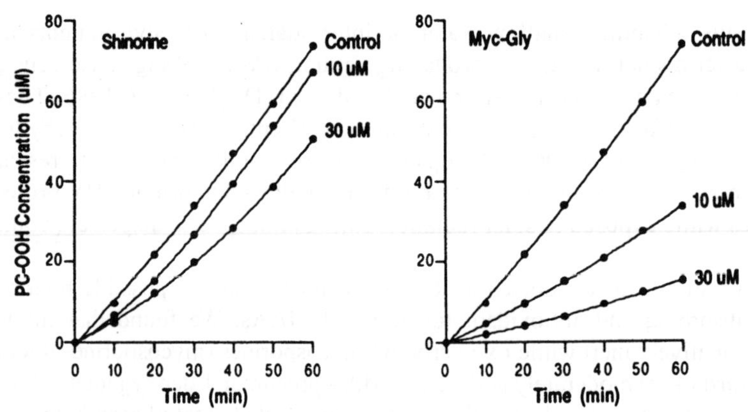

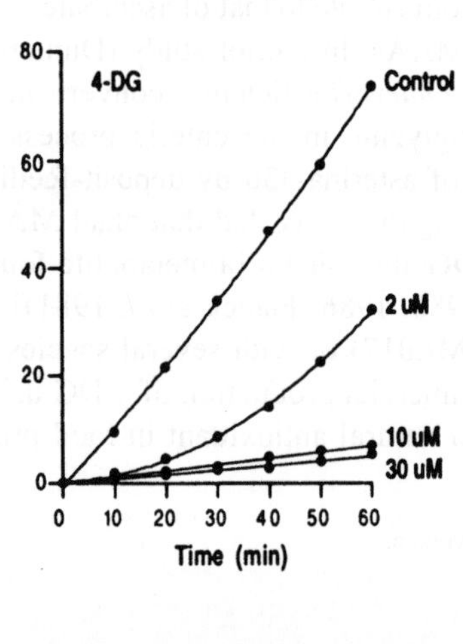

Figure 2. Phosphatidylcholine (PC) peroxidation inhibition assay for purified shinorine, mycosporine-glycine (Myc–Gly) and 4-deoxygadusol (4-DG). PC–OOH is the phosphatidylcholine hydroperoxide product of PC oxidation (data, in part, from Dunlap and Yamamoto 1995).

observed (Dunlap, pers. observation) in marine autotrophs and in autotrophic symbiosis together with MAAs, and has strong antioxidant properties. The antioxidant activities of mycosporine-glycine and 4-deoxygadusol are demonstrated by the phosphatidyl-choline (PC) peroxidation inhibition assay (Niki 1990) by which these antioxidant compounds retard the rate of chemically-initiated, free-radical hydroperoxidation of PC *in vitro* (Fig. 2). While the functional abilities, metabolic fate and possible cycling of these antioxidants in algal-invertebrate symbioses are yet unknown, the oxidative robustness of imino-mycopsorines is in keeping with their primary role as stable biological sunscreens in coral reef organisms (Dunlap and Yamamoto 1995).

REFERENCES

Bently, R., 1990, *Crit. Rev. Biochem.* **25**:307–384.

Chioccara, F, Della Gala, A., and De Rosa, M., 1980, *Bull. Soc. Chim. Belg.* **89**:1101–1106.

Chioccara, F, Zeuli, L., and Novellino, E.,1986, *Comp. Biochem. Physiol.* **80B**: 459–461.

Dunlap, W.C., Chalker, B.E., aand Oliver, J.K., 1986. *J. Exp. Mar. Biol. Ecol.*, **104**:239–248.

Dunlap, W.C., and Shick, J.M. (in press). *J. Phycol.* (invited review).

Dunlap, W.C., and Yamamoto, Y., 1995, *Comp. Biochem. Physiol.* **112B**:105–114.

Favre-Bonvin, J., Arpin, N., and Brevard, C., 1976, *Can J. Chem.* **54**:1105–1113.

Kuhl, M, Cohen, Y, Dalsgaard, T., Jorgensen, B.B., and Revsbech, N.P.,1995, *Mar. Ecol. Progr. Ser.* **117**:159–172.

Masaki, K., Dunlap, W.C., Yamamoto, Y., Karube, I., Larsen, R.M. and Matsukawa, R., 1996, A natural antioxidant and its production process. Toyo Suisan Kaisha Pty. Ltd. Japanese Patent Application No. 9604230, 23 September,

Niki, E., 1990, *Methods in Enzymology* **186**:100–108, Academic Press, New York.

Plank, P.A., Fraser, N.W.,, Grant, P.T., Middleton, C., Michell, A.I., and Thompson, R.H., 1981, *Biochem. J.* **199**:741–742.

Shick, J.M., Dunlap, W.C., Chalker, B.E., Banaszak, A.T. and Rosenzweig, T.K., 1992, *Mar. Ecol. Prog. Ser.*, **90**:139–148.

Young, H., and Patterson, V.J., 1982, *Phvtochem.* **21**:1075–1077.

8

UNUSUAL MARINE STEROLS MAY PROTECT CELLULAR MEMBRANES AGAINST ACTION OF SOME MARINE TOXINS

Tatiana N. Makarieva, Valentine A. Stonik, Ludmila P. Ponomarenko, and Dmitry L. Aminin

Pacific Institute of Bioorganic Chemistry
Russian Academy of Sciences
Vladivostok-22, Russia

In the course of our continued investigations on membranolytic toxins from marine organisms we have discovered that many toxin-producing animals contain unusual sterols instead of cholesterol in their free sterol fractions. We call as "biochemical coordination" the phenomenon, which involves the mutually conditioned presence of two different series of natural products in biological organisms.

We found several cases of "biochemical coordination" of the type "membranolytic toxins-unusual sterols". For instance, we studied toxins and free sterols from some holothurians (sea cucumbers), including *Pseudostichopus trachus*, *Holothuria* (*Microtele*) *nobilis, H. scabra*, the ascidian *Polycitor* sp., the sponges *Rhizochalina incrustata* and *Trachyopsis halichondroides*. We established that the holothurians contain triterpene glycosides (Stonik *et al.*, 1988), the ascidian contains varacin and its derivatives (Makarieva *et al.*, 1995), the sponges contain rhizochalin (Makarieva *et al.*, 1989) and sulfated polyhydroxylated steroids (Makarieva *et al.*, 1995), as toxins, respectively. All the animals have unusual sterols as dominant steroid constituents of their free sterol fractions instead of cholesterol.

In fact, 14α-methylsterols such as **5** were isolated as principal membrane constituents from holothurin B-containing sea cucumbers *H. nobilis* and *H. scabra*. The sea cucumber *P. trachus* contains other toxic triterpene glycosides, but cholesterol was a minor membrane sterol in this animal. Unusual sterol (**6**) was accumulated in sterol fraction of cellular membranes of *P. trachus* instead of cholesterol. The sea cucumber *B. natans* contains the predominant sterol as 24-ethylcholestanol.

Stanols were shown to be main sterols in the Far-Eastern ascidian *Polycitor* sp. which contains a polysulfide toxin varacin and its derivatives (Makarieva *et al.*, 1995).

New Developments in Marine Biotechnology, edited by Le Gal and Halvorson.
Plenum Press, New York, 1998.

Some marine toxins isolated

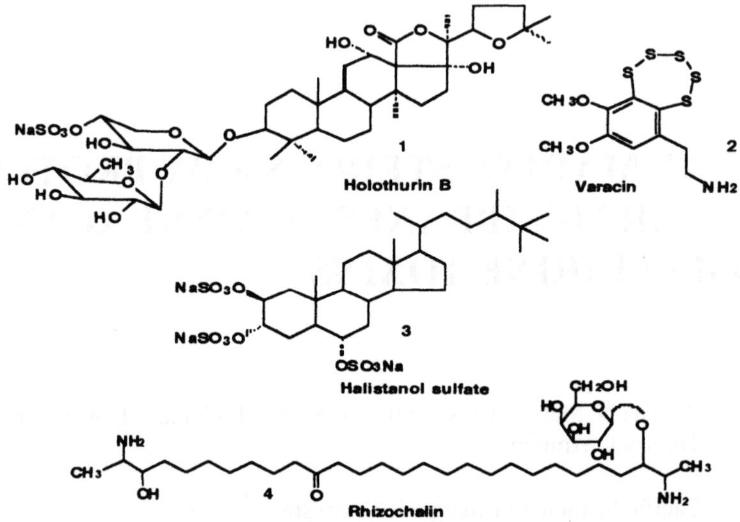

Trisulfated steroid toxin, halistanol sulfate (**3**), and some related toxins were isolated from the sponge *T. halichondroides*. A main sterol of this species was identified as 24-isopropylcholestadienol (**7**). In addition to sterol **7** other highly alkylated in the side chain sterols were found in this sponge (Shubina *et al.*, 1985).

Rhizochalin (**4**) was isolated as a new toxic secondary metabolite of mixed biogenesis from the sponge *R. incrustata*. Studies on sterol composition revealed that 24,25-methylenecholestanol (**8**) is a main sterol in this animal (Makarieva *et al.*, 1996). Sterols containing a cyclopropane ring in their side-chain have so far only been found in marine organisms. Sterol (**8**) belongs to this rare group of natural steroids.

We suggest that unusual sterols may protect cellular membranes of some toxic marine animals against action of their own toxins. To examine this suggestion, we compared effects of toxins (**3**) and (**4**) on cholesterol-containing cells with those on cells with unusual sterols (**7**) and (**8**), introduced into cellular membranes. Some obtained data are given in Figure 1.

Cytotoxic and hemolytic effects of marine toxins (**3**) and (**4**) on mouse Ehrlich carcinoma cells and erythrocytes were observed at the same ED_{50} of about 12.5 μg/ml for the both toxins. The preincubation of cell suspensions with cyclopropane-containing sterol (**8**) at concentration 10 μg/ml during 15 min decreased the cytotoxic and hemolytic

Some unusual sterols identified

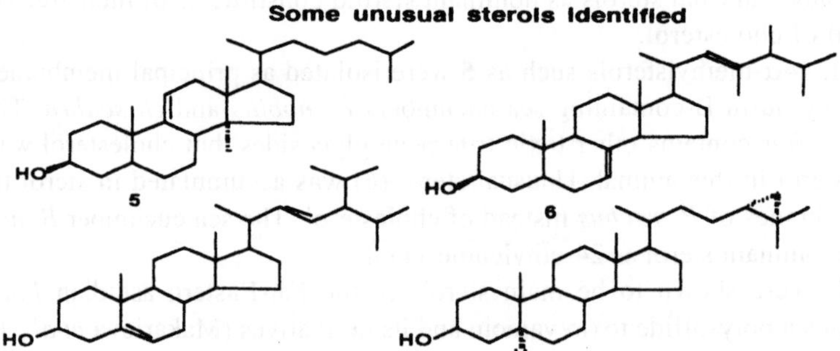

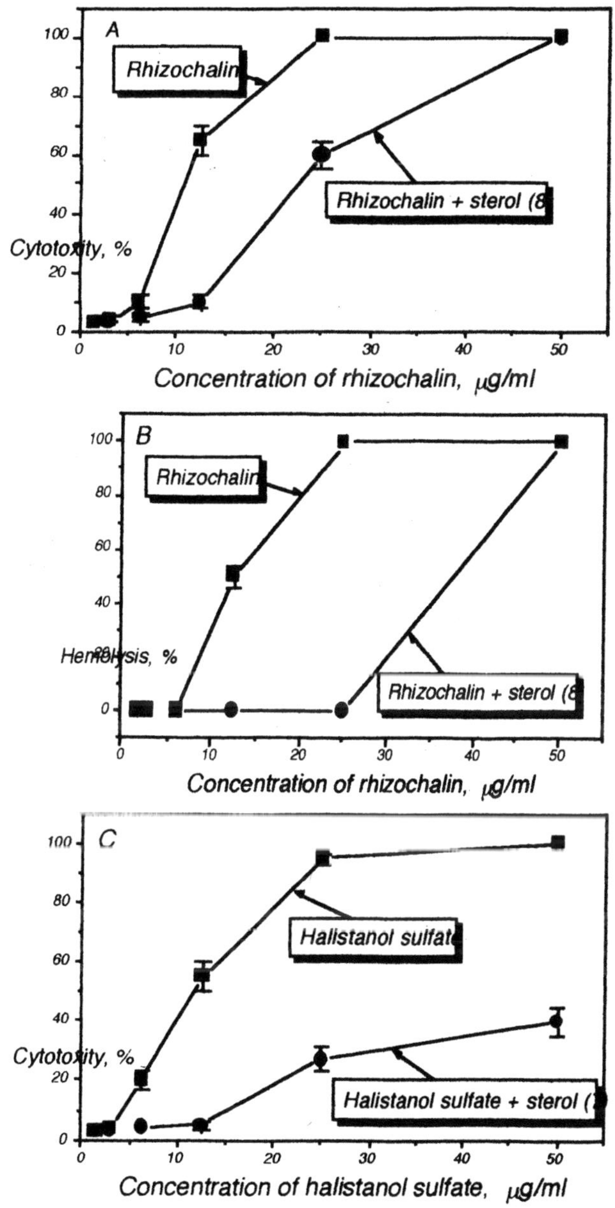

Figure 1. Protective action of unusual sterols against some marine toxins. **A:** Action of (**4**) on Ehrlich intact carcinoma cells and on those preincubated with sterol (**8**); **B:** Action of (**4**) on intact erythrocytes and on those, preincubated with sterol (**8**); **C:** Action of (**3**) on intact Ehrlich carcinoma cells and on those, containing sterol (**7**).

effects of rhizochalin more than two times (see Figure 1, A and B). The treatment of the carcinoma cells with lecithin liposomes followed by the treatment with the same liposomes, containing **7**, led to 43% change of cholesterol for sterol **7** (in accordance with the data of GLC analyses). Such alteration in sterol composition of tumor cells resulted in decrease of the cytotoxity of halistanol sulfate (**3**) against these cells more than five times (see Figure 1, C).

Thus, the data obtained showed that unusual sterols from marine invertebrates can protect cellular membranes against action of marine toxins, isolated from the same organisms. Earlier, a protective biological function of sterols was unknown.

REFERENCES

Makarieva, T.N., Denisenko, V.A., Stonik, V.A., Milgrom, Y.M., Rashkes, Y.W., 1989, *Tetrahedron Letters*, **30** (47): 6581–6584.

Makarieva, T.N., Ponomarenko, L.P., Stonik, V.A., 1995, *Khim. Prir. Soedin.*, **95**: 758–759.

Makarieva, T.N., Shubina, L.K., Kalinovsky, A.I., Stonik, V.A., Elyakov, G.B., 1983, *Steroids*, **42** (3): 267–281.

Makarieva, T.N., Stonik, V.A., Dmitrenok, A.S., Grebnev, B.B., Isakov, V.V., Rebachyk, N.M., 1995, *J. Nat. Prod.*, **58**: 254–258.

Makarieva, T.N., Stonik, V.A., Ponomarenko, L.P., Kalinovsky, A.I., 1996, *Chem. Research (S)*, **10**: 468–469.

Shubina, L.K., Makarieva, T.N., Kalinovsky, A.I., Stonik, V.A., 1985, *Khim. Prir. Soedin.*, **85**: 232–239.

Stonik, V.A., Elyakov, G.B., 1988, in Marine Biorganic Chemistry, Ed. P.J. Scheuer, Springer-Verlag, Berlin, Heidelberg, p. 43.

9

SECONDARY METABOLITES OF MARINE ORGANISMS

K. Mukesh, Miryam Z. Sahni, Valadmir Belenky Wahrman, and Gurdial M. Sharma*

Departments of Chemistry and Biology
William Paterson University
Wayne, New Jersey 07470

This paper describes several novel compounds which have been isolated from the blood plasma and tissue extracts of marine animals. Structures of these compounds are also presented.

In connection with another project we were monitoring temporal changes in the quaternary structure of the corrinoid binding protein present in the amebocyte free hemolymph (blood plasma) of the horseshoe crab, *Limulus polyphemous* by HPLC using a precaliberated gel filtration column (Sharma & Sahni, 1993). Blood samples used in the experiments were collected from horseshoe crabs, captured twice a week from New Jersey beaches during the breeding season (summer months, May–August). Also included in the analysis were blood samples that were taken from Limuli obtained from local fishermen once a month during the non-breeding winter months (September–April).

Qualitatively, HPLC profiles of all blood plasma samples were similar. The chromatograms invariably showed two protein peaks in the 400–200 Kd molecular weight region and several peaks in the region where salts, mono- and oligosaccharides, amino acids, nucleosides, modified nucleosides and other low molecular weight compounds (Mr < 1000) are expected to elute. Quantitatively, the chromatograms revealed that concentrations of low molecular weight compounds in the Limulus hemolymph changed markedly from week to week during the breeding season; but not in the non-breeding winter months. This observation (confirmed over a period of several years) suggested that some low molecular weight substances present in Limulus hemolymph during breeding season may be playing an important role in the reproductive biology of the horseshoe crab. We, therefore, decided to fractionate the chemicals present in blood plasma of the horseshoe crab with the hope of discovering new biologically active substances some of which might promote or inhibit growth.

Hemocyanin and other high molecular weight proteins present in the blood plasma of the horseshoe crab were removed by ultafiltration of the fluid through 4 packets of

New Developments in Marine Biotechnology, edited by Le Gal and Halvorson.
Plenum Press, New York, 1998.

PTGC OMT 05 polysulphone filters assembled in a Minitan Ultrafiltration System (Millipore Corporation). The filtrate (blood serum) from the ultrafiltration experiment was found to contain almost all of the small molecular weight compounds that were found to be present in the original blood plasma samples. Chromatography of the blood serum over a Sephadex G-25 column, using water as the eluent, resolved the small molecules into four fractions designated as E, F, G and H (Fig. 1). Bioassays revealed that fraction F contains compounds that destroy malignant melanoma cells, human colon carcinoma cells, stop the growth of marine phytoplankton cells and deforms them (Figs. 2 and 4). In addition fraction F also showed strong spermicidal activity (Fig. 3). None of the other three fractions showed these biological characteristics.

Encouraged by the discovery of anticancer, spermicidal and antiphytoplankton agents in Limulus blood plasma, we decided to search for similar types of chemicals in other marine organisms. Small molecular weight compounds present in the body fluids and/or tissue extracts of ten different marine animals (Table 1) were resolved into chromatographic fractions. Each fraction was tested for anticancer, spermicidal and antiphytoplankton activities. The data revealed that whereas substances with spermicidal and cytotoxic activities were present in four of the 11 animals examined, the compounds that inhibit the growth of marine phytoplankton were present in the tissue extracts and/or body fluids of all marine organisms used in the investigation. Another significant observation that we made was that although antiphytoplankton compounds occur at an

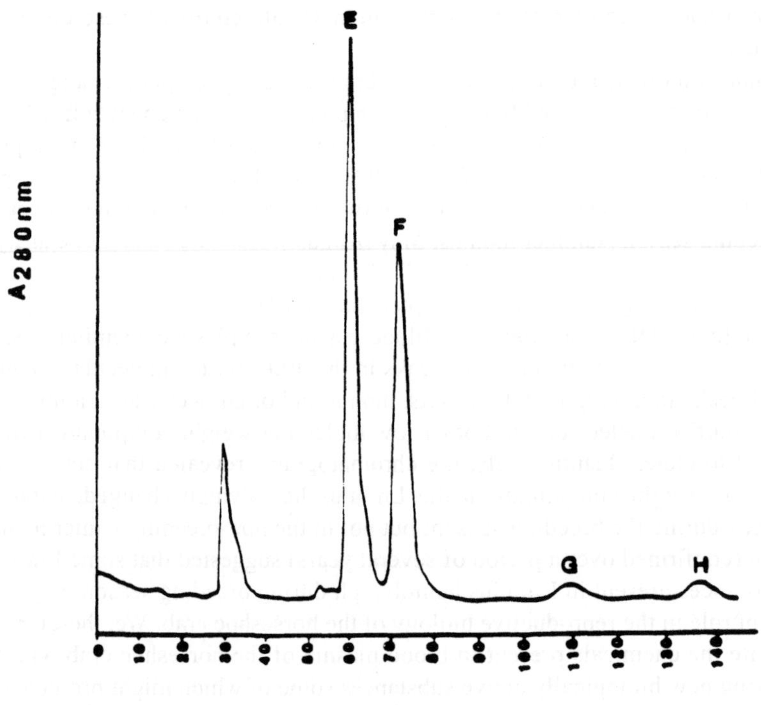

Figure 1. Sephadex G-25 chromatography of *Limulus* blood serum samples. Column: 3 × 90 cm; sample size: 10–30 mL; eluent: distilled water; flow rate: 40 mL/hr; fraction size: 90 drops (5 mL)/tube; total fractions: 140; detection: 280 nm; temp.: 4°C.

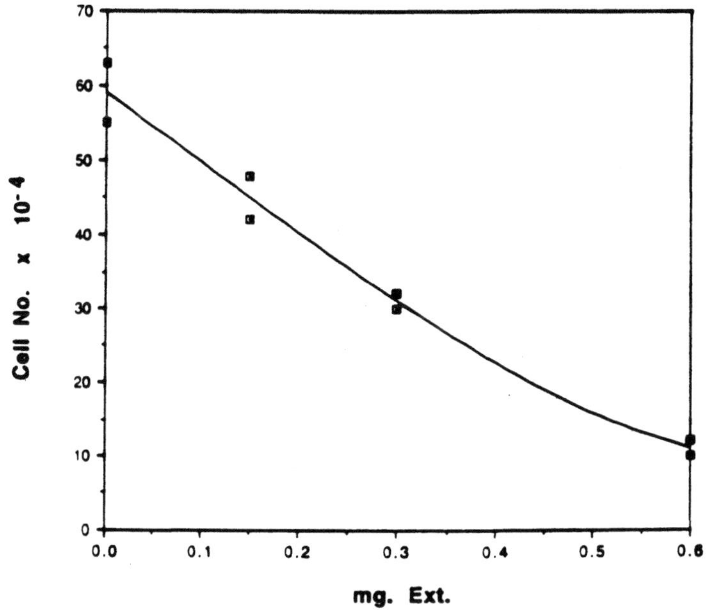

Figure 2. Effect of fraction F on human colon carcinoma cells. Fraction F (5 mg) was dissolved in 2.0 mL cell growth media, and aliquots were added to HT-29, human colon carcinoma cells.

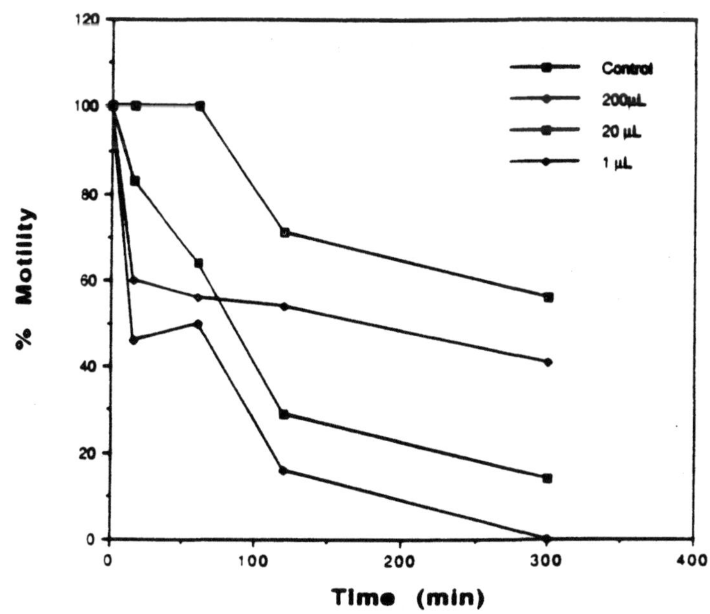

Figure 3. Effect of fraction F on mouse sperm motility. Fraction F (6 mg) was dissolved in 2.0 mL sperm growth media. Known aliquots (1, 20 and 200 μL) of this solution were used for testing spermicidal activity.

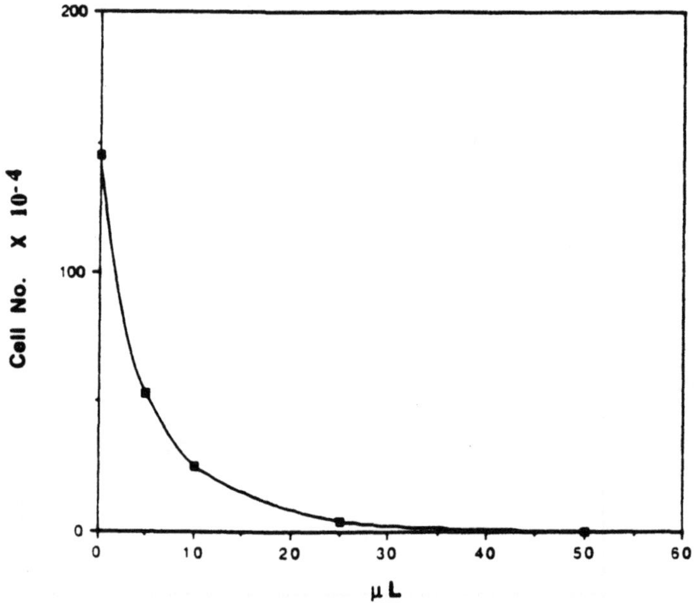

Figure 4. Effect of fraction F on the phytoplankton growth. Fraction F (5 mg) was dissolved in 2.0 mL growth media. Aliquots (5–50 µL) of this solution were used for testing antiphytoplankton activity.

extremely low concentration in marine animals, biologically they were extremely potent. As little as 0.1–0.2 mg material from appropriate crude chromatographic fraction was found to completely inhibit the growth of the marine phytoplankton *T. pseudonana*. It is possible that marine animals excrete these highly potent antiphtoplankton agents into the surrounding medium where they may keep primary productivity in proper ecological balance. Because of their high potency, the antiphytoplankton compounds may even be inhibiting the blooms of dinoflagellates which occasionally cause red tides in the marine environment.

Table 1. Biological activities of the body fluids and/or extracts of selected marine organisms

Marine organism	Biological specimen	Cytotoxic activity	Spermicidal activity	Antialgal activity
Horseshoe crab	Blood plasma	+++	+++	+++
	LAL	++	–	+++
Blue crab	Blood plasma	–	–	++
Rock crab	Blood plasma	–	–	++
Green crab	Blood plasma	–	–	+++
Barnacles	Animal extract	–	–	+++
Sea urchin	Animal extract	–	–	+++
Strombus gigas	Animal extract	+++	–	+++
Aplysia dactylomela	Animal extract	–	+++	–
Gorgonia ventalina	Animal extract	+++	+++	+++
Pseudoplexaura porosa	Animal extract	+++	+++	+++
Pseudopterogorgia americana	Animal extract	+++	+++	?

Abbreviations: +++, Very active (100% inhibition at a concentration of 300 µg/mL); ++, Active (50% inhibition at a concentration of 300 µg/mL); +, Moderately active (25% inhibition at a concentration of 300 µg/mL); –, No activity; ?, To be tested.

Having speculated on the significance of biological activities shown by the chromatographic fractions of the extracts and/or body fluids of marine organisms listed in Table 1, The next step was to identify the molecules responsible for these activities. We have isolated and identified secondary metabolites present in the Sephadex G-25 fractions E and F of Limulus blood serum (Fig. 1) and 50% methanol-water extracts of the queen conch, *Strombus gigas* and the sea fan, *Gorgonia ventalina*. The results of our investigations are summarized below.

Reverse phase chromatography using C-18 column revealed that the biologically inactive Sephadex fraction E of Limulus blood serum contains two compounds. These compounds were shown to be homarine **1** and the disaccharide **2** (Fig. 5). The biologically active Limulus fraction F was found to contain as much as 80% salt and over two dozen organic compounds at a very low concentration (analysis by HPLC using C-18 column). The aqueous solution of fraction F was fractionated by chromatography over an activated charcoal column. All organic matter was retained by charcoal; but salts washed out when the column was eluted with water. After the removal of salt, the column was eluted with methanol-ammonium hydroxide mixture (99:1 v/v). This solvent released all organic compounds from the charcoal column except the ones with biological activity. Charcoal was found to absorb the biologically active compounds irreversibly, as no solvent system was found to release them from the column, The mixture of organic compounds that came out of the charcoal column was found to contain uracil, uridine, the homarine derivatives **3** and **4** (Fig. 5), and many other compounds whose identity has not yet been established.

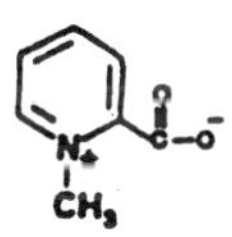

1

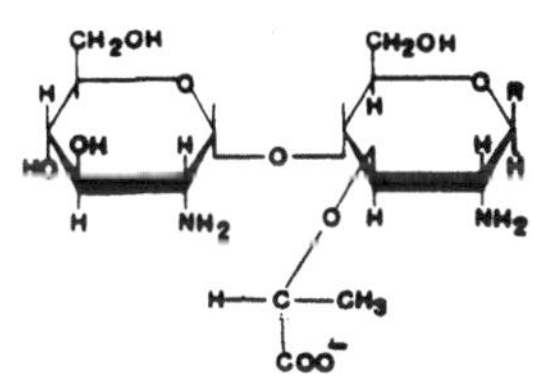

2

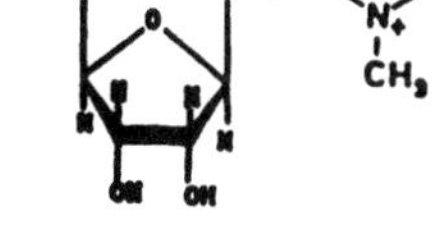

3

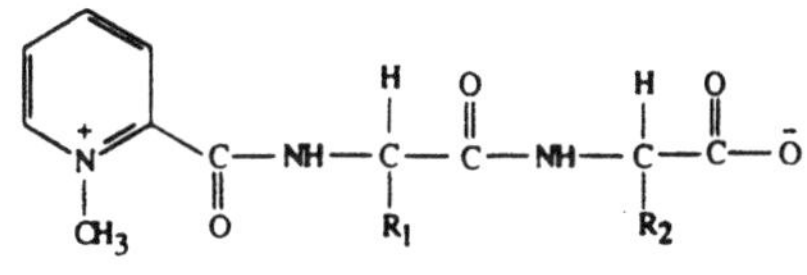

4

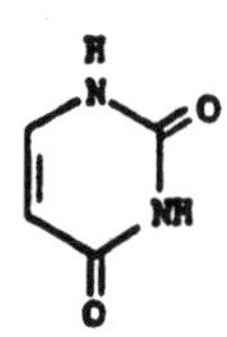

5

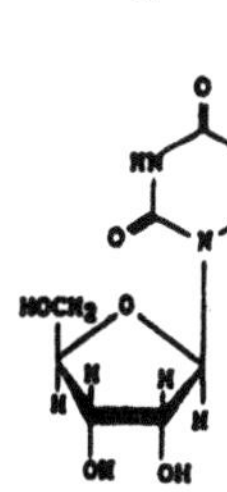

6

Figure 5. Compounds isolated from *Limulus* blood serum.

So far the anticancer, spermicidal, and antiphytoplankton agents of Limulus blood serum have defied all our other attempts at their isolation and identification.

For the isolation of biologically active secondary metabolites from the queen conch, *Strombus gigas*, the 50% methanol-water extract of the animal was partitioned between methylene chloride and water. The methylene chloride soluble organic material showed anticancer and spermicidal activities where as the antiphytoplankton agents were present in the aqueous layer.

The methylene chloride soluble material was partitioned between 90% methanol/water mixture and hexane. The 90% methanol layer upon rotary evaporation, gave a residue having pronounced anticancer and spermicidal activities. The hexane soluble material showed relatively weak anticancer and spermicidal properties.

The 90% methanol soluble material was resolved into three fractions by silica gel column chromatography. Solvents used for eluting the compounds from the column were

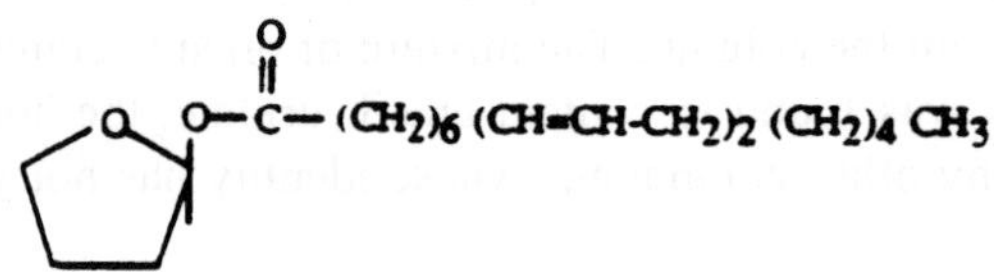

1

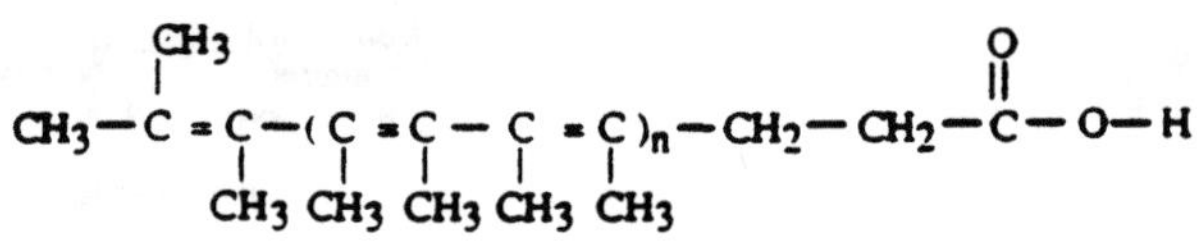

2

$$CH_3(CH_2)_4(CH{=}CH\,CH_2)_2(CH_2)_6{-}C(=O){-}O{-}C_{16}H_{33}$$

3

$$CH_3(CH_2)_4(CH{=}CH\,CH_2)_2(CH_2)_6COOH$$

4

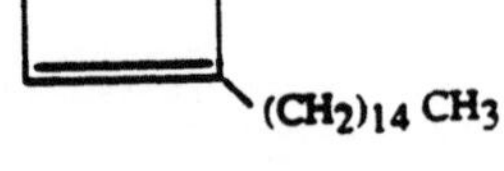

5

Figure 6. Compounds isolated from *Strombus gigas*.

benzene (fraction 1), methylene chloride (fraction 2) and methylene chloride:methanol mixture (fraction 3). Fraction 3, a liquid, showed anticancer properties. GC/MS analysis revealed that fraction 3 is a complex mixture in which fatty acids predominate. By using TLC and reverse phase HPLC four compounds were isolated from the mixture. On the basis of their high resolution mass spectral and NMR data these compounds were assigned structures shown in (Fig. 6).

Surprisingly, the four anticancer compounds isolated from the methanol extract of *S. gigas* were also present in the extracts of the sea fan, *Gorgonia ventalina*. It appears these compounds are synthesized by the algae upon which the animal feed. In fact, most of the secondary metabolites present in the extracts of *S. gigas* and *G. ventalina* seem to be of dietary origin rather than being synthesized by the animals themselves.

ACKNOWLEDGMENTS

This research is a result of work funded by the NOAA office of Sea Grant, under grant # NA 36-RG0505. New Jersey Sea Grant # NJSG-97366. Additional funds were provided by the Research Corporation, Tuscon Arizona and Center for research of William Paterson University.

REFERENCES

Sharma, G. M. and Sahni, M. K, 1993, Marine Proteins in Clinical Chemistry in Marine Biotechnology, Volume 1: Pharmaceutical and Bioactive Natural Products, edited by David H. Attaway and Oskar R. Zaborsky. Plenum Press, New York, pp. 153–180.

10

BIOSYNTHETIC STUDIES ON THE SALINAMIDES, DEPSIPEPTIDES FROM A MARINE *STREPTOMYCES*

Bradley S. Moore

Department of Chemistry
Box 351700
University of Washington
Seattle, Washington 98195

In recent years, the field of natural products research has erupted with the emergence of combinatorial biosynthesis as a means to generate large numbers of molecules with great structural diversity (Hutchinson, 1997). These studies have provided diverse chemical structures which are required for high throughput bioassays in the discovery of new lead drug candidates. Much of the work in combinatorial biochemistry has centered on the Actinomycetes, particularly the common soil genus *Streptomyces*, as these microorganisms are a proven source of structurally diverse secondary metabolites possessing broad ranges of biological activities. As the frequency of novel bioactive compounds identified from terrestrial actinomycetes decreases with time, however, academic and industrial programs are increasingly screening actinomycetes from varied environments for their ability to generate novel metabolites. These studies have not only supplied new potential lead compounds, but have also revealed new biochemical pathways that can be tapped in a combinatorial fashion to produce new molecular structures.

We have recently initiated a pilot study to engineer biochemical diversity in marine streptomycete natural products, as these marine strains present a vast untapped resource of novel metabolites (Jensen et Fenical, 1996). In this paper we present our current progress on the biosynthesis of the salinamides for future combinatorial manipulations to engineer new analogs.

Salinamides A (**1**) and B (**2**) are novel, bicyclic depsipeptides produced by *Streptomyces* sp. CNB-091 that are related to each other as the respective epoxide and chlorohydrin (Trischman et al., 1994). The marine streptomycete, isolated from the surface of the jellyfish *Cassiopeia xamachana* in the Florida Keys, requires saltwater for optimal growth and salinamide production. Significantly, **1** contains structural units that are either unknown or of limited distribution in other natural products.

New Developments in Marine Biotechnology, edited by Le Gal and Halvorson.
Plenum Press, New York, 1998.

The salinamides exhibit potent topical anti-inflammatory activities as shown by the phorbol ester-induced mouse ear edema assay in which they inhibit edema by nearly 90% at the standard testing dose of 50 μg/ear (Trischman et al., 1994). In addition, these compounds exhibit moderate antibiotic activity against Gram-positive bacteria. Computer modeling indicates that the molecules contain a cleft between the disubstituted benzene ring and the adjacent glycine loop that is large enough to accommodate a two amino acid residue,[3] raising the possibility that their selectivity against Gram-positive bacteria could be related to the well characterized mode of action of the vancomycin group of glycopeptide antibiotics.

The basic building blocks of **1** are reasonably evident from the structure and from general biochemical knowledge. The salinamide A core peptide ring system is likely derived from the hexapeptide Thr–DIle–Hpg–MePhe–DaThr–Ser, which is synthesized by a nonribosomal peptide synthetase (Figure 1). The proposed pathway proceeds through the desmethyl analogs of the naturally occurring minor salinamides E and C (Trischman et al., in preparation). The seven-carbon fragment connecting *p*-hydroxyphenylglycine (hpg) to glycine in **1** is expected to be biosynthetically derived from (2*E*,4*E*)-4-methylhexa-2,4-dieneoyl coenzyme A (CoA) (**3**), as evident in the structure of salinamide C.

Feeding experiments with stable isotope-labeled precursors were conducted to establish the biosynthetic building blocks of the proposed **3**-derived moiety in **1**. Two hypothetical biosynthetic pathways are conceivable. If polyketide in origin, **3** should be derived from one propionate and two acetate units. Alternatively, **3** may originate via a pathway involving isoleucine catabolism to 2-methylbutyryl-CoA, followed by dehydrogenation to (*E*)-2-methyl-2-butenoyl-CoA (**4**), a common intermediate of both pathways.

Results from numerous feeding experiments support the isoleucine catabolic pathway and are summarized below in Figure 2. C4-C5, but not C8-C42 of **1**, was derived intact from [1,2-$^{13}C_2$]acetate, while C50, but not C6, was enriched by [1-^{13}C]propionate, thus eliminating the polyketide scenario. The single incorporation of the methyl carbon of acetate at C6 further implied that the remaining 5-carbon fragment (C6-C8, C40, C42) was derived from 2-methylbutyrate, as the reference isoleucine moiety in **1** was also enriched from acetate to the same extent. This conclusion was corroborated by the high specific incorporation (52%) of 2-methyl[1-^{13}C]butyrate at C6. The propionate-derived moiety in **1** was also enriched at C50 in this feeding experiment via the β-oxidation product [1-^{13}C]propionate.

The unsaturated [1-^{13}C]**4** as the sodium salt likewise enriched C6, but at a much lower relative incorporation (13%). We therefore synthesized the corresponding *N*-acetylcysteamine (NAC) thioester, as these CoA mimics have been shown to be more highly incorporated than the free acids due to their presumed more facile transesterification with a thiol group on an enzyme. Incorporation in this case was considerably higher at 23%. Feeding experiments with [2,3-$^{13}C_2$]**3**-NAC thioester and the ^{13}C-^{15}N dilabeled glycyl ester of **3** are currently underway.

Re-examination of the absolute configuration of **1** indicated that the published structure is the enantiomer. The absolute stereochemistry had originally been determined by chiral GC analysis of the hydrolytic threonine units providing D-threonine and L-*allo*-threonine, since single crystal X-ray diffraction analysis of **2** established only the relative stereochemistry (Trischman et al., 1994). We investigated the absolute configurations of the salinamide hydrolytic amino acids derivatized as their *N*-dansylates by capillary electrophoresis using β-cyclodextrin in the running buffer as the chiral recognition agent (Trischman et al., in preparation). This analysis conclusively determined that the published structure was the enantiomer; the two threonine residues we identified were L-threonine and D-*allo*-

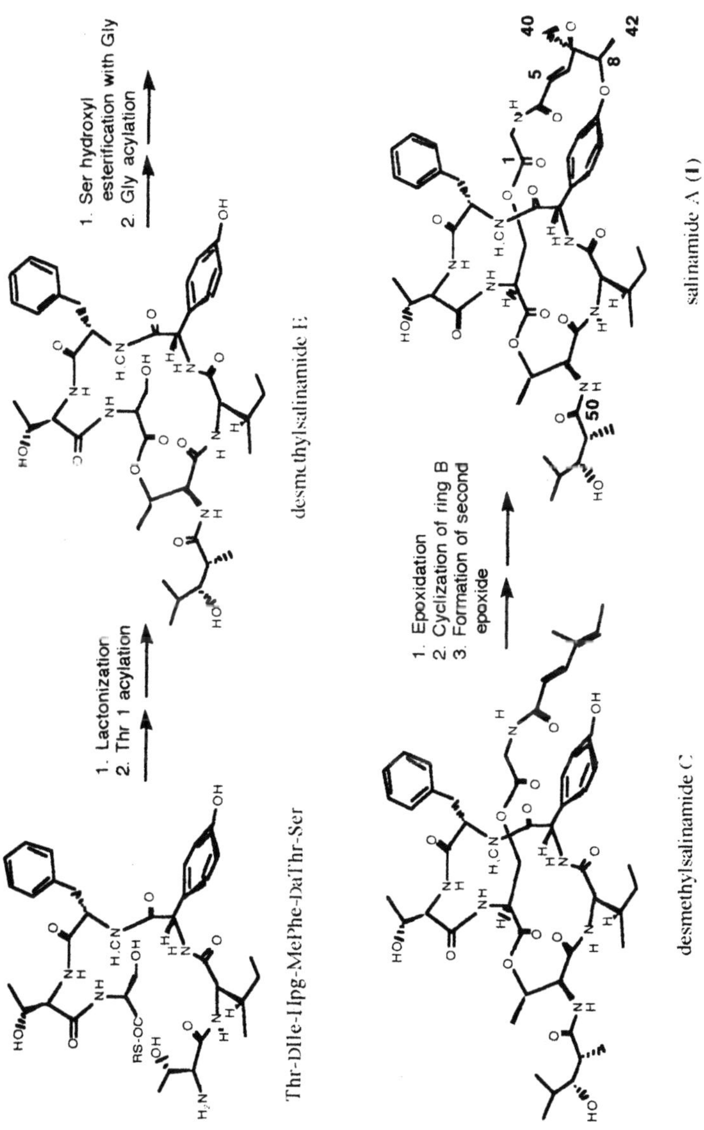

Figure 1. Proposed biosynthesis of salinamide A (**1**).

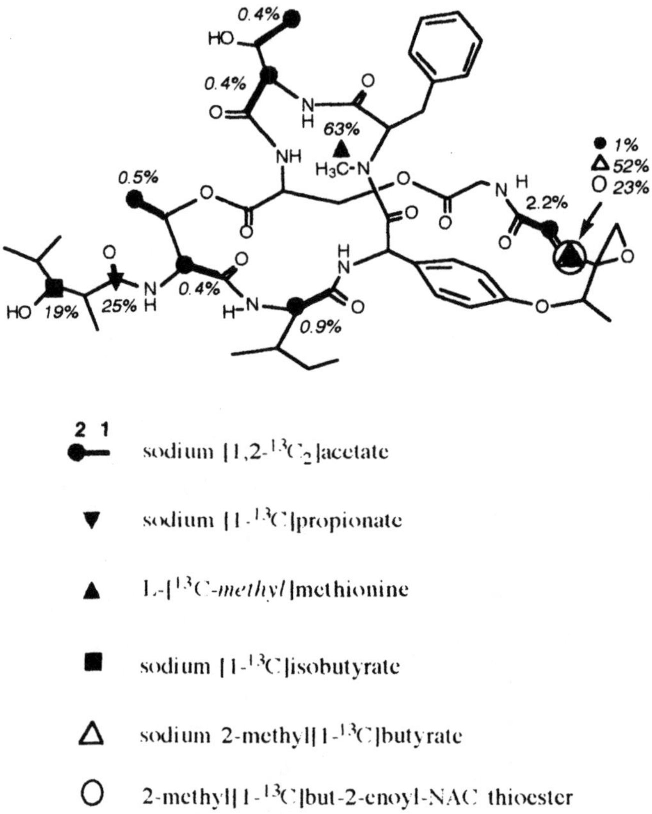

Figure 2. Origin of the carbons in **1**.

threonine. This was corroborated by the identification of L-serine, which is D configured in the original structure.

The majority of peptide antibiotics are synthesized by large, often multisubunit, enzyme complexes called nonribosomal peptide synthetases (Kleinkauf & Von Döhren, 1996). These multifunctional enzymes employ a thio-template mechanism for peptide synthesis. The order of the aligned domains determines the sequence of the amino acids in the product. Each domain is composed of several defined modules responsible for distinct functions of the enzymes. The mode of synthesis of the salinamides is suggestive of a nonribosomal pathway because the hexapeptide core in **1** contains two D-amino acids, an *N*-methyl amino acid, and the nonproteinogenic hpg. This hypothesis was supported by the fact that by supplementing the growth media with L-valine, salinamide D, in which a D-valine moiety replaces the D-isoleucine residue in **1**, and its chlorohydrin analog were the major salinamides produced (Trischman et al., in preparation). We elucidated their structures by 2D-NMR (HMQC and HMBC), MS and CE analyses and are currently evaluating their anti-inflammatory activity.

A PCR approach (Turgay & Marahiel, 1994) with degenerate oligodeoxynucleotide primers derived from highly conserved amino acid sequences in peptide synthetase domains was used to amplify peptide synthetase fragments from *Streptomyces* sp. CNB-091 genomic DNA. Partial sequencing of these fragments showed that one contained the conserved motifs of peptide synthetases. This PCR product encoding part of a peptide synthetase will be used to screen a cosmid library of *Streptomyces* sp. CNB-091 in order to clone the salinamide biosynthetic gene cluster. The characterization of the biosynthetic gene cluster will allow for future genetic manipulations.

ABSTRACT

Salinamides are novel, bicyclic depsipeptides produced by *Streptomyces* sp. CNB-091 that exhibit potent topical anti-inflammatory activities. As biosynthetic knowledge of bioacive metabolites from marine bacteria is sparse, we have focused on salinamide biosynthesis at the biochemical and genetic levels to serve as a model for future work with marine bacterial secondary metabolic pathways. Specifically, we have (1) demonstrated that the published structures for salinamides A and B are the enantiomers of the correct structures based on chiral capillary electrophoresis analysis of the hydrolyzed amino acids, (2) determined the basic biosynthetic building blocks for the two novel non-amino acid moieties through feeding experiments with ^{13}C-labeled precursors, (3) generated two structural analogs through precursor-directed biosynthesis that will later be tested for their anti-inflammatory activities, and (4) cloned and sequenced a non-ribosomal peptide synthetase gene from the salinamide producing organism by PCR that we will use to probe our cosmid library for the salinamide biosynthetic gene cluster.

ACKNOWLEDGMENTS

I am indepted to my coworkers Dr. Dieter Seng, Mr. David Kho, Mr. Michael Dahl and Mr. Ronald Pangilinan for their contributions and to Prof. William Fenical and Mr. Paul Jensen, Scripps Institution of Oceanography, for providing a culture of *Streptomyces* sp. CNB-091 and reference compounds. This work was funded in part by a grant from the Washington Sea Grant Program, University of Washington, pursuant to National Oceanographic and Atmospheric Administration Award No. NA36RG0071, Project No. R/B-20. D.S. is a recipient of a Deutsche Forschungsgemeinshaft postdoctoral fellowship.

REFERENCES

Hutchinson, C. R. In Biotechnology of Antibiotics: Second Edition, Revised and Expanded; W. R. Strohl, Ed.; Marcel Dekker, Inc: New York, 1997; pp 683–702.

Jensen, P. R., Fenical, W., 1996, *J. Indust. Microbiol.*, **17**: 346–351.

Kleinkauf, H.; Von Döhren, H, 1996, *Eur. J. Biochem.*, **236**:335–351.

Trischman, J. A.; Moore, B. S.; Seng, D.; Jensen, P. R.; Fenical, W., manuscript in preparation.

Trischman, J. A.; Tapiolas, D. M.; Jensen, P. R.; Fenical, W.; McKee, T. C.; Ireland, C. M.; Stout, T. J.; Clardy, J., 1994, *J. Am. Chem. Soc.*, **116**:757–758.

Turgay, K.; Marahiel, M. A., 1994, *Pept. Res.*, **7**:238–241.

11

DEREPLICATION AND PROFILING OF MARINE BACTERIA BY FATTY ACID ANALYSIS OF CRUDE EXTRACTS USING FOURIER TRANSFORM MASS SPECTROMETRY

David J. Bourne, Eliane Abou-Mansour, Russell T. Hill, and Peter T. Murphy

Australian Institute of Marine Science
PMB 3, Townsville
Queensland 4810, Australia

1. INTRODUCTION

Methods for grouping of bacteria include analysis of fatty acids by pyrolysis mass spcctromctry of frcc fatty acids (Sanglicr ct al., 1992), their methyl esters (Basile et al., 1995), phospholipids and glycolipids by fast atom bombardment mass spectrometry (Heller et al., 1987), measurement of fatty acid content by methylation and gas chromatographic analysis of methyl esters (Bertone et al., 1996), GCMS (Dzierzewicz et al., 1996), and mass spectral fingerprints generated by matrix assisted laser desorption of intact bacterial cells (Claydon et al., 1995). Methods developed for grouping terrestrial bacteria can be transferred to marine species (Basile et al., 1995). There has been considerable recent interest in methods for phylogenetic grouping and identification of marine bacteria, where traditional biochemical tests are not effective in grouping isolates.

For this study 35 bacterial isolates were analysed for free fatty acids (FFAs).

2. METHODS

A cubic centimetre of sponge with the trivial name of Very White Fan (undescribed species) was macerated in sterile seawater (10 mls), streaked onto marine agar plates and incubated at 26°C for seven days. Colonies were isolated and regrown in pure culture, plates scraped to harvest cells, cells suspended in sterile seawater and centrifuged. The pellet (0.1–0.2 gm) was resuspended, washed twice with distilled water pelleted by centrifugation. The pellet was extracted with methanol (5 mls) and sonicated for 30 minutes. This methanol extract was directly analysed by negative ion electrospray.

New Developments in Marine Biotechnology, edited by Le Gal and Halvorson.
Plenum Press, New York, 1998.

The mass spectrometer used in this study was a Bruker 47e BioApex equipped with an Analytica ion source. The crude extracts were infused at a rate of 100 μl/hr. Sixteen scans were acquired over the mass range of 147 to 4000 amu at a rate of 1.2 secs per scan. Resolution varied from about 40,000 at 150 amu to 8000 at 1000 amu.

The ratio of fatty acids was determined along with the total fatty acid content with reference to other components.

3. RESULTS AND DISCUSSION

Figure 1 shows the spectra of the crude extracts from four of the isolates.

Fourier transform ion cyclotron resonance mass spectrometry (FT-ICR MS) is a instrumental technique which offers ultrahigh resolution, mass accuracy and high sensitivity. These attributes are clearly demonstrated in Figure 1 where the ions generated have a resolution greater than 30,000. Resolution is a measure of ability to discriminate between neighbouring peaks and is equal to peak width at half peak height in atomic mass units (amu) divided by peak mass (amu). An instrument that operates at high resolution in its normal mode can distinguish between ions at the same nominal mass. Results are not shown here but an expansion of the spectrum of microbe #25 between m/z 266 and m/z 270 shows clearly that there are two ions of almost equivalent intensity at m/z 267; m/z 267.0736 and 267.2341. Instruments operating at lower resolution can not distinguish between these ions and would consequently give an exaggerated value for C17:1. Figure 1 shows examples of very simmilar and very different fatty acid patterns in the crude microbial extracts.

High mass accuracy also adds another level of specificity. This feature is a consequence of high resolution and stability and linearity of mass scale. The accurate mass table

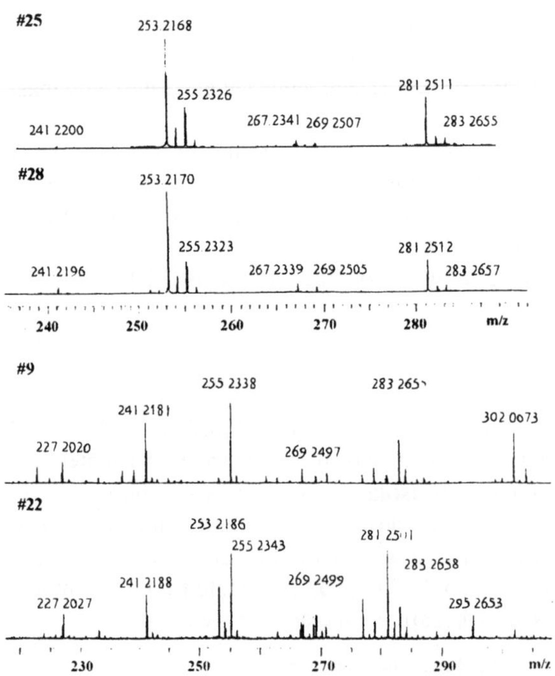

Figure 1. Negative ion electrospray mass spectra of four bacterial extracts. The two extracts on the left show very clearly that they possess a similar profile of fatty acids. The other two extracts demonstrate different populations of FFAs.

Table 1. Masses of eight FFA anions from the spectrum of microbe #28. This high mass accuracy and its maintenance over a day's repetitive analysis is typical of the instrument

Fatty acid	Ion formula	Theoretical	Experimental	Error (ppm)
C14:0	$C_{14}H_{27}O_2^-$	227.2016	227.2027	4.8
C15:0	$C_{15}H_{29}O_2^-$	241.2172	241.2188	6.6
C16:0	$C_{16}H_{31}O_2^-$	255.2330	255.2343	5.1
C16:1	$C_{16}H_{29}O_2^-$	253.2172	253.2186	5.5
C17:0	$C_{17}H_{33}O_2^-$	269.2486	269.2499	4.8
C18:0	$C_{18}H_{35}O_2^-$	283.2642	283.2658	5.6
C18:1	$C_{18}H_{33}O_2^-$	281.2485	281.2501	5.7
C19:cyp[a]	$C_{19}H_{35}O_2^-$	295.2642	295.2653	3.7

[a]cyp = cyclopropane

(Table 1) illustrates the precision of the analysis. Ions can be assigned to fatty acids with errors less than 10 ppm and often lower than 5 ppm.

Fourteen FFAs were analysed in each extract. These were the saturated FFAs from C12:0 to C18:0 inclusive, the mono-unsaturates from C14:1 to C18:1, and C18:2 and C19:cyp. Several extracts were tested by GCMS after methylation along with authentic lactobacillic acid to confirm that the C19 FFA contained the cyclopropane moiety.

Examination of Table 2 reveals that there are some major differences and some similarities in fatty acid content. Extracts #2 and #6 have very large percentages of C19:cyp, extracts #8 and #12 are very low in C19:cyp but high in C16:0, C16:1 and C18:1. A statistical analysis of the entire 35 extracts involving the derivation of Euclidean distances by unweighted pair group averages generated the tree diagram shown in Figure 2. Raw data for this analysis include the fourteen FFAs and total fatty acid content as a percentage of the total ion population in the mass spectra.

Table 2. A representative portion of FFAs and crude microbial extracts. Results are expressed as percentages of total fatty acid content. Not all FFAs or extracts are shown here

Microbe #	C12:0	C14:0	C15:0	C16:0	C16:1	C18:0	C18:1	C19:cyp
1	0.8	9.8	2.4	33.3	13.6	24.6	14.1	0
2	0	0	0	8.2	0	8.1	15.9	64.1
3	0	1.6	31.3	8.8	11.3	0	0	0
4	0.2	2.6	9.2	20.3	5.9	10.8	12.9	3.4
5	0.2	2	0.4	16	27.7	11.5	40	0.02
6	0	0	0	4.9	0	4.7	4.4	86.1
7	0.4	3.6	2.3	18.9	2.2	18.7	10.4	22.4
8	0.4	4.6	5.7	30	32.6	4.1	15.8	0
9	0.6	7.1	21	28.2	3.1	29.3	3.1	0.4
10	0	9.3	18.9	26.2	14.8	12.3	1.8	0
11	0	4	7	25	24.2	13.7	0	0
12	0.1	3.9	5.6	23	29.1	4.9	20.7	4.5
13	0.7	0	11.4	19.7	20.5	3.9	1.7	6.9
14	0	7.5	0	31.4	5	32.8	9.4	6
15	0	7.4	11.1	25.4	3	32.1	0	0
16	0.2	3.9	5.8	24.5	35.3	6.2	16.7	0.4
17	0	3.4	10.7	34.1	0	37.9	0	0
18	0	5.4	24	0	0	12	0	0
19	0	4	8.9	34.5	0	41.1	0	0
20	3.5	15.5	8.1	3.6	30.6	19.7	0	0

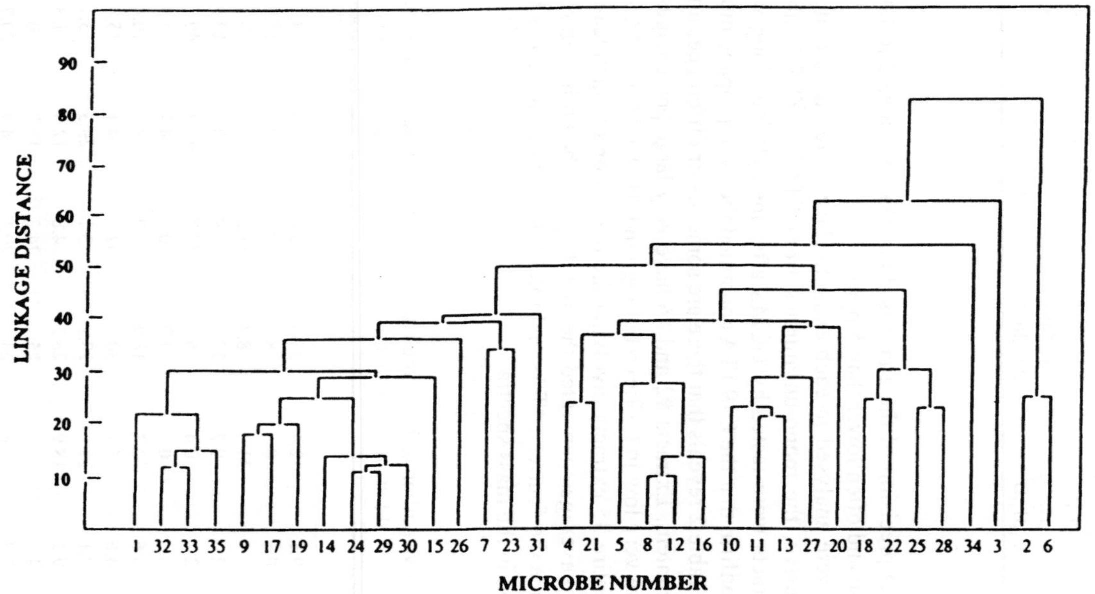

Figure 2. Tree diagram generated by determination of Euclidean distances by unweighted pair group averages of FFAs and the total fatty acid content as a percentage of the total ion population. Note that there are several clusters with linkage distances less than 30%.

Thirty percent is becoming the level below which isolates can be regarded as belonging to the same species or at least the same genus. There are several microbes which can be grouped at this level. Some support for our cluster analysis can be gained by phenotypic characterisation of certain isolates and bioactivity profile. Microbes #32, #33 and #35 are actinomycetes, #8, #12 and #16 are vibrios while #7 and #23 produce related bioactive compounds. Full validation of this clustering procedure awaits 16S RNA sequence analysis.

Inclusion of other metabolites or secondary products other than fatty acids deserves serious consideration. Many of the extracts contain similar populations of compounds. The mass spectra of two isolates (#2 and #6, linkage distance of about 23%) show considerable homology in chemical fingerprint as determined by negative ion electrospray. Major ions include m/z 295 (C19:cyp), m/z 331, 408, 434, 686 and 712. Variation in exact mass occurs in the third or fourth decimal place, indicating that they are identical.

4. CONCLUSIONS

We have shown that FTMS offers a very specific method of FFA analysis for the phylogenetic grouping and dereplication of marine bacteria. This methodology may be transferable to other types of mass spectrometer but at a cost of specificity and mass accuracy. No derivatisation is required and the determinations on the crude extracts can be carried out at about 70 samples in an eight hour period. Very small amounts (less than 50 μl) of extract is consumed (mainly dead volume and needle flushing. A system with a low dead volume injector would use much less (5–10 μl). This procedure is very amenable to automation and the only disadvantage is the high cost of instrumentation.

5. ABSTRACT

Fatty acid determination and cluster analysis is a commonly used method for phylogenetic grouping of bacteria. We have developed a method for analysis of free fatty acids using an electrospray source coupled to a Fourier Transform Ion Cyclotron mass spectrometer. This method offers several advantages over more traditional techniques. It is very rapid, specific and does not require any derivatisation or chromatography. All analyses were conducted on crude extracts.

REFERENCES

Basile F., Voorhees K. J. and Hadfield T. L. 1995, *Appl. Environ. Microbiol.* **61**:1534–1539

Bertone S., Giacomini M., Ruggiero C., Piccarolo C. and Calegari L. 1996, Appl. *Environ. Microbiol.* **62**:2122–2132

Claydon M. A., Davey S. N., Edward-Jones V. and Gordon D. B. 1996, *Nature Biotech.* **14**(11) 1584–1586

Dzierzewicz Z., Cwalina B., Kurkiewicz S., Chodurek E. and Wilczok T. 1996, *Appl. Environ. Microbiol.* **62**:3360–3365

Heller D. N., Cotter R. J., Fenselau C., Platt J. A, and Uy O. M. 1987, *Anal. Chem.* **60** :1415–1419

Sanglier Jean-J., Whitehead D., Saddler G. S., Ferguson E. V. and Goodfellow M. 1992, ***Gene*** **115** :235–242

12

BIOCOMPATIBLE ALGINATES FOR USE IN BIOHYBRID ORGANS

Gerd Klöck,[1,*] Patrik Gröhn,[1] Christan Hasse,[2] and Ulrich Zimmermann[1]

[1]Department of Biotechnology
University of Würzburg
Biocenter—Am Hubland
D-97074 Würzburg, Germany
[2]Department of Surgery
University of Marburg
Baldinger Strasse, D-35043 Marburg, Germany

Using artificial, semi-permeable membranes, transplanted tissue can be physically isolated from the immune system of the recipient (immunoisolation). This allows, in principle, transplantation of cells of different genera, without the need for pharmacological immunosuppression (Clayton et al., 1993; Colton, 1995; Lanza & Chick, 1997). In the past, a whole range of different concepts have been employed to achieve this goal (see Clayton et al., 1993; Colton, 1995; Geisen et al., 1990; Lanza & Chick, 1997). It is now generally accepted that micocapsules composed of natuaral polysacccharides, especially of alginates, are the material of choice (Clayton et al., 1993; Geisen et al., 1990; Lanza & Chick, 1997; Smidsrod & Skjak-Braerk, 1990).

Alginates (extracted from brown seaweeds) are anionic polysaccharides composed of variable homopolymeric regions of D-mannuronic acid and L-guluronic acid interspaced with regions of alternating blocks. Alginates have hydrogel-forming properties as di-, tri- or multivalent cations (e.g. Ca^{2+}, Ba^{2+} or Fe^{3+}) bind to the G-blocks (Smidsrod & Skjak-Braerk, 1990).

Entrapment of living cells in alginate gels can be accomplished under very mild conditions, and is therefore widely used for the immobilization of microorganisms, as well as of many eucaryotic cells (Smidsrod & Skjak-Braerk, 1990). Thus, microencapsulation of tissues is an intriguing alternative to postoperative immunosuppression in transplantation. In recent years, several laboratories have used alginates for microencapsulation of islets of Langerhans (Clayton et al., 1993; De Vos et al., 1993; Geisen et al., 1990; Lanza & Chick,

* Correspondence: kloeck@biozentrum.uni-wuerzburg.de

New Developments in Marine Biotechnology, edited by Le Gal and Halvorson.
Plenum Press, New York, 1998.

1997; Lim & Sun, 1980; Zkorn et al., 1992) and parathyroid tissue (Hasse et al., 1997; Hasse et al., 1996).

In order to be applied as implantable devices, the alginate gels must meet several requirements. Besides mechanical and chemical stability, which can be accomplished by using barium-alginate gels (Geisen et al., 1990; Hasse et al., 1997; Klöck et al., 1997; Zeborn et al., 1992; Zimmermann et al., 1992), a narrow pore size distribution and a high biocompatibility are prerequisites for any therapeutic application of alginate encapsulated material in the human body.

In the past, most studies with alginates encapsulated cells were performed using commercial alginates which are manufactured mainly for use in food industry and textile printing. In most of these studies, inflammatory reactions and fibrotic overgrowth of the beads result if Langerhans islets were encapsulated in raw alginate hydrogel beads (Clayton et al., 1993; De Vos et al., 1993). The consequence was random failure of the islets some time after transplantation. Analysis of raw alginates by using the free flow electrophoresis technique demonstrated that commercial alginates contained at least 10–20 fractions (characterised by different electrophoretic mobilities) which showed mitogenic activity (Zimmermann et al., 1992).

Highly purified algin preparations free of these adverse contaminations with endotoxins and other mitogens became recently available by a new purification process (Klöck et al., 1994).

Animal studies with these new, purified algin formulations revealed the absence of a mitogen-induced foreign body reaction, even when the purified material (after cross-linking with Ba^{2+}-ions) is implanted into animal models with elevated macrophage activity (diabetes-prone BB rat) (Klöck et al., 1994; Klöck et al., 1997; Pfeffermann et al., 1996).

To establish alternative in vitro models to predict the biocompatibility of alginate materials, we compared the response of different macrophage systems as well as of murine splenocytes to these alginates *in vitro* with their performance *in vivo*. The human macrophage based assays (measuring alginate-induced release of proinflammatory cytokines) were not predictive of the outcome of implantation studies in the BB/OK rat (Klöck et al., 1997; Pfeffermann et al., 1996). In murine J774A1 macrophages alginate induced a weaker proinflammatory cytokine response in biocompatible alginate samples than in many of the fibrogenic samples (Pfeffermann et al., 1996). However, fibrogenic alginates always showed a strong co-stimulatory activity towards the LPS-induced mitogenic proliferation of splenocytes prepared from Balb-c mice (Pfeffermann et al., 1996), whereas non-fibrogenic samples did not influence the response of splenocytes to LPS (Fig. 1). It is concluded, that the co-stimulatory *in vitro* assay presented here will allowes the prediction of biocompatibility *in vivo* (Pfeffermann et al., 1996).

Murine splenocytes (prepared from 8 weeks old male Balb-c mice) were cultured for 3 days with varying concentrations of LPS (E.coli serotype 055:B12) in the presence of 1 mg ml^{-1} of the different alginate samples (Pfeffermann et al., 1996). After 3 days, splenocyte activation was measured by the XTT reduction (Pfeffermann et al., 1996). Dotted line (o): activation induced by LPS alone; closed cycles: addition of 1 mg ml^{-1} of commercial alginate (Manugel GMB, Kelco Inc.), triangles: addition of 1 mg ml^{-1} of Manugel GMB after purification by our new process (Klöck et al., 1994).

Cross linkage of alginates with barium ions produces a much more resistant matrix than with calcium. Barium alginate gels are stable *in vitro* as well as *in vivo* for several months (Geisen et al., 1990; Hasse et al., 1997; Zekorn et al., 1992). In contrast to cross-linkage by Ca^{2+}-ions, Ba^{2+}-ions lead to gel matrices which are chemically very stable under both *in vitro* and *in vivo* conditions. Several studies demonstrated, that barium alginate

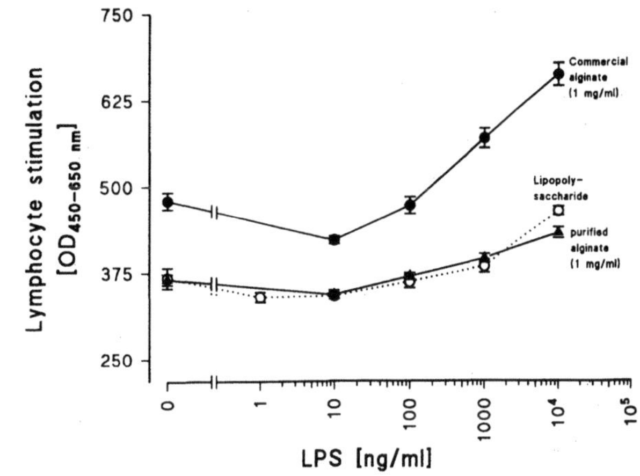

Figure 1. In vitro assay of the biocompatibility of alginates: costimulatory effect of alginate on LPS induced mitogenic activation of murine splenocytes.

encapsulated islets were able to restore normoglycemia in different animal models for long periods of time (Clayton et al., 1993; Geisen et al., 1990; Zeborn et al., 1992). Parathyroid tissue (microencapsuled in barium alginate) was also sucessfully iso-, allo-and xenotransplanted into parathyroidectomized rats (Hasse et al., 1997; Hasse et al., 1996). In vivo, the microencapsuled grafts preserved parathyroid function over a six month period resulting in complete reversal of hypoparathyroidism. Correspondingly, histologic examination revealed vital parathyroid tissue in intact barium alginate microcapsules (Hasse et al., 1997; Hasse et al., 1996).

In addition to the utilization as implantable cell reactors in therapy and biotechnology, highly purified, biocompatible alginates have broad application potential as ocular fillings, tissue replacements, microencapsulated growth factors and/or interleukins or slow-release dosage forms of antibodies, surface coatings of sensors and other invasive medical devices, and in encapsulation of genetically engineered cells for gene therapy.

REFERENCES

Clayton, H.A., James, R.F., London, N.J., 1993, Islet microencapsulation: a review. *Acta Diabetol* **30**:181–189

Colton, C.K., 1995, Implantable biohybrid artificial organs. *Cell Transplantation* **4**:415–436

De Vos, P, Wolters, G.H., Fritschy, W.M., Van Schilfgaarde, R., 1993, Obstacles in the application of microencapsulation in islet transplantation. *Int J Artif Organs* **16**:205–212

Geisen, K, Deutschländer, H., Gorbach, S., Klenke, C., Zimmermann, U., 1990, Function of barium alginate microencapsulated xenogenic islets in different diabetic mouse models. Frontiers in Diabetes Research Lessons from animal diabetes III:142–148 (Editor: Shafir, E.)

Hasse, C., Klöck, G., Bartsch, D., Zielke, A., Zimmermann, U., Rothmund, M., 1997, First successful xenotransplantation of microencapsulated human parathyroid tissue in experimental hypoparathyroidism: Long term in vivo function without immunosuppression. *J. Microencapsulation* (in press).

Hasse, C., Klöck, G., Zielke, A., Schlosser, A., Barth, P., Zimmermann, U., and Rothmund, M., 1996, Transplantation of parathyroid tissue in experimental hypoparathyroidism: in vitro and in vivo function of parathyroid tissue microencapsulated with a novel alginate. *Int. J. Artif. Organs* **19**:1–7.

Klöck, G., Frank, H., Houben, R., Zekorn, T., Horcher, A., Siebers, U., Wöhrle, M., Federlin, K., Zimmermann, U., 1994, Production of purified alginates suitable for use in immunoisolated transplantation. *Applied Microbiol. Biotechnol.* 40:639–643.

Klöck, G., Pfeffermann, A., Ryser, C., Gröhn, P., Kuttler, B., Hahn, H.J., and Zimmermann, U. Biocompatibility of mannuronic-acid rich alginates. 1997, *Biomaterials* 18:707–713.

Lanza, R.P., Chick, W.L. 1997, Transplantation of encapsulated cells and tissues. *Surgery* **21**:1–9.

Lim, F, Sun, A.M., 1980, Microencapsulated islets as a bioartificial pancreas. *Science* **210**:908–910

Pfeffermann, A., Klöck, G., Gröhn, P., Kuttler, B., Hahn, H.-J., and Zimmermann. U., 1996, Assay procedures for the assessment of the biocompatibility of alginate implants. *Cell Engineering* **4**:167–173

Smidsrod, O., Skjak-Braerk, G., 1990, Alginate as immobilization matrix for cells. *Trends Biotechnol* **8**:71–78

Zekorn, T., Horcher, A., Siebers, U., Schnettler, R., Klöck, G., Hering, B., Zimmermann, U., Bretzel, R.G., and Federlin, K. 1992, Barium-cross-linked alginate beads: a simple, one-step method for successful immunoisolated transplantation of islets of Langerhans. *Acta Diabetologia* **29**:99–106.

Zimmermann, U., Klöck, G., Federlin, K., Hannig, K., Kowalski, M., Brezel, R.G., Horcher, A., Entenmann, H., Siebers, U. and Zekorn, T.,1992, Production of mitogen contamination free alginates with variable rations of mannuronic to guluronic acid by free flow electrophoresis. *Electrophoresis* **13**:269–274.

13

PRODUCTION OF BIOACTIVE COMPOUNDS BY CELL AND TISSUE CULTURES OF MARINE SEAWEEDS IN BIOREACTOR SYSTEM

Gregory L. Rorrer,[1] William H. Gerwick,[2] and Donald P. Cheney[3]

[1]Department of Chemical Engineering
[2]College of Pharmacy
Oregon State University
Corvallis, Oregon 97331
[3]Marine Science Center
Nahant, Massachusetts 01908

SUMMARY

Macrophytic marine algae, commonly known as seaweeds, are a source of unique natural products with medicinal properties. Of particular interest are bioactive compounds produced through the selective oxidation of fatty acids to eicosanoids and related oxylipins, or the selective halogenation of terpenoids to brominated/chlorinated monoterpenes [Carte, 1996]. Cell and tissue cultures established from anatomically-complex marine seaweeds have the potential to biologically synthesize these compounds in a controlled environment at a scale required for continued drug development or commercial production. In pursuit of this goal, over the past six years we have developed several unique phototrophic cell and tissue culture systems representing brown, green, and red seaweeds. These include a gametophyte cell suspension culture from the brown alga *Laminaria saccharina* [Qi & Rorrer, 1995; Rorrer et al., 1995; Zhi & Rorrer, 1996], a freely suspended tissue culture of the green alga *Acrosiphonia coalita* [Rorrer et al., 1996], an differentiated filament clump culture of the red alga *Agardhiella subulata* established by a callus-induction technique, and a freely suspended microplantlet culture established by controlled regeneration of filament clumps [Huang et al., 1997]. For example, the *Agardhiella subulata* cell and tissue culture system is illustrated in Figure 1.

The controlled cultivation of these culture systems has been demonstrated in stirred tank [Qi & Rorrer, 1995; Rorrer et al., 1996; Ramanan et Rorrer, unpublished], bubble column [Zhi & Rorrer, 1996], immobilized mesh [Huang et al., 1997], and tubular recycle photobioreactor systems [Mullikin & Rorrer, 1997]. Representative growth curves for the *L.*

New Developments in Marine Biotechnology, edited by Le Gal and Halvorson.
Plenum Press, New York, 1998.

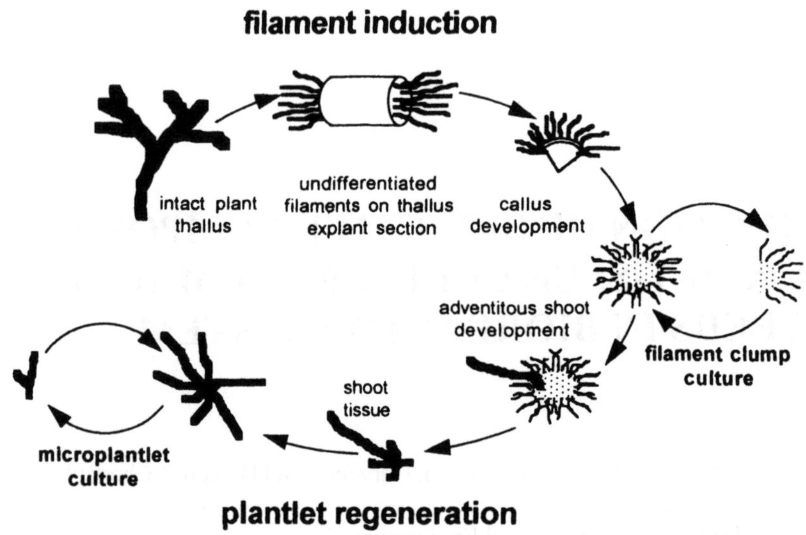

Figure 1. Development of the filament clump and regenerated microplantlet suspension culture systems for the red *Agardhiella subulata*.

saccharina female gametophyte cell suspension culture system in stirred tank vs. tubular recycle photobioreactors are shown in Figure 2. Cell densities exceeding 2 g of dry cell weight per liter of culture were achieved with specific growth rates up to 0.2 day^{-1} under selected conditions of light, mixing intensity, aeration, and mode of nutrient delivery. The *L. saccharina* gametophyte cell suspension culture system was further studied for bioactive compound production. This culture biologically synthesized three bioactive hydroxy fatty acids derived from regio- and stereoselective ω-*6* lipoxygenase oxidation: 15-hydroxy-5,8,11,13-eicosatetranenoic acid (15-HETE) from arachidonic acid, 13-hydroxy-9,11-octadecadienoic acid (13-HODE) from linoleic acid, and 13-hydroxy-9,11-octadecatetraenoic acid (13-HOTDA) from stearidonic acid [Rorrer et al., 1997]. Feeding the culture with the precursors linoleic acid and γ-linolenic acid stimulated the production of all three hydroxy fatty acids up to 400% of the controls, as shown in Figure 3. All of our studies suggest that bioprocess technology for the controlled production of bioactive compounds from cell and tissue cultures of marine seaweeds in bioreactor systems holds significant promise for securing future supplies of novel drugs derived from marine seaweeds.

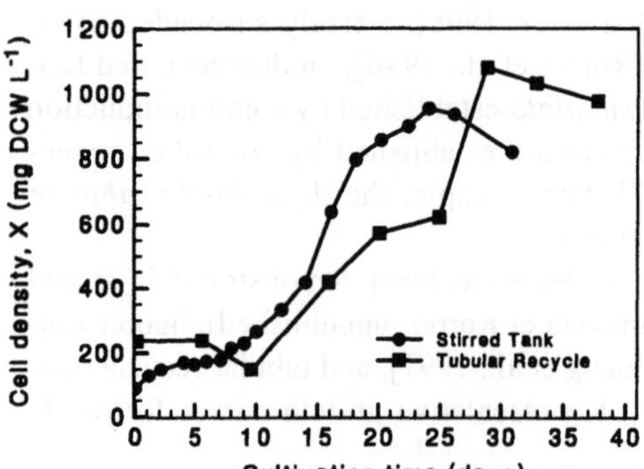

Figure 2. Representative growth curves *L. saccharina* female gametophyte cell suspension cultures in stirred-tank and tubular-photobioreactors at 12°C. Stirred tank bioreactor: 250 rpm, 0.4 aeration rate, 77 μE/m^2-sec incident light intensity. Tubular recycle bioreactor: 15 min residence time in tubular section, 6 min residence time in aeration section, 18 μE/m^2-sec incident light intensity.

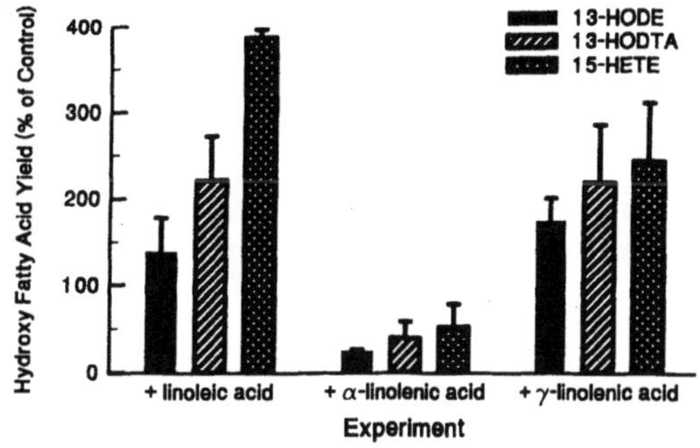

Figure 3. Hydroxy fatty acid yields *L. saccharina* female gametophyte cell suspension cultures incubated with 200 mg/L of a given fatty acid for 10 days, normalized against the yield of the control culture.

ACKNOWLEDGMENTS

This work was supported by grant no. NA36RG0451 (project no. R/BT-8) from the National Oceanic and Atmospheric Administration to the Oregon State University Sea Grant College Program under the National Sea Grant Program Marine Biotechnology Initiative.

REFERENCES

Carte, B.K., 1996, *BioSci.*, **46**, 271–286.

Huang, Y.M., Maliakal, S., Cheney, D.P., Rorrer, G.L., 1997 (Submitted to *J. Phycol.*).

Mullikin, R.K., Rorrer, G.L., 1997, *BioHydrogen '97 Proceedings*, O.R. Zaborsky, Ed., Plenum Publishing Corporation, New York, NY (in press).

Qi, H., Rorrer, G.L., 1995, *Biotech. Bioeng.*, **45**, 251–260.

Ramanan, S., Rorrer, G.L. (manuscript in preparation).

Rorrer, G.L., Polne-Fuller, M., Zhi, C., 1996, *Biotech. Bioeng.*, **49**, 559–567.

Rorrer, G.L., Yoo, H.D., Huang, Y.M., Hayden, C., Gerwick, W.H., 1997, *Phytochem.* **46**:871–877).

Rorrer, G.L., Zhi, C., Modrell, J., Gerwick, W.H., 1995, *J. Applied Phycol.*, **7**, 187–198.

Zhi, C., Rorrer, G.L., 1996*Enz. Microb. Tech.*, **18**, 291–299.

14

THE MERMAID'S PURSE, OR WHAT THE SKATE CAN TELL US ABOUT KEEPING EGGS SAFE IN ONE BASKET

Thomas J. Koob,[1,*] David P. Knight,[2] Marina Paolucci,[3] Bradley Noren,[4] and Ian P. Callard[4]

[1]Skeletal Biology Section
Shriners Hospital for Children
Tampa, Florida 33612
[2]Department of Biological Sciences
King Alfred's College
Winchester, Hants, United Kingdom
[3]Department of Biology
University of Naples Federico II
Naples, Italy
[4]Department of Biology
Boston University
Boston, Massachusetts 02215

Egg laying vertebrates with relatively small clutches of large yolked eggs (elasmobranchs, reptiles and birds) have developed a diversity of tertiary membranes to contain and protect the egg/embryo from environmental assaults. The eggs of oviparous elasmobranchs (some sharks and all skates) are left to fend for themselves in the marine environment for up to a year or longer. Once released, the egg capsule is moored to stationary objects where it must resist attacks of molluscs, bacteria, and other predatory organisms. Elasmobranchs have solved this problem quite successfully by the invention of the tough, leathery and extremely durable egg capsule, the product of a specialised region of the reproductive tract, the oviducal, shell or nidamental gland. This material is able to withstand the corrosive action of sea water for remarkably long periods, remarkable especially given that it is composed almost entirely of proteins.

Elasmobranch egg capsules have attracted the attention of biologists for over a hundred years with regard to the morphological, synthetic, secretory and regulatory questions

* Correspondence: Thomas J. Koob, Ph.D., Skeletal Biology Section, Shriners Hospital for Children, 12502 North Pine Drive, Tampa, FL 33612, USA. Tel: 813-972-2250, Fax : 813-975-7127; E-mail: tkoob@earthlink.net

New Developments in Marine Biotechnology, edited by Le Gal and Halvorson.
Plenum Press, New York, 1998.

posed by the complex assembly process. In recent years, an understanding of the polymerization mechanism and the physicochemical properties of the mature secretory product has lead us to speculate as to its potential adaptation or utility for biomedical and marine applications. We here summarize the current status of the basic research and report on some preliminary studies exploring these potential applications.

THE REPRODUCTIVE CYCLE OF THE LITTLE SKATE, *RAJA ERINACEA*

In captive egg-laying little skates (*Raja erinacea*), eggs are released in pairs every 8–10 days. Capsular proteins begin to be secreted by the nidamental gland about 12 hours prior to ovulation, corresponding to the end of the periovulatory progesterone peak, when estrogen levels are quite low (Koob *et al.*, 1986). The ovulated egg enters the partially formed egg case which is subsequently completed. The fully formed egg case is held in the reproductive tract for 24–48 hours prior to oviposition. During this period polymerization of the assembled structural proteins is completed (via a quinone tanning process: see below). Although the endocrine regulation of this process has not been delineated as yet, the gland of the immature animal is quite small and shell gland growth is positively correlated with plasma estradiol levels as the animal advances into the active egg-laying phase (Koob *et al.*, 1986). Both estrogen (Reese and Callard, 1991) and progesterone (Paolucci and Callard, 1997) receptors have been characterized in the little skate reproductive tract, which confirms that the glandular processes are under endocrine control.

MORPHOLOGY OF THE NIDAMENTAL GLAND IN RELATION TO THE SECRETORY PROCESS

In *Raja erinacea* the shell gland is a symmetrical kidney-shaped structure, flattened in the dorso-ventral plan, located in the upper oviduct. The two halves of the gland are arranged on the dorsal and ventral sides of the upper oviduct. The principal secretory unit of the gland consists of tubules bordered by a simple epithelium, the cells of which contain abundant, morphologically distinct cytoplasmic granules. The epithelial cells secrete the products into the lumen of the tubules from whence it issues into the lamellae emptying into the oviducal lumen. Contained within the complex secretory material, but compartmentalised in some way, are the six major structural proteins and two enzymes which catalyze the tanning or cross-linking process. The white untanned proteinaceous mixture is extruded through the lamellae into the oviduct lumen, the material from the more rostral tubuluar glands forming the innermost layer of the shell, with successive layers added by the extrusion of material from more caudal glandular zones. Successive layers are annealed together to form the exceedingly tough, resistant laminate of 35 layers, which is dark brown in color. The process of tanning continues for some time after laying as can be determined by further deepening of the color after eggs are laid (Koob, 1987).

BIOCHEMISTRY OF THE GLANDULAR PROTEINS AND THE TANNING PROCESS

Analysis by SDS/PAGE gel electrophoresis of the assembled protein complex isolated from untanned capsular material yielded six distinct major structural proteins ranging from 20 to 90 kDa in size (Koob & Cox, 1993). These structural proteins are co-secreted with the tanning enzymes, tyrosine hydroxylase and catechol oxidase, involved in the stabilization of the secreted protein complex into the tough, sclerotized egg case. The catalysts are secreted in a manner such that the elements of the final mixture are not activated until the assembled proteins have exited the gland. The composite then begins to harden and darken within the tract even before the oocyte reaches the lumen of the nidamental gland and enters the partially formed capsule. Measurement of one enzyme of the tanning complex (catechol oxidase) indicates very low levels of activity in glands from immature animals, reaching 10 fold higher once the gland becomes active in mature animals. No changes in enzyme activity were observed during the ovulatory cycle.

POSSIBLE BIOMEDICAL, MARINE, AND BIOMARKER APPLICATIONS

Possible use of the native and fully matured capsule material as an adjunct in skeletal repair is under investigation by the senior author (patent pending). Preliminary trials indicate that the native (untanned, insoluble) protein precursors provoke an immune response coupled with extensive osteolysis when integrated into experimental cortical defects in rabbit tibia. However, the fully tanned shell material is non-antigenic, directs bone deposition, and shows evidence of integration into identical cortical bone lesions. Thus, the polymerization process per se obviates antigenicity of the capsule proteins in this rabbit model. The use of this material for bridging congenital, trauma related, or surgically created bone defects appears promising.

The process of shell gland synthesis and secretion may also be of utility as a biomarker of the effects of xenoestrogens and other endocrine disrupting agents in the marine environment. We have noted, for example that the little skate concentrates heavy metals (e.g. cadmium) within the shell gland. Further, after a single low level exposure to cadmium *in vivo*, the ability of the shell gland to secrete the shell protein complex is progressively inhibited over time. This suggests that shell gland activity could be a useful marker for endocrine disruptors in the coastal environment.

In addition to these possible clinical and ecotoxicological applications, the potential for a natural product which is converted from a malleable protein mix to a tough, resistant, fibrous material on site under controlled conditions in an aqueous medium may have applications as non-toxic marine coatings, replacing the highly toxic tributyl tin (TBT). Oviparous elasmobranchs solved the problem of marine corrosion in producing a material which perseveres for years sea water. We aim to capture the relevant chemistry in order ultimately to design superior materials for use in the marine environment.

ACKNOWLEDGMENTS

Supported by the Shriners of North America (#9610) to TJK and NSF IBN-9407348 to IPC.

REFERENCES

Koob, T.J., 1987, *J. Exp. Mar. Biol. Ecol.* **113**: 155–166.
Koob, T.J. and Cox, D.L., 1993, *Environ. Biol. Fishes* **38**: 151–157.
Koob, T.J., Tsang, P. and Callard, I.P., 1986, *Biol. Reprod.* **35:** 267–275.
Paolucci, M., and Callard, I.P., 1997, *Gen.Comp.Endocrinol.*, in press.
Reese, J.C., and Callard, Ian P., 1991, *Gen .Comp.Endocrinol.***84**:170–181.

15

IN VITRO PRODUCTION OF MARINE-DERIVED ANTITUMOR COMPOUNDS

Shirley A. Pomponi, Robin Willoughby, Amy E. Wright, Claudia Pecorella, Susan H. Sennett, Jose Lopez, and Gail Samples

Harbor Branch Oceanographic Institution, Inc.
Fort Pierce, Florida 34946

The sustainable supply of marine-derived bioactive compounds to meet demands for pre-clinical and clinical evaluation is a major consideration in drug development strategies. It is also a concern to environmental resource managers, particularly since conservation of genetic resources may preclude harvesting as a bulk supply strategy. Our research addresses this need by focusing on in vitro production of antitumor compounds through cell culture of the source organisms. Our objectives are: to establish cultures of bioactive marine invertebrates that can be used as models to study in vitro production of antitumor compounds and the factors which control expression of their production; to provide bulk supplies of these compounds through in vitro production; and to produce new structural analogs via manipulation of culture conditions.

We have established primary cell cultures of two bioactive marine invertebrates: the sponge Teichaxinella morchella which produces the antitumor compound, stevensine (Albitzi and Faulkner, 1985), and the tunicate Ecteinascidia turbinata from which the antitumor ecteinascidins have been isolated (Rinehart et al., 1990; Wright et al., 1990). Cultures were initiated using techniques developed in our laboratory for the dissociation and selective enrichment of cells known to contain the bioactive metabolites: archaeocytes in T. morchella (Pomponi and Willoughby, 1994; Pomponi et al., 1997), and stem cells in the embryos and larvae of E. turbinata. Cells are inoculated into microwell plates (10^5 to 10^6 cells/ml) in the appropriate nutrient medium. Sponge cell culture media are based on commercially available media (Medium 199 or Iscove's MDM). Osmolality is increased to 1000 mOsm by addition of NaCl. The pH is adjusted to 8.1 and stabilized by addition of 5 mM Trizma buffer. Tunicate medium is based on a formula by Kawamura and Fujiwara (1995). Rifampicin (1.16 mM), streptomycin, or penicillin are used to control bacterial contamination, and ketoconazole or amphotericin B (Fungizone) is used to inhibit fungal growth in long-term maintenance of cultures.

We have developed multiwell plate assays to measure effects of growth-regulating compounds on DNA and protein synthesis. To quantify and monitor synthesis of DNA in

New Developments in Marine Biotechnology, edited by Le Gal and Halvorson.
Plenum Press, New York, 1998.

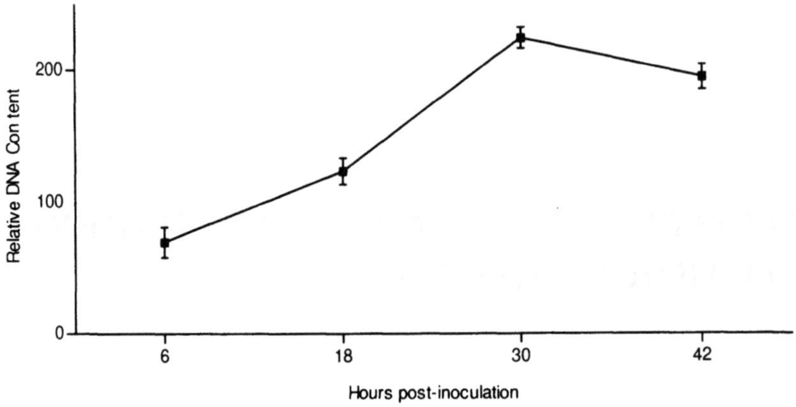

Figure 1. PHA-induced DNA synthesis in Teichaxinella morchella cells, expressed as percent of control. Cultures were incubated in Medium 199 with 1.5% PHA. DNA was measured by the Hoechst 33342 microwell plate assay. At 18 hours post-inoculation, the DNA content of PHA-stimulated cultures increased. The relative DNA content of PHA-stimulated cultures reached a plateau at 30 hours, at greater than 200% of the DNA content of the untreated control. (Mean ± S.E., n = 5). (From Pomponi et al., 1997).

cultures, we have developed a multiwell assay using the nucleic acid probe, Hoechst 33342 (Pomponi et al., 1997). Protein synthesis is measured using a modification of the sulphorhodamine B (SRB) multiwell plate assay (Skehan et al., 1989).

Since a cell line has not yet been developed, our research is focused on optimization of culture conditions for sustained growth, cell division, and bioactive metabolite production by normal cells. A medium supplement consisting of several growth factors, organic nutrients, a lectin (phytohemagglutinin [PHA]), a transport factor (transferrin), and hormones was formulated (Pomponi et al., 1997). Phytohemagglutinin stimulated DNA synthesis (Fig. 1) and cell division (Fig. 2) in T. morchella cultures. The supplement, as well as some of its constituents (epidermal growth factor, bovine pituitary extract, PHA, insulin, and prostaglandin E2) stimulated protein synthesis (Fig. 3A) in T. morchella cultures. Epidermal growth factor and insulin stimulated DNA synthesis in E. turbinata cultures, however, in the presence of fetal bovine serum (3%), DNA synthesis is inhibited.

To determine if Teichaxinella morchella cultures which underwent cell division in response to phytohemagglutinin (PHA) continued to produce the bioactive metabolite stevensine, PHA-stimulated and control (unstimulated) cultures were subsampled for analysis of stevensine concentration after 36 hours and 8 days in culture. Microanalytical

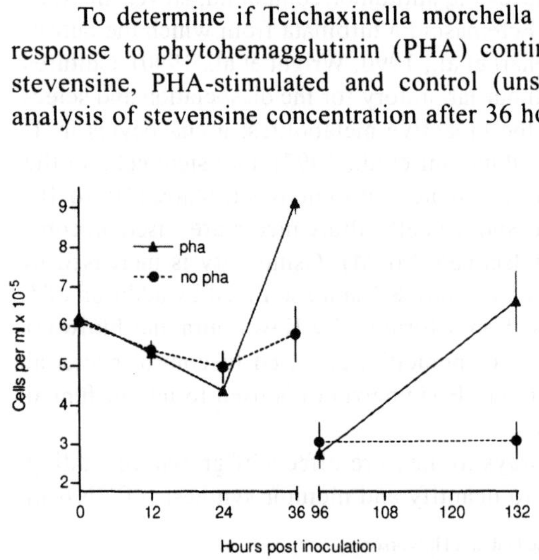

Figure 2. PHA-induced growth of Teichaxinella morchella cells. Cultures were treated as described in Fig. 1. PHA-stimulated cultures (solid line) doubled after 36 hours, as measured by direct cell counts. Cell numbers in unstimulated controls (dashed line) did not increase. After 96 hours, subcultures were stimulated with PHA, and doubled after 36 hours. Control cultures showed no increase. (Mean ± S.E., n= 4). (From Pomponi et al., 1997).

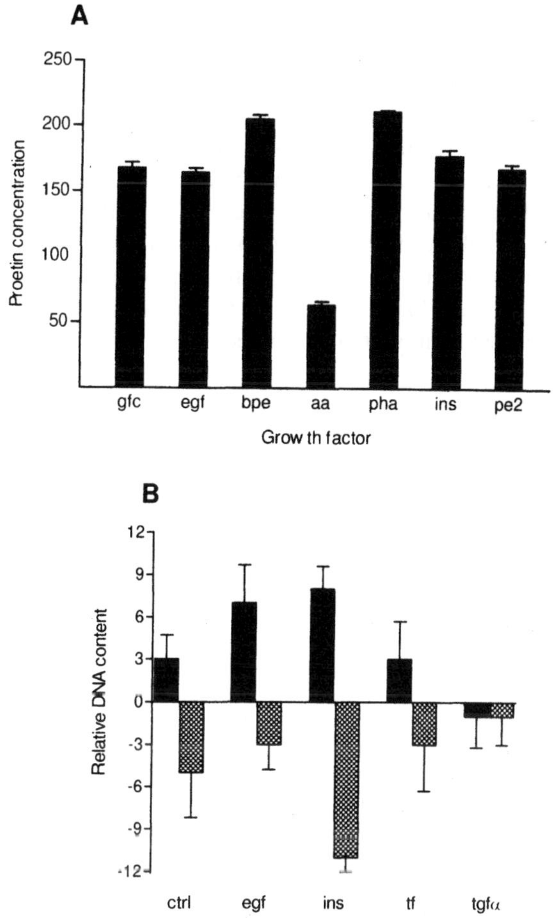

Figure 3. (A) Effect of growth factors on protein synthesis in Teichaxinella morchella cultures after 48 hours incubation in Iscove's MDM. Data are plotted as percent of control. Control cultures (horizontal dashed line) received no growth factors. gfc, growth factor cocktail (consists of egf, epidermal growth factor, 10 ng/ml; bpe, bovine pituitary extract, 25 μg/ml; aa, arachidonic acid, 10 μM; linoleic acid, 0.2 μM; cholesterol, 5 μM; phosphoethanolamine, 10 μg/ml; pha, phytohemagglutinin, 1.5%; tf, transferrin, 5 μg/ml; ins, insulin, 10 μg/ml; pe2, prostaglandin E2, 50 ng/ml; hydrocortisone, 50 nM, retinol acetate, 0.3 μM). Concentrations of components of gfc tested individually are the same as in the complete gfc. (Mean ± S.E., n = 5). (B) Effect of growth factors on DNA synthesis in Ecteinascidia turbinata cultures after 48 hours incubation in Kawamura medium, with (black bars) and without (patterned bars) 3% fetal bovine serum. Cell-free controls were subtracted from data shown. Concentrations are as listed for Fig. 3A. tgf-a, transforming growth factor-a, 10 ng/ml. (From Pomponi et al., 1997).

HPLC techniques were developed to monitor in vitro production of stevensine (Pomponi et al., 1997). A total of 6×10^7 cells from each culture were analyzed. After 36 hours, both PHA-stimulated and control cultures contained the same amount of stevensine per cell, indicating that cells which divided continued to produce the bioactive metabolite (Fig. 4).

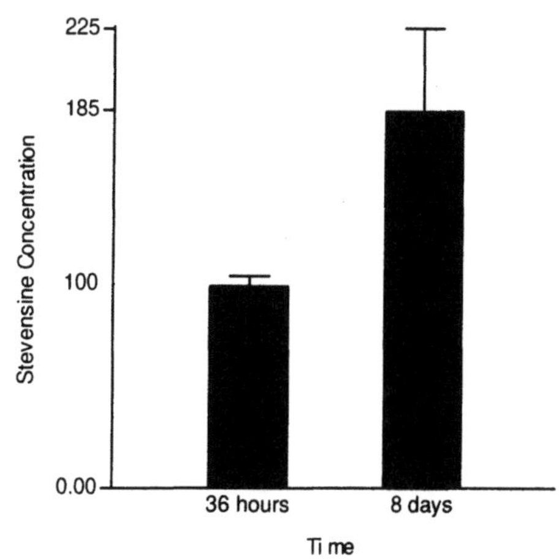

Figure 4. Concentration of stevensine per cell in cultures of T. morchella incubated in PHA, expressed as percent of control. One population doubling occurred within 36 hours in PHA-stimulated cultures; control cultures did not divide. After 8 days, PHA-treated cultures showed an increase in stevensine concentration per cell. (From Pomponi et al., 1997).

After eight days in culture, the PHA-stimulated cells contained more stevensine per cell than at the beginning of the experiment (Fig. 4). These results indicate that the cells retain their ability to synthesize stevensine after doubling, and demonstrate that sponge cells will respond to mitogenic stimuli without disruption of biosynthetic pathways.

In summary, methods have been developed for the primary culture of sponge and tunicate cells stimulated with vertebrate growth-regulating compounds and the lectin phytohemagglutinin. Development of a continuous sponge or tunicate cell line has not yet been achieved. In vitro production of the sponge-derived antitumor compound stevensine demonstrates, however, that this approach is feasible for further development as a biological model for studying the factors which control expression of production of marine-derived antitumor compounds.

ACKNOWLEDGMENTS

This research was funded by grants from the National Institutes of Health (CA55871, CA56191) and from the National Sea Grant Marine Biotechnology Program (NA3 6RG-0070).

REFERENCES

Albitzi, K.F., Faulkner, D.J., 1985, *J. Org. Chem.* **50**:4163–4164.

Kawamura, K., and Fujiwara, S., 1995, *Cell Struc. Func.* **20**: 97–106.

Pomponi, S.A., Willoughby, R., 1994, In: van Soest, R.W.M., Th.M.G. van Kempen and J.C. Braekman (Eds.). Sponges in Time and Space. Proc. 4th Intl. Porifera Congress, Amsterdam, Rotterdam, Netherlands, pp. 395–400.

Pomponi, S.A., Willoughby, R., Kaighn, M.E., Wright, A.E., 1997, In: Maramorosch, K. and J. Mitsuhashi (Eds.). Invertebrate Cell Culture: Novel Directions and Biotechnology Applications. Science Publishers, Inc., U.S., pp. 231–237.

Rinehart, K.L., Holt, T.G., Fregeau, N.L., Stroh, J.G., Keifer, P.A., Sun, F., Li, L.H., Martin, D.G., 1990, J. Org. Chem. **55**:4512–2415.

Wright, A.E., Forleo, D.A., Gunawardana, G.P., Gunasekera, S.P., Koehn, F.E., McConnell, O.J., 1990, J. *Org. Chem.* **55**:4508–4512.

16

STRUCTURE AND FUNCTION OF BARNACLE CEMENT PROTEINS

Kei Kamino and Yoshikazu Shizuri

Marine Biotechnology Institute
1900 Sodeshi, Shimizu, Shizuoka 424, Japan

1. UNDERWATER ADHESIVE PROTEINS PRODUCED BY MARINE ORGANISMS

We can find various sessile organisms in the marine environment. The barnacle is typical of sessile animals which firmly attach to various substrata in water. The barnacle secretes proteinaceous cement from a cement gland to the space between the calcareous barnacle base and substratum for the adhesion (Saroyan et al., 1970; Dougherty, 1990). While amino acid compositions of cements in some species have been reported (Walker, 1972; Barnes & Blacktock, 1974), protein constituents have remained unstudied because of the cement insolubility. We faced at least two difficulties in our studies on underwater adhesive proteins. First problem was analysis of the insoluble protein, because the underwater adhesive is generally insolubilized after secretion from the animal (Waite, 1987). The second difficulty was measurement of specific adhesive activity.

Only adhesive proteins containing peptidyl-3,4-dihydroxyphenylalanine (DOPA) have been biochemically studied well in mussels (Rzepecki & Waite, 1995) and sabellariid polychaetes (Waite et al., 1992), although their role in underwater adhesion is still unclear. The DOPA moiety in these adhesive proteins is used as a marker for purification of their soluble precursors from the secretory organ. Proteins in other underwater adhesives lacking DOPA, including barnacle cement (Naldrett, 1993), remain to be purified as well as characterized.

We have previously shown which material of the barnacle should be analyzed for understanding barnacle underwater adhesion (Kamino et al., 1996). Several partial amino acid sequences were also determined by using aqueous formic acid as a solvent following cyanogen bromide cleavagc for solubilization of cement. Although reducing agent treatmemt was used in this previous method, the reductive was poor. In this study, we successfully modified the previous method to solubilize and identify all major constituents in the barnacle cement by using the reductive reagent, dithiothreitol (DTT).

New Developments in Marine Biotechnology, edited by Le Gal and Halvorson.
Plenum Press, New York, 1998.

2. SOLUBILIZATION OF INSOLUBLE CEMENT

For preliminary separation of insoluble cement, we employed centrifugation with guanidium hydrochloride (Gdm) solution as a solvent. By this method, cement was separated into three fractions: Gdm-soluble fraction (GSFI), Gdm-soluble fraction after treatment by excess amount of DTT, and Gdm-insoluble fraction after treatment by DTT (GIF). Weight percentages of GSFI, GSF2 and GIF in the cement were 24%, 70% and 6%, respectively. SDS-PAGE analysis showed that GSFI was composed of proteins near 60kDa, named Mrcp-60k, and some minor proteins with molecular weights of less than 20kDa. The GSF2 gave further bands in SDS-PAGE near l00kDa and 50kD, named Mrcp-l00k and Mrcp-50k, respectively. Mrcp-60k was also solubilized in GSF2. Mrcp-l00k and Mrcp-50k in GSF2 were not solubilized by treatment in 4.2% 2-mercaptoethanol, 15 mM Tris-HCl buffer (pH 6.8), 2% SDS at 100°C for 3min which is used in pretreatment of SDS-PAGE of Laemmli system (Laemmli, 1970). After solubilization by DTT-treatment, these proteins became soluble in 5% aqueous acetate solution or SDS-containing solution. Although GIF was also composed of a proteinaceous substance, the method for solubilization is not known. Chemical cleavage by cyanogen bromide was not effective for solubilization of GIF. This study indicates that cement is composed of three major proteins named Mrcp-l00k, -60k and -50k, and some minor proteins.

3. DETERMINATION OF PRIMARY STRUCTURES

To reveal entire structures of the cement components, cDNA cloning technique were employed using DNA probe generated from a partial amino acid sequence of each protein. So far, cDNAs corresponding to three major proteins, Mrcp-100k, -60k and -50k, and two minor proteins, Mrcp-20k and Mrcp-16k, were cloned and sequenced. Four of these five proteins (not Mrcp-16k) were novel in structure, and SF1-I6k was a lysozyme homologue. Each cloned cDNA contained 2979, 1302, 1692, 606 and 471 bp as open reading frames respectively, including putative signal peptides. Isoelectric points were calculated as 9.84, 6.92, 10.53, 4.72 and 8.83, respectively.

A brief profile of each proteins follows. Mrcp-l00k and Mrcp-50k are similar in the some properties, i.e. effect of reductant for solubilization from cement, isoelectric point, high hydrophobicity and amino acid composition. However, sequence similarity was not found. Some repetitive sequence motifs were found in the primary structure of Mrcp-50k. Mrcp-60k was characterized by an abundance (60% of total residues) of four amino acids: Ser, Thr, Gly and Ala. Also, a high amount of carbohydrate was attached in this protein. Mrcp-60k is relatively hydrophilic compared with the previous two major proteins. Mrcp-20k had some unique characteristics. The primary structure was composed of repetitive sequences containing many Cys residues. Also, Mrcp-20k had many charged amino acids and was extremely hydrophilic. The fifth protein in this study, Mrcp-16k, was a Iysozyme homologue. Interestingly, the crude cement had lytic activity of bacteria, although origin of the activity is not confirmed.

4. EXPRESSION OF GENES

Northern blotting analysis revealed that all genes were only expressed at the basal portion of adult barnacles where the histologically distinguished cement gland is located, and that these genes were not expressed at the stage of nauplius and cypris swimming

Table 1.

	Mrcp-100k	*Mrcp*-50k	*Mrcp*-60k	*Mrcp*-20k	*Mrcp*-16k
M.W. and pI	111768, 9.84	62359, 10.53	40166, 6.92	20403, 4.72	15679, 8.83
Similar proteins	novel	novel	novel (Mucine)	novel (sp185, fp-2)	Lysozyme
Characters of primary structure		Repetitive motif		Repetitive sequence	
Amino acid composition			S, T, G, A-rich	Cys-rich	
Property of polypeptide	Hydrophobic	Hydrophobic		Hydrophilic	
known modification			Glycosylation		
Comment	Similar each other.		OH-rich	Charge-rich	

larva. The latter result indicates that tentative underwater adhesive of cypris larva used for searching out habitat is not same as adult barnacle permanent adhesive.

5. PUTATIVE FUNCTION OF CONSTITUENTS

Although the function of each protein in barnacle underwater adhesion is not known, we propose several purposes. Mrcp-16k may be present to remove bacterial film on substratum surfaces for attachment, or to protect cement from bacterial degradation. Mrcp-60k is rich in hydroxy-group donated by Ser, Thr and carbohydrates. In the case of DOPA-containing proteins in mussels (Rzepecki & Waite, 1995), proteins were always rich in hydroxylated amino acids such as DOPA, hydroxyl-Pro and hydroxyl-Arg. These abundant hydroxy-groups may be useful to remove a water-boundary layer at the attachment site. Mrcp-l00k and -50k are insoluble proteins in the cement, and were difficult to solubilize. These proteins may be available to immobilize other components, and/or to cure whole cemenl for strong cohesion. Mrcp-20k, which is difficult to speculate on the functional role, has abundant charged amino acids. A high amount of charged amino acids may play an important role in interaction with the substratum.

6. FUTURE EXPECTATIONS

The underwater adhesive proteins produce by marine organisms function extracellularly. These proteins should require functional spontaneous assembly in water. Understanding the characteristics of these adhesive proteins may lead to a stronger understanding of barnacle settlement, and also to biomimetic designs for biological polymers. Genes cloned in this study may eventually be available to produce underwater adhesive. Moreover, these genes and antibodies corresponding to each protein will be effective tools for understanding the mechanism of barnacle underwater adhesion.

REFERENCES

Barnes, H., and Blackstock, J., 1974, *J. exp. mar. Biol. Ecol.* **16**:87–91
Dougherty, W.J., 1990, *J. crusta. Biol.* **10**(3):469–478
Kamino, K., Odo, S., and Maruyama, T., 1996, *Biol. Bull.* **190**:403–409

Laemmli, U.K., 1970, *Nature* **227**:680–685
Naldrett, M.J., 1993, *J. Mar. Biol. Ass. U.K.* **73**(3):689–702
Rzepecki, L.M., and Waite, J.H., 1995, *Mol. Mar. Biol. Biotech.* **4**(4):313–322
Saroyan, J.R.,Lindner, E. and Doolcy, C.A., 1970, *Biol. Bull.* **139**:333–350
Waite, J.H., 1987, *Int. J. Adhesion and Adhesives* **7**:9–14
Waite, J.H., Jensen, R.A., and Morse, D.A., 1992, *Biochemistry* **31**, 5733–5738
Walker, G., 1972, *J. Mar. Biol. Ass. U.K.* **52**:429–43

17

THE DEVELOPMENT AND COMMERCIALIZATION OF TETRAPLOID TECHNOLOGY FOR OYSTERS

Standish K. Allen, Jr.* and Ximing Guo

Rutgers University
Haskin Shellfish Research Laboratory
Port Norris, New Jersey 08349

The first use of biotechnology in shellfish aquaculture was induced triploidy which has realized commercial success in the USA (Allen et al., 1989) and is of great interest in most other parts of the world. There are limitations to producing triploids in the hatchery that include toxicity of the treatment, reduced survival of the progeny, and often unpredictable results. Our lab has pioneered the use of tetraploid oysters for crossing with diploids to yield all-triploid progeny (Guo et al., 1996).

For all practical purposes, it can be concluded that tetraploids produce 100% triploids when mated with diploids. The production of all-triploid progenies also suggests that tetraploid oysters produce strictly diploid gametes. Triploid production by mating tetraploids and diploids is as simple as producing normal diploids, and no artificial treatment is needed (Guo et al., 1996). The mated triploids (C. gigas) were as viable as normal diploid controls in the lab. The survival of CB induced triploids, on the other hand, is always greatly reduced, in most cases by at least half (Allen et al., 1989). In other studies, the survival of triploids was mostly about 10–60% of normal diploids (Stanley et al., 1981; Tabarini, 1984; Downing and Allen, 1987; Allen and Bushek, 1992). Synchronization of meiotic events is critical for the inhibition of PB II by CB treatment (Allen et al., 1989). The CB treatment can be lethal to eggs if its misses a critical window of meiosis II (Allen et al., 1989; Guo et al., 1992a,b). Unsynchronized meiosis is not a problem for mated triploids.

Guo et al. (1996) examined the reproductive potential of tetraploids and the early data are encouraging. In studies in August 1994, they found that fecundity of tetraploid

* Current address: College of William and Mary, School of Marine Science, Virginia Institute of Marine Science, Gloucester Point, Virginia 23062-1346

New Developments in Marine Biotechnology, edited by Le Gal and Halvorson.
Plenum Press, New York, 1998.

Table 1. Fecundity of diploid and tetraploid females, in millions of eggs, enumerated after dissection from the gonad. Diploids came from a randomly mated population of about the same age

Group	n	Mean ($\times 10^6$)	SEM*
Diploid	11	5.18	0.99
Tetraploid	14	3.46	0.55

*Standard error of the mean.

females was relatively high, only slightly lower than diploids. Table 1 shows fecundity for 14 tetraploid females compared to 11 diploids of about the same size, although the diploids were not produced contemporaneously with the 4n group.

These same authors also learned that the breeding potential of tetraploids for creating triploids is high. In experiments conducted to examine survival of 2 × 2 crosses between diploids and tetraploids, survival of [4n × 2n] and the reciprocal were at least as high as the diploids control, and two orders of magnitude higher than triploids produced by standard induction procedures (3nCB in Table 2 below).

Natural triploids exhibited significant differences in growth. The larger eggs from tetraploids led to larger D-stage larvae at 24 hrs after fertilization. However, the reciprocal cross of diploid female × tetraploid male also yielded triploid D-stage larvae that were larger than their normal diploid counterparts. Size advantage of natural triploid oysters was also obvious at 8 months after spawning, about two months before first sexual maturity. At this time, triploid oysters were 35–51% larger than normal diploids. The size advantage of natural triploids decreased somewhat between eight and ten months after spawning, when sexual maturation occurred.

The development of tetraploids themselves was a protracted process littered with failures until we tested a rather unconventional approach. This approach calls for the use of eggs from triploid females crossed with sperm from diploids, followed by inhibition of the first polar body (PBI). Survival of the progeny to metamorphosis and beyond is hypothetically due to the increased volume of the triploid egg. We have documented the chromosome divisions giving rise to tetraploid embryos following the inhibition of PBI (of triploid eggs) and they are of the same nature as those seen in diploid eggs when PBI is inhibited (Que et al., 1997). Three year classes of tetraploid × diploid crosses in our lab have yielded > 99.95% (n~2000) triploidy, substantiating the practical value of this cross for both production and possibly, population control. For the latter, all-triploid populations of non-native oysters may be possible for aquaculture where the native species are not doing well.

Table 2. Percent survival of fertilized eggs to spat (30 days post-fertilization) of crosses between diploid (2n) and tetraploid (4n) Pacific oysters. Female ploidy is listed first

Group (f × m)	n	Mean (%)	SEM
[2n × 2n]	3	1.66	1.17
[2n × 4n]	4	5.41	2.7
[4n × 2n]	3	1.63	0.59
[4n × 4n]	3	0.002	0.001
3nCB[a]	1	0.01	—

[a]3nCB refers to triploids induced by standard techniques of CB treatment.

REFERENCES

Allen, Jr., S.K. and D. Bushek. 1992, *Aquaculture* **103**:241–251.
Allen, S.K., Jr., S.L. Downing, and K.K. Chew, 1989, University of Washington Press, Seattle, Washington.
Downing, S.L. and S.K. Allen, Jr. 1987, *Aquaculture* **60**:1–15.
Guo, X., Cooper, K., Hershberger, W.K., and Chew, K.K. 1992a, *Biol Bull* **183**:381–386.
Guo, X., Hershberger, W.K., Cooper, K., and Chew, K.K. 1992b, *Biol Bull* **183**:387–393.
Guo, X., G.A. DeBrosse and S.K. Allen, Jr. 1996, *Aquaculture* **142**:149–161.
Que, Huayong, X. Guo, F. Zhang, and S.K. Allen, Jr. 1997, *Biol. Bull.* **193**:14–19.
Stanley, J.G., S.K. Allen Jr., and H. Hidu. 1981, *Aquaculture* **23**:1–10.
Tabarini, C.L. 1984, *Aquaculture* **42**:151–160.

18

NEW TECHNOLOGY FOR THE ACCELERATION OF REPRODUCTIVE MATURATION IN ECONOMICALLY IMPORTANT CRUSTACEANS

Milton Fingerman, Rachakonda Sarojini, and Rachakonda Nagabhushanam

Tulane University

The ready availability of females that are about to spawn, they having fully developed ovaries, would help assure the successful operation of commercial facilities designed for aquaculture of economically important crustaceans. Consequently, finding ways to stimulate ovarian maturation by non-surgical intervention is a major goal of several laboratories. The approach in this laboratory is to identify endogenous neurotransmitters that control release of the neurohormones that regulate gonadal maturation, in particular the gonad-stimulating hormone (GSH) from the brain and thoracic ganglia. A second important neurohormone involved in reproduction is the gonad-inhibiting hormone (GIH) that is present in the sinus gland in each eyestalk. GIH (Panouse, 1943) was actually discovered earlier than was GSH (Otsu, 1963). Both GSH and GIH are present in males and females. The fact that GIH is produced in the eyestalk is the basis for the commonly used technique in shrimp hatcheries of eyestalk ablation to induce ovarian maturation. Removing this inhibitor leads to precocious gonadal maturation. However, eyestalk ablation occasionally results in a high incidence of mortality. Consequently, discovery of techniques to induce maturation without surgery has a high priority. Once the neurotransmitters that regulate release of GSH and GIH have been identified, these neurotransmitters and their agonists or antagonists may provide a means by using them to induce gonadal maturation as needed by the hatcheries.

Experiments done in this laboratory showed 5-hydroxytryptamine (5-HT) stimulates gonadal maturation in male and female sand fiddler crabs, *Uca pugilator* (Richardson *et al.*, 1991), and red swamp crayfish, *Procambarus clarkii* (Kulkarni *et al.*, 1992). This action of 5-HT is indirect, 5-HT stimulating release of GSH from the brain and thoracic ganglia. Studies with ovarian explants showed 5-HT has no direct effect on the ovary. But, when ovarian explants were incubated with 5-HT and brain or thoracic ganglia greater ovarian maturation occurred than when ovarian explants were incubated with brain or thoracic ganglia alone, 5-HT presumably enhancing GSH release. 5-HT in males, as in females, triggers GSH release; but in males GSH in turn stimulates the androgenic glands

New Developments in Marine Biotechnology, edited by Le Gal and Halvorson.
Plenum Press, New York, 1998.

which release the androgenic gland hormone, and it is this hormone that directly produces testicular maturation. The androgenic gland hormone is not a neurohormone; it is instead a classical type hormone, like the molting hormone. In male crustaceans GSH and GIH do not act directly on the testes, only indirectly regulating the testes by directly affecting the androgenic glands, and it is the androgenic gland hormone that directly controls the testes. Removal of both eyestalks from a male results in hypertrophy of the androgenic glands with subsequent testicular maturation. In contrast, in females GSH and GIH directly affect the ovaries. Support for this GSH-releasing role of 5-HT was obtained by use of 5-HT agonists and antagonists. The 5-HT receptor blocker LY53857 inhibited ovarian maturation in females whose ovaries were undergoing maturation because of their usual annual reproductive cycle. On the other hand, the 5-HT agonists fenfluramine, which is a 5-HT releaser, and fluoxetine, which is a 5-HT potentiator, induced accelerated ovarian maturation. Similar studies have also been done with males. Crustacean testes do not produce a hormone. The androgenic gland hormone is responsible not only for differentiation and maturation of the testes, but also for the development of the accessory reproductive structures. However, crustacean ovaries produce a hormone that induces development of the ovipositing setae and brood chamber.

Like 5-HT, red pigment-concentrating hormone (RPCH) stimulates ovarian maturation by stimulating GSH release (Sarojini *et al.*, 1995b). The role of RPCH as a hemolymph-borne neurohormone involved in crustacean color changes is well established (Brown, 1950). The presence of RPCH in all the major portions of the crustacean central nervous system, in neurons that appear not to be neurosecretory as well as in neurosecretory cells, had raised the question as to whether this neurohormone might have some other role in addition to that in color changes (Mangerich *et al.*, 1986). So we see interestingly that one of the earliest-known crustacean neurohormones has another role, namely as a neurotransmitter, which helps to explain its wide distribution in the central nervous system. The calcium ionophore A23187 mimics the action of RPCH on the ovaries, which suggests that calcium acts here as a second messenger for RPCH. In contrast to 5-HT and RPCH, dopamine (DA) inhibits gonadal maturation in both sexes (Sarojini *et al.*, 1995a,c). At least one of the ways DA exerts this effect is by inhibiting GSH release. However, it is possible that DA exerts part of this inhibitory action by triggering GIH release from the sinus glands also. A dopaminergic agonist and an antagonist were used on male fiddler crabs. The DA agonist ADTN inhibited testicular maturation, like DA, but the DA antagonist spiperone induced maturation, presumably blocking the antigonadal activity of endogenous DA.

By various techniques, including immunocytochemistry and HPLC, the presence of opioids in the crustacean central nervous system is now well established (Fingerman and Nagabhushanam, 1992). The opioid methionine enkephalin, like DA, was found to inhibit ovarian maturation in a dose related manner whereas the opioid antagonist naloxone induces ovarian maturation (Sarojini *et al.*, 1997). Methionine enkephalin and naloxone directly affect GSH release from the brain and thoracic ganglia. This inhibitory action of methionine enkephalin is due at least in part to inhibition of GSH release, but stimulation of GIH release may in addition also occur. Use of the highly specific delta opioid receptor agonist DADLE and the highly specific delta opioid receptor antagonist ICI-174,864 revealed that this action of methionine enkephalin is most likely mediated by delta opioid receptors. 5-HT and DA, like RPCH and methionine enkephalin, are indeed present in the crustacean central nervous system (Fingerman *et al.*, 1994). It is interesting that already two neurotransmitters have been identified, 5-HT and RPCH, that induce ovarian maturation and two, DA and methionine enkephalin, that inhibit this process. The results described herein provide the first evidence for crustaceans that not only classical aminergic neurotransmitters but also peptidergic neurotransmitters can have important roles in the regulation of ovarian maturation. Because both 5-HT and naloxone stimulate ovarian maturation, supplementing the diet with 5-HT and/or naloxone may be one way to accelerate reproductive activity, and thereby provide a new tool for crustacean mariculture.

REFERENCES

Brown, F.A., Jr., 1950, *Biol. Bull.*, **98**:218–226.
Fingerman, M. and Nagabhushanam, R., 1992, *Comp. Biochem. Physiol.*, **102C**:343–352.
Fingerman, M., Nagabhushanam, R., Sarojini, R. and Reddy, P.S.,1994, *J. Crust. Biol.*, **14**:413–437.
Kulkarni, G.K., Nagabhushanam, R., Amaldoss, G., Jaiswal, R.G. and Fingerman, M., 1992, *Invert. Reprod. Develop.*, **21**:231–240.
Mangerich, S., Keller, R. and Dircksen, H., 1986, *Cell Tiss. Res.*, **245**:377–386.
Otsu, T.,1963, Embryologia, **8**:1–20.
Panouse, J., 1943, *C.R. Acad. Sci. Paris*, **217**:553–555.
Richardson, H.G., Deecaraman, M. and Fingerman, M.,1991, *Comp. Biochem. Physiol.*, **99C**:53–56.
Sarojini, R., Nagabhushanam, R., Devi, M. and Fingerman, M.,1995a, *Comp. Biochem. Physiol.*, **111C**:287–292.
Sarojini, R., Nagabhushanam, R. and Fingerman, M., 1995b, *J. Exp. Biol.*, **198**:1253–1257.
Sarojini, R., Nagabhushanam, R. and Fingerman, M., 1995c, *Experientia*, **51**:156–158.
Sarojini, R., Nagabhushanam, R. and Fingerman, M.,1997, *Comp. Biochem. Physiol.*, **117C**:207–210.

19

ENDOCRINE FACTORS REGULATING CRUSTACEAN REPRODUCTIVE MATURATION

Lei Liu and Hans Laufer*

Department of Molecular and Cell Biology
University of Connecticut

1. INTRODUCTION

Crustaceans including shrimp, crab and lobster are major marine resources with a high commercial value. The U.S. alone imports more than $2 billion worth of shrimp annually. Aquaculture will serve as an important supplement to commercial shrimp fishing for domestic shrimp consumption. Actually, shrimp farming has grown into a $4 billion industry worldwide.

A major impediment for successful shrimp aquaculture is an inadequate knowledge of the controls of maturation and reproduction. We have undertaken an assessment of several endocrine substances and their metabolism that appear to play major functions in regulating reproduction in crustaceans. Methyl farnesoate (MF), an isoprenoid related to insect juvenile hormone, appears to be a key substance acting as a gonadotropin not only in commercially important crustaceans such as shrimp, crabs and crayfish but in model crustaceans (Laufer et al., 1987). The hormone is synthesized by the mandibular organs in highest concentrations by reproductively active males as well as in vitllogenic females in our model system the spider crab, *Libinia emarginata* (Laufer et al., 1997a). We found that MF has two major effects. In juveniles, MF can promote molting and regulates morphogenesis and metamorphosis during growth. In adult, MF appears to control reproduction. Thus, it can promote reproduction in shrimp by increasing fertility and hatchability (Laufer et al., 1997b).

A control on the synthesis of MF is an inhibitory neuropeptide hormone, mandibular organ inhibiting hormone (MOIH), from the eyestalk sinus gland-X organ. The sinus gland is considered to be the most important neuroendocrine regulation and integration center in crustaceans. It acts as a storage pool and release site for neuropeptides which are

* Correspondence: U-125, Department of Molecular and Cell Biology, University of Connecticut, Storrs, CT 06268 (USA) Tel: 860-486-4117, Fax: 860-486-4331

New Developments in Marine Biotechnology, edited by Le Gal and Halvorson.
Plenum Press, New York, 1998.

synthesized in peripheral neurons in the eyestalk. These peptides account for various physiological effects such as circadian and tidal rhythmicity, chromatic adaptation, glycogen metabolism, molting, gonadal development and cardiac activity (Beltz, 1988). Among the neuropeptides from the sinus gland, a novel family of peptides which is relatively large in size has been found. This family includes crustacean hyperglycemic hormone (CHH), molting inhibiting hormone (MIH), and vitellogenesis inhibiting hormone (VIH) (Keller, 1992). The members of this family share a high percentage of homology (60–90%) even among different species. Along with the structural similarities, overlapping biological effects have been observed. MIH of *Homarus americanus* causes hyperglycemia (Chang, 1990). CHH of *Carcinus maenas* inhibit ecdysteroid synthesis in isolated Y-organs (Webster and Keller, 1986). We found that certain CHH isomers of the crayfish *Procambarus clarkii* inhibit methyl farnesoate synthesis in MOs in culture and one of the isomers has been identified as MIH (Laufer et al., 1993).

Peptide isolation, functional bioassays and molecular biology methods for the crustacean SG neuropeptides have been well established in our lab.We have isolated several MOIHs from the crab *Libinia emarginata* (Liu and Laufer, 1996) and elucidated an amino acid sequence of one of the MOIHs (Liu et al., 1997). We also have cloned cDNA for MOIHs (Liu et al., 1997).

2. RESULTS AND DISCUSSION

Using a two-step HPLC procedure, we have identified and isolated 5 peptides with MO-IH activity. Three relatively large peaks: P21, P22 and P25 gave the highest inhibition. The molecular masses of these were determined by mass spectrometry as P21, 8,398 Dalton, P22, 8,474 Dalton and P25 8,439 Dalton. Amino acid analysis shows similar amino acid composition for P22 and P25 with 72 and 73 residues respectively, including 5 cysteines. These features are very similar to those of the CHH isomers identified by others. When the peptides which we purified were injected into de-eyestalked fiddler crabs, *Uca pugillator*, they caused dramatic increases of glucose level in the hemolyph—the expected hyperglycemic effect. These peptides appear therefore to be multi-functional.

Because the direct amino acid sequencing of the intact peptides did not give any results, the peptides appear to be blocked at their N-termini. One of the common modifications at the N-terminus of neuropeptides is pyroglutamic acid. It can be removed by pyroglutamate aminopeptidase. We found the behavior of the peptide on HPLC changed after treated with pyroglutamate aminopeptidase. This confirmed the first amino acid residue which was determined to be pyroglutamate. The peptide was digested by Lys-C and Glu-C endo-peptidase to generate short peptides with free N-termini. The digested fragments were collected and sequenced. The complete sequence was resolved by mapping the fragments together (for details, see Liu et al., 1997). P22 has an amino acids sequence of 72 residues and resembles the CHH of *C. maenus* by 70%.

A cDNA library was constructed using the λZiplox cDNA cloning kit (GibcoBRL) according to the manufactory instruction manual. Screening 500,000 clones using the homologous probe generated by PCR resulted in 9 positive clones. One of the clones contained the longest insert (about 1 kb). The full-length cDNA (972 bp) encoding the preproMOIH was elucidated. This cDNA sequence encodes a preproMOIH with 137 amino acid residues with a 78 bp 5′ untranslated region and a 482 bp 3′ untranslated region. The preprohormone peptide is proceeded by a 26-amino acid long signal peptide, followed by a 34-amino acid long precursor related peptide. The dibasic peptide Lys–Arg between the precursor related peptide and the MOIH indicates the cleavage site of the precursor related peptide during the post-translational modification.

In summary, the research presented here provides us important information on the isolation, characterization, structure and functions of the MOIHs in crustaceans. This information may be used in practical applications for aquaculture. Major evidence that MF functions as a gonadotropin during maturation is that it can be administered to crustaceans, including economically important species, such as *Procambarus clarkii* and *Penaeus vannamei* to enhance egg maturation and larva production.

ACKNOWLEDGMENTS

The research reported here was supported in part by grants from the Sea Grant College Program, NOAA, whose support is gratefully acknowledged.

REFERENCES

Beltz, Barbara S., 1988, Crustacean Neurohormones, in Laufer, H. and Downer, R. (eds): Endocrinology of Selected Invertebrate Types, New York, Alan R. Liss, Inc., pp. 235–258.

Chang, Ernest S., Prestwich, Glenn D., and Bruce, Marilyn J., 1990, *Biochemical and Biophysical Research Communication*, 171(2), pp. 818–826.

Keller, R., 1992, *Experientia*, **48**:439–448.

Laufer, H., P. Takac, J.S.B. Ahl, and M.R. Laufer, 1997a, *Invert. Reprod. Devel.*, **31**:63–68.

Laufer, H., J. Paddon, and M. Paddon, 1997b, A hormone enhancing larva production in the pacific white shrimp, *Penaeus vannamei.* In: IV symposium on aquaculture in Central America: focusing on shrimp and tilapia, D.E. Alston, B.W. Green and H.C. Clifford, (eds.), the Latin American Chapter of the World Aquaculture Society, pp. 161–162.

Laufer H., L. Liu, and F. Van Herp, 1993, A neuropeptide family that inhibits the mandibular organ of crustacean and may regulate reproduction, Insect Neurochemistry and Neurophysiology, ed. Alexej B. Bovkovec and Marcia J. Loeb, CRC press, Boca Ruton, FL, p. 203–206.

Laufer, H., Borst, D., Baker, F. C., Carrasco, C., Sinkus, M., Reuter, C. C., Tsai, L., W., and Schooley, D. A., 1987, *Science*, **235**:202–205.

Lei Liu, Hans Laufer, Peter J. Gogarten, and Minhua Wang, 1997, cDNA cloning of a mandibular organ inhibiting hormone from the spider crab Libinia emarginata, in 5th International Conference of Invertebrate Neurochemistry and Neurophysiology, Sept.7–11, (in press).

Lei Liu, Hans Laufer, Yajun Wang, and Timothy Hayes, 1997, *Biochem. Biophy. Res. Comm.*, **237**:694–701.

Lei Liu and Hans Laufer, 1996, *The Archive of Insect Biochemistry and Physiology*, **32**:376–386

Webster, S.G. and Keller, R., 1986, *J. Comp. Physiol. B* **156**:617–624.

20

STUDIES ON THE SEA BASS *DICENTRARCHUS LABRAX* L. IMMUNE SYSTEM FOR ITS CONTROL IN AQUACULTURE

G. Scapigliati, L. Abelli, N. Romano, L. Mastrolia, and M. Mazzini

Dipartimento di Scienze Ambientali
Università della Tuscia
I-01100 Viterbo, Italy

To achieve a more comprehensive analysis of the anatomy and physiology of the sea bass immune system, and to address important questions concerning the immune response of this fish species reared in many countries because of its high economical relevance, we report in the present work on the data obtained by using the anti-thymocytes and T-cells monoclonal antibody (Mab) DLT15, and the anti-immunoglobulin and B cells Mab DLIg3.

The Mab DLT15, obtained by immunizing mice with paraformaldehyde-fixed thymocytes, was previously described as recognizing specifically in immunofluorescence and FACS analysis thymocytes and peripheral T-cells of the sea bass. For the first time in fish immunology the distribution of T-cells in lymphoid and non-lymphoid organs of a teleost fish was shown (Scapigliati et al., 1995; Romano et al., 1997). The Mab DLT15 was employed in immunocytochemistry to study the ontogeny of T-cells, the results have shown that thymus develops in sea bass at day 30 post hatching and it's a primary organs for T-cells, and the appearance of DLT15-immunoreactive cells in the lymphoid organs was in the sequence thymus, head kidney and spleen (Abelli et al., 1996). The presence of a mucosa-associated immune system in teleosts has been established (Peleteiro and Richards, 1985; Rombout et al., 1993; Joosten, 1997), and current research focuses on this lymphoid compartment to improve methods for immunizing fish against disease. The studies on the teleost gut-associated lymphoid tissue (GALT) have shown that like in mammals the majority of GALT lymphoid cells lack surface immunoglobulins (Rombout et al., 1993) and have been consequently regarded as T-cells. The Mab DLT15 was employed to show by immunocytochemistry that the majority of GALT lymphocytes are T-cells (Abelli et al., 1997), and to assess their ontogeny (Picchietti et al., 1997). GALT T-cells have been purified from percoll-enriched cells preparations by using the Mab DLT15 and immunomagnetic sorting, and this protocol gave pure fractions of T-cells

New Developments in Marine Biotechnology, edited by Le Gal and Halvorson.
Plenum Press, New York, 1998.

Table 1. Percentages of Mab-positive cells in lymphoid and non-lymphoid organs of sea bass[a]

	DLIg3	DLT15
Thymus	2.5 ± 2.1	79.5 ± 10
PBL	21.1 ± 2.9	3.1 ± 1.2
Spleen	29.9 ± 5.8	7.1 ± 2.9
Head kidney	33.1 ± 4.6	6.5 ± 3.1
GALT	2.7 ± 1.8	55.3 ± 2.9
MALT (gills)	3.1 ± 2.1	60.4 ± 5.3

[a]Cells were enriched in leucocyte fractions by centrifugation on Percoll gradients at a density of 1.07 g/cm^3 and analyzed for their immunofluorescence by FACS analysis. Numbers represent the mean ± SD of diverse experiments (positive values minus the background staining). PBL = peripheral blood leucocytes; GALT = gut-associated lymphoid tissue; MALT = mucosa-associated lymphoid tissue.

(98% by FACS). DLT15-purified cells have been used to prepare total RNA which was copied in cDNA. By RT-PCR assay using primers obtained from the sequence of trout T-cell receptor (TcRβ) (Partula et al., 1995), a 104 bp DNA was amplified whose sequence was very similar to that of trout TCRβ and also similar, at a lesser extent, to the sequence of other vertebrate TcR's.

The Mab DLIg3, obtained immunizing mice with purified sea bass immunoglobulins (Ig), was previously described as recognizing in western blotting the light chain of Ig, and in immunofluorescence and FACS analysis Ig-bearing cells from lymphoid and non-lymphoid organs (Scapigliati et al., 1996; Romano et al., 1997). The Mab DLIg3 has been also employed to setup ELISA assays for the quantitation of total and antigen-specific Ig from sea bass serum (Scapigliati et al., 1996). By this assay it was shown that the total Ig content raised with the age of the fish, being 0.4 ± 0.18 mg/ml at 7 months, 5.6 ± 3.8 mg/ml at two years, and 11.3 ± 4.8 mg/ml at ten years. The seasonal variations of Ig content were analyzed, and during a two-years monitoring, the immunoglobulin levels were slightly different in spring (6.7 ± 3 mg/ml), in summer (3.83 ± 1.25 mg/ml), and in winter (7.8 ± 4.9 mg/ml). This seasonal trend could also be related to the winter reproductive cycle of sea bass, since detectable amounts of Ig were determined by ELISA assays of mature egg content. The Mab DLIg3 was employed in immunocytochemistry to study the "in situ" distribution of B-cells in lymphoid and non-lymphoid organs (Scapigliati et al., 1995; Abelli et al., 1996; Romano et al., 1997), and to show the presence of immunoreactive Ig in early embryos.

To reassume, the percentage of T-cells and B-cells in sea bass lymphoid and non-lymphoid organs determined by immunofluorescence and FACS analysis is shown in Table 1.

Taken together, we have produced and characterized two Mab that are useful tools to study the morpho-functional organization of the sea bass immune system. At the present, the sea bass is the only teleost fish for which anti T-cells and anti B-cells Mab are available, and these reagents will be indispensible for the monitoring of humoral and cellular immune responses after antigen challenge and/or vaccination of sea bass in aquaculture.

REFERENCES

L. Abelli, S. Picchietti, N. Romano, L. Mastrolia and G. Scapigliati.1996, *Fish Shellfish Immunol.*, **6**:493–505,.
L. Abelli, S. Picchietti, N. Romano, L. Mastrolia and G. Scapigliati, 1997, *Fish Shellfish Immunol.*, **7**:235–246.

Joosten, E. P. H. M., 1997, Immunological aspects of oral vaccination in fish. PhD thesis, Wageningen Agricultural University, Wageningen, The Netherlands

Peleteiro, M. C. and Richards, R. H., 1985, *J. Fish Dis.* **8**:161.

S. Picchietti, F. R. Terribili, L. Mastrolia, G. Scapigliati and L. Abelli, 1997, Expression of lymphocyte antigenic determinants in developing GALT of the sea bass *Dicentrarchus labrax* (L.). *Anat. Embriol.*, in press.

Romano, N., Abelli, L., Mastrolia, L. and Scapigliati, G1997, Immunocytochemical detection and cytomorphology of lymphocyte subpopulations in a teleost fish *Dicentrarchus labrax* (L). *Cell Tissue Res.*, **289**:163–171.

Rombout, J. H. V. M., Taverne-Thiele, J. J. and Villena, M. I. 1993, *Dev. Comp. Immunol.* **17**:55.

G. Scapigliati, M. Mazzini, L. Mastrolia, N. Romano and L. Abelli, 1995, *Fish Shellfish Immunol.*, **5**:393–405.

G. Scapigliati, N. Romano, S. Picchietti, M. Mazzini, L. Mastrolia, D. Scalia and L. Abelli, 1996, *Fish Shellfish Immunol.*, **6**:383–401.

21

DEVELOPMENT OF DNA VACCINES FOR AQUACULTURE

Joël Heppell,[1] Tong Wu,[1] Niels Lorenzen,[2] Anthony E. Ellis,[3] Susan M. Efler,[1] Neil K. Armstrong,[1] Joachim Schorr,[4] and Heather L. Davis[1]

[1]Ottawa Civic Hospital Loeb Research Institute
725 Parkdale Avenue
Ottawa K1Y 4E9, Canada
[2]Danish Veterinary Laboratory
Hangøvej 2
DK8200 Århus, Denmark
[3]SOAFD Marine Laboratory
P.O. Box 101, Victoria Road
Aberdeen AB1 9DB, Scotland
[4]Qiagen GmbH
Max-Volmer-Strasse 4
4010, Hilden, Germany

Aquaculture is a rapidly growing industry in many countries (Meyer, 1991; Leong & Fryer, 1993). Its development is driven by an increasing demand for fish and shellfish products, and the continuous decline of wild caught harvests (Meyer, 1991; Hanfman, 1993). However, loss due to infectious disease is a major concern for fish farmers. Approximately 10% of all aquatic animals are lost annually due to pathogens (Leong & Fryer, 1993). For various reasons, such as inefficacy, high cost and/or environmental concerns, traditional antigen-based vaccines (*i.e.*, whole killed, live attenuated and subunits) have only been used successfully for the prevention of a limited number of diseases, mainly those of bacterial origin. DNA vaccination could circumvent many of the disadvantages associated with classical methods of immunization. In mammals and birds, they have been shown to induce very potent, long-lasting and protective humoral and cell-mediated immune responses against numerous viral, bacterial or parasitic diseases (Wahren, 1996; Donnelly *et al.*, 1997; Davis & Brazolot Millan, 1997).

To further investigate the potential of DNA vaccines for aquaculture, plasmid DNA encoding the luciferase or *lacZ* reporter genes under the control of the cytomegalovirus (CMV) immediate early promoter (pCMV-luc and pCMV-lacZ, respectively) were injected intramuscularly (IM) into adult zebra fish, *Brachydanio rerio* (0.2–0.9 g), or young rainbow

New Developments in Marine Biotechnology, edited by Le Gal and Halvorson.
Plenum Press, New York, 1998.

trout, *Oncorhynchus mykiss* (0.5–2.0 g). High luciferase activity was detected in the injected muscle with all doses of DNA. Expression of the reporter gene was dose- dependent, and when compared to values obtained with Balb/C mice injected IM with similar doses of pCMV-luc, luciferase activity in fish muscle was significantly higher (Table 1). Since effective immunization of mice is possible with low doses of DNA, this boded well for application of DNA vaccines to fish.

To study the longevity of foreign gene expression in fish, other groups of animals were injected with 0.1 μg (zebra fish) or 1 μg (rainbow trout) of pCMV-luc and sacrificed at different time points. Luciferase activity in injected muscle increased between 0 and 60 h after direct gene transfer, then stayed constant for at least 12 weeks.

Other rainbow trout were injected with plasmid DNA encoding the *lacZ* gene, and killed at different time points for histological evaluation of the injected muscle. Transverse sections of the muscle tissue were cut with a cryostat microtome, stained with eosin and X-gal (for β-galactosidase activity) and compared with sections from fish injected with saline only (stained with haematoxylin and eosin). Cells expressing the *lacZ* gene (stained in blue) appeared completely normal at all times, and other than the needle-stick injury, which had disappeared within 28 days, no tissue damage was caused by the DNA vaccine per se.

Thus, experiments with reporter genes showed that DNA injection in fish muscle induce high expression levels of foreign genes for a prolonged period of time, without causing permanent tissue damage.

To demonstrate that expression of plasmid-encoded antigen in fish can induce protective immune responses against infectious diseases, two fish pathogens were used as models. The first one, viral haemorrhagic septicaemia virus (VHSV) is a rhabdovirus responsible for a very important disease in European trout farms (Lorenzen *et al.*, 1993). It has also been isolated in marine fish such as turbot (Meier *et al.*, 1994; Stone *et al.*, 1997).

Table 1. Luciferase activity in fish epaxial muscle, and mouse tibialis anterior (TA) muscle, after IM injection with different doses of plasmid DNA encoding the luciferase reporter gene

DNA injected (μg)[a]	Luciferase activity (RLU/sec)b		
	Rainbow trout[c]	Zebra fish[c]	Mouse[d]
0.01	65 898 (± 30 774)	17 089 (± 10 520)	N.D.
0.1	327 724 (± 177 583)	365 481 (± 91 636)	1649 (± 542)
1	1 100 347 (± 669 634)	23 694 854 (± 10 115 959)	5466 (± 1536)
10	3 225 068 (± 1 869 474)	4 995 132 (± 2 583 771)	43 082 (± 5419)
50	4 520 741 (± 1 609 457)	3 879 339 (± 2 236 777)	70 713 (± 15 921)

[a]DNA was diluted in 10 μl of saline (0.15 M NaCl) for fish, or 50 μl of PBS for mice.
[b]Luciferase activity was measured with a luminometer and total relative light unit (RLU) per second for the whole homogenized tissue section was calculated.
[c]Mean luciferase activity (RLU/sec) ± standard error of the mean, detected 60 h after injection. For each dose, n ≥ 5.
[d]Mean luciferase activity (RLU/sec) ± standard error of the mean, detected 72 h after injection. For each dose, n = 10.
N.D. = not done

Genes coding for the glycoprotein G and nucleocapsid N protein of VHSV were cloned individually into an expression vector for DNA vaccination (pCMV-vhsG and pCMV-vhsN constructs, respectively).

Rainbow trout (average weight 13 g) were injected with recombinant plasmids pCMV-vhsG and/or pCMV-vhsN. For each vaccination group, 66 to 70 animals divided among three aquaria, were challenged by bath inoculation with live VHSV 52 days after vaccination. Mortalities were recorded for the next 28 days and relative percent of survival (RPS) was calculated (Table 2).

In the control group injected with a non-expressing plasmid (vector without insert), 92.5% (62/67) of the fish died compared to less than 6% in groups injected with pCMV-vhsG alone or in combination with pCMV-vhsN. No significant difference was observed between fish injected with 10 and 50 μg pCMV-vhsG, suggesting that smaller doses would still provide the same high level of protection. Compared to the G construct, the RPS value obtained with N alone was lower. Nevertheless, all DNA vaccine formulations were superior to the killed vaccine made of inactivated VHSV particles.

With a plaque neutralization assay, anti-VHSV neutralizing antibodies were detected in sera collected from non-challenged fish injected with pCMV-vhsG, but not with pCMV-vhsN. Since pCMV-vhsN alone provided protection to fish, cell-mediated immunity could play a relatively more important role in the case of the N gene.

The second pathogen used as model, infectious pancreatic necrosis virus (IPNV), is a birnavirus isolated worldwide in a large variety of marine and freshwater fish species. Sickness has been traditionally associated with very young salmonids, in which it is sometimes responsible for very high mortality rates (Wolf, 1988), but more recently it has also been observed in older fish (Smail *et al.*, 1992, 1995; Stangeland *et al.*, 1996). IPNV genome segments coding for the VP2 and VP3 proteins, were cloned individually into an eukaryotic expression plasmid for DNA vaccination (pCMV-VP2t and pCMV-VP3 constructs, respectively). These recombinant plasmids were injected IM, alone or in combination, into 50 g Atlantic salmon (*Salmo salar*) kept in fresh water. Sera were collected at different time points after injection, and evaluated for an IPNV-specific antibody response. Preliminary ELISA data showed that salmons raised antibodies to IPNV. More experiments are underway to determine the neutralization titer of the fish sera.

Results reported herein show that DNA vaccines can provide an inexpensive, safe and efficacious alternative to the classical antigen-based approaches for the aquaculture industry. They should also allow the development of effective vaccines against diseases,

Table 2. Relative percent of survival (RPS) of rainbow trout vaccinated with different DNA- or antigen-based vaccines after challenge with live VHSV

Vaccine used	DNA or protein injected per fish (μg)[a]	RPS[b]
pCMV-vhsG	50	96.8
pCMV-vhsG	10	93.8
pCMV-vhsN	5	65.6
pCMV-vhsG + pCMV-vhsN	5 + 5	93.7
Non-expressing plasmid (control)	50	—
Inactivated VHSV (antigen-based vaccine)	1.3[c]	55.4

[a]DNA was injected IM, in the epaxial muscle. Inactivated VHSV was injected intraperitoneally. In both cases, the total volume injected per fish was 25 μl.

[b]RPS was calculated as follows: 100 × {1 – (% mortality in vaccinated group / % mortality in control group)}.

[c]Total amount of protein.

such as viral haemorrhagic septicaemia, for which it has not been possible heretofore to develop an effective or sufficiently inexpensive vaccine.

REFERENCES

Davis, H.L. and Brazolot Millan, C.L. 1997, *Blood Cell Biochemistry* (in press).

Donnelly, J.J., Ulmer, J.B., Shiver, J.W. and Liu, M.A. 1997, *Annual Review of Immunology* **15**: 617–648.

Hanfman, D.T. 1993, The status and potential of aquaculture in the United States: an overview and bibliography. National Agricultural Library, Beltsville, MD, USA.

Leong , J.C. and Fryer, J.L. 1993, *Annual Review of Fish Diseases* **3**: 225–240.

Lorenzen, N., Olesen, N.J., Vestergård Jørgensen, P.E., Etzerodt, M., Holtet, T.L. and Thøgersen, H.C. 1993, *Journal of General Virology* **74**: 623–630.

Meier, W., Schmitt, M. and Wahli, T. 1994,. *Annual Review of Fish Diseases* **4**: 359–373.

Meyer, F.P. 1991, *Journal of Animal Sciences* **69**: 4201–4203.

Smail, D.A., Bruno, D.W., Dear, G., McFarlane, L.A. and Ross, K. 1992, *Journal of Fish Diseases* **15**: 77–83.

Smail, D.A., McFarlane, L., Bruno, D.W. and McVicar, A.H. 1995, *Journal of Fish Diseases* **18**: 631–638.

Stangeland, K., Høle, S. and Taksdal, T. 1996, *Journal of Fish Diseases* **19**: 323–327.

Stone, D.M., Way, K. and Dixon, P.F. 1997, *Journal of General Virology* **78**: 1319–1326.

Wahren, B. 1996, *Immunotechnology* **2**: 77–83.

Wolf, K. 1988, Fish Viruses and Fish Viral Diseases. Cornell University Press, Ithaca, NY, USA.

22

GENETIC MANIPULATION AND STRAIN IMPROVEMENT IN COMMERCIALLY VALUABLE RED SEAWEEDS

D. Cheney,[1] B. Rudolph,[2] L. Z. Wang,[1] B. Metz,[1] K. Watson,[1] K. Roberts,[1] and I. Levine[3]

[1]Seaweed Biotechnology Laboratory
Marine Science Center
Northeastern University
Nahant, Massachusetts 01908
[2]Copenhagen Pectin
DK-4623 Lille Skensved, Denmark
[3]Coastal Plantations International, Inc.
Poland, Maine 04274

1. INTRODUCTION

Over the past twenty years, seaweed cultivation has advanced to a level where there is hardly a single commercially valuable species that isn't currently being cultivated. The largest cultivation industries for red seaweeds include the production of carrageenans from *Eucheuma* and *Kappaphycus*, of agar from *Gracilaria* and, the largest of all, the production of *Porphyra* or nori. In most of these cases, future improvements in production will most likely depend more on improving the suitability of the species being cultured than improvements in culture methods. Seaweed strain improvement efforts have traditionally relied upon classical plant breeding techniques (for reviews, see Cheney, 1990; Patwary and van der Meer, 1992), with the single most commonly used method being simple strain selection of wild plants or of new genetic variants produced by mutagenesis, colchicine treatment, and to a lesser extent, the use sexual hybridization. Sexual hybridization hasn't had much impact with commercially valuable red algae because it has been generally impractical or impossible in most commercially valuable red seaweed (see Cheney, 1990).

Until recently there has not been a practical and generally applicable approach to cultivar improvement in commercially valuable red seaweeds. However, we believe that has now changed. Over the past 15 years, several new techniques have been developed at the Seaweed Biotechnology Laboratory of Northeastern University which have made the production of genetically improved cultivars of red seaweeds not only possible but practi-

New Developments in Marine Biotechnology, edited by Le Gal and Halvorson.
Plenum Press, New York, 1998.

cal. Through advancements in protoplast fusion, mutagenesis, and a new technique called cell-cell fusion, we have produced genetically altered and improved new strains (including hybrids) of the carrageenophytes *Eucheuma, Kappaphycus* and *Chondrus,* the agarophyte *Gracilaria*, and the world's most widely eaten seaweed nori or *Porphyra*. The purpose of this short report is to present a brief overview of our approach to strain improvement and to provide some examples of its success. Further details of our work will be presented elsewhere.

2. METHODS AND RESULTS

One of the key ingredients of any modern plant strain improvement program is an understanding of tissue culture and protoplast technology of the species of interest. We started over twelve years ago to do this by developing tissue culture and protoplast isolation techniques in *Agardhiella* and *Gracilaria,* respectively (Bradley and Cheney, 1990; Cheney, et al., 1986). At that time, protoplast fusion was just beginning to be applied to seaweeds. Several reports of successful protoplast fusion in seaweeds appeared in the late 1980s and early 1990s (e.g. Cheney, 1990; Fujita and Migita, 1987; Fujita and Saito, 1990; Reddy et al., 1992), however, few studies provided proof of having produced true hybrids. In addition, fusion techniques were very limited in the species to which they could be applied, primarily because of the inability to regenerate protoplasts in most anatomically-complex seaweeds.

We have developed a protoplast fusion technique that gets around the regeneration problem in seaweeds which we have called "spore-protoplast fusion". Because of the high regeneration rate of spore-protoplasts, the spore-protoplast fusion technique has been demonstrated to work with species of seaweeds in which the traditional source of protoplasts (ie. vegetative tissue) does not produce protoplasts capable of regeneration. So far, we have successfully used the technique with the carrageenophytes *Eucheuma*, *Kappaphycus* and *Chondrus*, as well as the agarophyte *Gracilaria*. In *Gracilaria,* spore-protoplasts were fused between two different species, *G. chilensis* and *G. tikvahiae*, using the fusagen PEG (see Cheney, 1990) and hybridity was determined using isoenzyme electrophoretic techniques. The hybrid plant possessed a combination of phosphosglucose isomerase (PGI) patterns that served as genetic markers of the parental species (Cheney and Duke, 1995). Similarly, we have also fused spore-protoplasts of the different life history phases of *Chondrus crispus*, which produce different types of carrageenan in the hope of producing novel combinations. The haploid phase of *C. crispus* produces *kappa* carrageenan (plus a small amount of *iota)* while the diploid phase produces *lambda* carrageenan. One plant we've produced so far from a N&2N fusion experiment appears to produce a novel type of *kappa* carrageenan. Specifically, it exhibits a reduced level of 3,6 anhydrogalactose based upon IR analyses conducted by us and molar ratio determinations conducted by a collaborator, Thierry Chopin of the University of New Brunswick. The spore-protoplast fusion technique is protected by a US patent (Cheney and Duke, 1995), as well as several foreign patents.

A second technique has also been developed by us to produce somatic hybrids in which protoplasts are not involved but tissue culture is utilized; we have called this technique "cell-cell fusion". In this technique, tissue cultured-generated filaments from two different species are co-cultured in such a manner that nuclei are exchanged between them via the formation of connector cells. This technique has been utilized to produce some interesting and unique plants between fusions of the *iota* carrageenan-producing *Eucheuma*

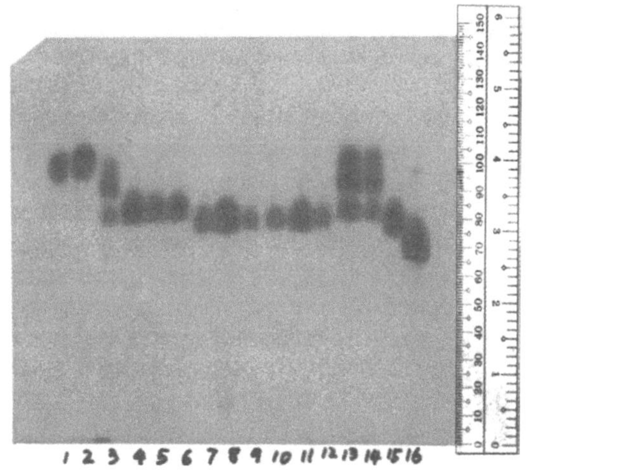

Figure 1. Phosphoglucose isomerase (PGI) patterns of *Eucheuma denticulatum* (lanes #1 & 2), *Kappaphycus alvarezii* (lanes #4–11) and putative hybrids (lanes #3, 13 & 14).

denticulatum (previously called *E. spinosum*) and the *kappa* carrageenan-producing *Kappaphycus alvarezii* (previously called *E. cottonii*). An example of the unique PGI isoenzyme pattern exhibited by a hybrid produced by cell-cell fusion is shown below (Fig. 1). More interesting from a commercial view point is that some plants produced in this manner produce altered and novel carrageenan compositions. The cell-cell fusion technique is protected by a US patent (Cheney and Wang, 1996), as well as several foreign patents.

Lastly, we have recently begun developing new techniques for fusing protoplasts of the edible seaweed *Porphyra* or nori. To date, we have concentrated on fusions between the plant currently being commercially farmed in Maine, *Porphyra yezoensis*, and a local species, *P. umbilicalis*. *Prophyra* is a particular challenge because others have tried and apparently so far failed to produce a true hybrid plant (see Fujita and Migita, 1987; Fujita and Saito, 1990). Although our results are just preliminary at this point, they suggest that we may have developed a new approach for producing polyploids and/or hybrids in *Porphyra* (see Levine and Cheney, this volume).

In conclusion, we believe that the techniques that we have developed so far and are continuing to develop hold great promise for strain improvement of red seaweeds, and that they are an example of what can be accomplished using biotechnological techniques.

ACKNOWLEDGMENTS

Support from Copenhagen Pectin, a division of Hercules, Inc., Coastal Plantations International, and the National and MIT Sea Grant Programs is gratefully acknowledged.

REFERENCES

Bradley, P. and D. Cheney, 1990, *Hydrobiologia* **204/205**:353–360

Cheney, D., 1990, In: *Economically important marine plants of the Atlantic*, C. Yarish et al. (eds), Conn. Sea Grant College Program, pp 15–25.

Cheney, D. and C. Duke, 1995, US Patent No. 5,426,040. "Methods for producing improved strains of seaweed by fusion of spore-protoplasts, and resultant seaweeds·ans phycocolloids".

Cheney, D. and LZ Wang, 1996, US Patent No. 5,585,544. "Method of causing somatic hybridization between two species of algae".

Fujiita, Y. and S. Migita, 1987, *Jpn. J. Phycol.* **35**:201–208.

Fujita, Y. and M. Saito, 1990, *Hydobiologia* **204/205**:161–166.
Levine, I. and D. Cheney, North American *Porphyra* cultivation—from molecules to markets (this volume).
Patwary, M. and J. van der Meer, 1992, *Korean J. Phycology* **7**:281–318.
Reddy, C. et. al., 1992, *J. of Applied Phycol.* **4**:57–65

23

EXPRESSION OF THYROID HORMONE RECEPTOR-α IN THE GROWTH AND DEVELOPMENT OF THE SEA BREAM (*SPARUS AURATA*)

Lynda Llewellyn,[1,2] Vimi P. Ramsurn,[1,2] Trevor Wigham,[2] Deborah M. Power,[3] and Glen E. Sweeney[1]

[1]School of Molecular and Medical Biosciences
University of Wales
P.O. Box 911
Cardiff CF1 3US, United Kingdom
[2]School of Pure and Applied Biology
University of Wales
P.O. Box 915
Cardiff CF1 3TL, United Kingdom
[3]Unidade de Ciencias e Tecnologias Agrarias
Universidade do Algarve
Campus de Gambelas
Faro 8000, Portugal

1. INTRODUCTION

Thyroid hormones (THs) play a number of roles in vertebrate organisms. In mammals, in addition to influencing basic metabolic rate, THs affect prenatal and post-natal development, with transgenic mice lacking the thyroid hormone receptor (TR)-α dying within a few weeks of birth, apparently due to an inability to complete some of the maturation processes required for survival in the post-weaning period (Fraichard *et al.*, 1997). Additionally, TH deficiency during differentiation causes a number of abnormalities including neurological and skeletal defects.

Metamorphosis in amphibia is induced by THs, which have also been implicated in fish growth and development. Peaks of TH have been observed during Japanese flounder metamorphosis (de Jesus *et al.*, 1993) and salmon smoltification (Folmar and Dickhoff, 1980). Treatment of larvae with TH has been shown to promote the survival of threadfin larvae (Brown and Kim, 1995) but has also produced skeletal abnormalities in other species (Higgs *et al.*, 1982), however the role of THs in fish is still unclear.

New Developments in Marine Biotechnology, edited by Le Gal and Halvorson.
Plenum Press, New York, 1998.

The sea bream (*Sparus aurata*) is a major aquaculture species in Southern Europe, in 1996 sea bream accounted for at least 50% of farmed fish in the Mediterranean and Europe. Cultivated sea bream however show a higher incidence of skeletal abnormalities which reduce viability and commercial success (Santamaria *et al.*, 1994). In spite of the potential importance of the role of hormones during the growth and development of the sea bream there is little information available as to their precise function. THs mediate their action by binding to thyroid hormone receptors (TRs), of which there are two types TR-α and TR-β, which act directly on target genes. During metamorphosis in amphibia there is a rapid increase in the concentration of TRs which coincides with the response to T_3 (triiodothyronine) (Yaoita and Brown, 1990). The aim of this investigation was to study the expression of thyroid hormone receptor mRNAs during the growth and development of the sea bream to ascertain whether they play a role in this process.

2. RESULTS

Total RNA was extracted from liver of a sea bream 24 h after treatment with T_3 (1 μg/g) and used as a template for cDNA synthesis. PCR was carried out using degenerate oligonucleotide primers, designed from well conserved regions of published TR sequences. A 387 bp PCR product was generated which was subcloned and sequenced and showed homology to TRs. The alignment of the amino acid sequence of sea bream TR with other TRs is shown in Figure 1. The high amino acid identity with flounder TR-α (93%) and *xenopus* TR-α (82%) compared to *Xenopus* TR-β (70%) indicated that the sea bream TR was the α form of the receptor.

This fragment of sea bream TR-α was subsequently used in expression studies using RT-PCR. β-actin and EF1-α were used as controls for the tissue and larval experiments respectively. Juvenile sea bream (approx. 100 g) were treated with T_4 (1 μg/g) and tissues were extracted 24 h following treatment. The highest level of TR-α mRNA expression was

```
Sea bream TR   ---------------------------IEENRQRRKREEMVRTLQTRPEPNTAEWELIRM
Flounder α     RFKKCISVGMAMDLVLDDSKRVAKRRLIEENREKRKREEMVRTLQIRPEPDTAEWELIRM
Flounder β     RFKKCIAVGMATDLVLDDSKRLAKRKLIEENRERRRKEELQKTVWDRLEPTQEEWDLIRM
Xenopus α      RFKKCIAVGMAMDLVLDDSKRVAKRKLIEENRQRRRKEEMIKTLQQRPEPSSEEWELIRI
                                          *****  *  **   *   * **   ** ***

Sea bream TR   VTEAHRHTNAQGSSWKQKRKFLS---------DDIGQGPVVPTSEGDKVDLEAFSEFTKI
Flounder α     ATDAHRHTNAQGSSWKQKRKFLS---------DDIGQSPMVPTSDGDKVDLEAFSEFTKI
Flounder β     VTEAHMATNAQGNHWKQKRKFLSAAGVKEDKPEEIGQASMANTPEGNKVDIEAFSQFTKI
Xenopus α      VTEAHRSTNAQGSHWKQRRKFLP---------EDIGQSPMASMPDGDKVDLEAFSEFTKI
                * **  *****  *** ****            ***        * *** **** ****

Sea bream TR   MTPAITRVVDFAKKLPMFSELPCEDQIILLKGCCMEIMSLRAAVR---------------
Flounder α     MTPAITRVVDFAKKLPMFSELPCEDQIILLKGCCMEIMSLRAAVRYDPDSETLTLNSEMA
Flounder β     ITPAITRVVDFAKKLPMFCELPCEDQIILLKGCCMEIMSLRAAVRYDPESETLTLNGEMA
Xenopus α      ITPAITRVVDFAKKLPMFSELTCEDQIILLKGCCMEIMSLRAAVRYDPDSETLTLSGEMA
               ***************** ** ***********************
```

Figure 1. Alignment of sea bream TR with flounder TR-α and TR-β and *Xenopus* TR-β. Residues conserved in all sequences are indicated with a *.

observed in brain, the lowest level in muscle and the other tissues were fairly constant (Figure 2a). Treatment with T_4 did not alter the level of expression in any tissues.

Tissues extracted from sea bream 6 h following treatment with T_3 and T_4 (1 μg/g) showed a similar expression pattern, with the highest levels of TR-α mRNA in brain, but the expression of TR-α increased significantly in liver 6 h following treatment with T_3 and to a lesser extent T_4.

There were trace amounts of TR-α observed in pre-hatch embryos and levels of tr-α did not appear until 1–2 days post hatch (dph) when they increased steadily until 67 dph. The increase in TR-α mRNA expression in larvae from 25–64 dph is shown in Figure 2b. There is a localisation of TR-α in the head of 64 dph larvae.

3. CONCLUSIONS

A 387 bp fragment of Tr-α was isolated and used to analyse the level of expression of TR in sea bream larvae and tissues. Trace amounts of TR-α transcripts, presumably of maternal origin, were observed in early stage embryos which suggests that thyroid

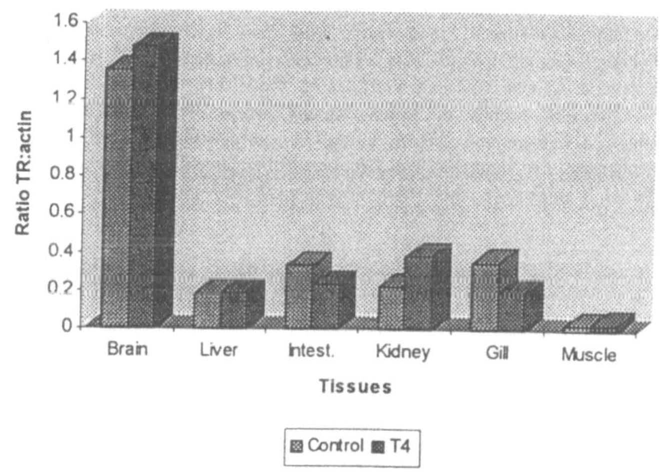

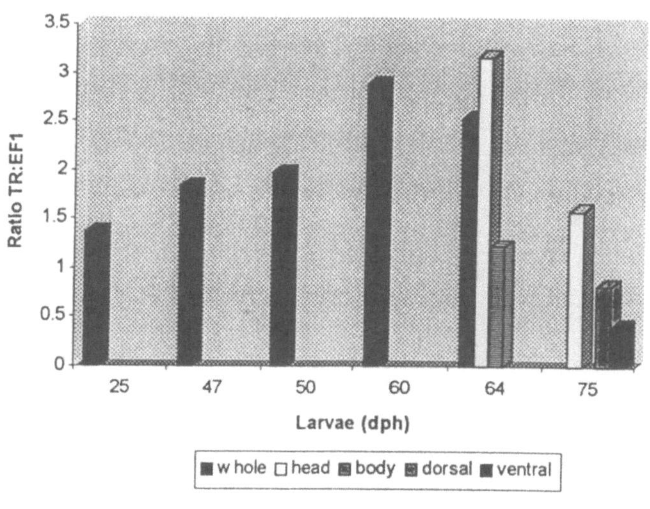

Figure 2. a. The ratio of TR:actin mRNA expression in untreated tissues and in tissues 24 h after treatment with T4. **b.** The expression of TR-α and EF1-α in larvae.

hormones may have no role during this stage of development. Levels of Tr-α appeared at 1–2 dph and increased steadily to a peak at 67 dph which may reflect major changes occurring at this stage, possibly maturation processes. Treatment with T_3 or T_4 did not alter the expression of TR-α. High levels of Tr-α mRNA expression were seen in the brain of juvenile sea bream and localised in the head of 64 dph larvae. This may indicate a role for TH in the adult brain and during its development. Treatment with t_3 and t_4 had no effect on sea bream tissues except in the liver 6 h following hormone treatment, although 24 h after treatment no increase was observed. This indicates that a negative feedback mechanism for regulation of Tr-α exists.

ACKNOWLEDGMENT

This work was financed by an EU grant AIR2-CT93-1483.

REFERENCES

Brown, C.L. and Kim, B.G., 1995, *Aquaculture* **135**:79–86
De Jesus, E.G., Hirano, T. and Inui, Y., 1993, *Fish Physiol. Biochem.* **11**:323–328
Folmar, L.C. and Dickhoff, W.W.,1980, *Aquaculture* **21**:1–37
Fraichard, A., Chassande, O., Plateroti, M., Roux, J.P., Trouillas, J., Dehay, C., Legrand, C., Gauthier, K., Kedinger, M., Malaval, L., Rousset, B. and Samarut, J., 1997, *EMBO J.* **16**: 4412–4420
Higgs, D.A., Fagerland, U.H.M., Eales, J.G. and McBride, J.R., 1982, *Comp. Biochem. Physiol.* **73B**:143–176
Santamaria, J.A., Andrades, J.A., Herraez, P., Fernandez-Llebrez, P and Becerra, J.,1994, *Anat. Rec.* **240**:248–254
Yaoita, Y. and Brown, D.D.,1990, *Genes Dev.* **4**:1917–1924

24

REGULATION OF *Dlx* HOMEOBOX GENE EXPRESSION DURING DEVELOPMENT OF THE ZEBRAFISH EMBRYO

The Potential Importance of Their Genomic Organization

Marc Ekker,[1,2] Genny Giroux,[1] Ted Zerucha,[1,2] Alison Lewis,[1,2]
Adriana A. Gambarotta,[3] and Joshua R. Schultz[3]

[1]Loeb Inst. for Medical Research
Ottawa, Canada
[2]Department of Medicine, Anatomy, and Neurobiology
University of Ottawa
[3]University of Ottawa Heart Institute

The comparison of gene structure, function, and regulation of gene expression in distant species constitute powerful tools to enhance our knowledge of the evolution of multigene families and to shed light on fundamental mechanisms of development. In our studies of the *dlx* family of homeobox genes in the zebrafish, we have compared the sequences, embryonic expression patterns and genomic organization of the zebrafish *dlx* genes with those of other vertebrates and suggest that conservation of shared *cis*-acting regulatory sequences in pairs of *dlx* genes constitute a key element of their coordinated expression during development. Our studies show the importance of analyzing developmental genes in teleost fish to understand the evolution of gene families and elucidate developmental strategies.

1. THE HOMEOBOX

The homeobox is a 180 bp sequence first identified in two homeotic genes of *Drosophila*, *Antennapedia* and *Ultrabithorax* (Gehring, 1987; McGinnis et al., 1984; Scott et al., 1989; Scott and Weiner, 1984). The homeobox encodes for a protein domain, the homeodomain, that mediates DNA recognition. This motif has been well conserved through evolution and is found in a wide variety of eukaryotic genes, including invertebrate, vertebrate and yeast genes that encode transcriptional regulators (Laughon and Scott,

New Developments in Marine Biotechnology, edited by Le Gal and Halvorson.
Plenum Press, New York, 1998.

1984; Pabo and Sauer, 1984; Scott et al., 1989). Homeobox-genes have been classified into families and sub-families based upon sequence similarities of their homeodomains.

2. THE VERTEBRATE *Dlx* GENES

Vertebrate genes with homeoboxes similar in sequence to that of the *Drosophila distal-less* gene constitute the *Dlx* family. Eight *dlx* genes have been isolated in the zebrafish (Akimenko et al., 1994; Ekker et al., 1992; Ellies et al., 1997; Ellies et al., 1997; Stock et al., 1996). *Dlx* genes in other vertebrates include those of the mouse (Porteus et al., 1991; Price et al., 1991; Robinson et al., 1991; Simeone et al., 1994), the newt (Beauchemin and Savard, 1992), the chick (Ferrari et al., 1995), human (Scherer et al., 1995; Simeone et al., 1994), the frogs *Xenopus laevis* (Asano et al., 1992; Dirksen et al., 1993; Papalopulu and Kintner, 1993), and *Eleutherodactylus coqui* (Fang and Elinson, 1996) and the axolotl (Mullen et al., 1996). Phylogenetic analyses (Akimenko et al., 1994; Stock et al., 1996) have shown that: i) vertebrate *Dlx* genes can be organized into six orthologous groups (shown in Table 1); ii) all vertebrate *Dlx* genes descend from a pair of ancestral genes (Fig. 1) and the six orthologous groups can be organized into two clades: one that would include the *Dlx1*, *Dlx6* and *Dlx7* genes and one that would include the *Dlx2*, *Dlx3* and *Dlx5*.

3. EXPRESSION AND FUNCTION OF *Dlx* GENES DURING EMBRYONIC DEVELOPMENT

The patterns of embryonic expression of *Dlx* genes suggest that they play important roles in the development of the forebrain (Akimenko et al., 1994; Papalopulu and Kintner, 1993; Porteus et al., 1991; Price et al., 1991; Robinson et al., 1991), olfactory placodes (Akimenko et al., 1994), branchial (visceral) arches and craniofacial derivatives (Akimenko et al., 1994; Dollé et al., 1992; Papalopulu and Kintner, 1993; Robinson and Mahon, 1994; Zhao et al., 1994), otic vesicle and inner ear (Akimenko et al., 1994; Ekker et al., 1992; Papalopulu and Kintner, 1993; Robinson and Mahon, 1994) and limbs/fins (Akimenko et al., 1994; Dollé et al., 1992; Morasso et al., 1995). Evidence for a role of *Dlx* genes in branchial arch development was strengthened by our observation that treatment of zebrafish embryos with exogenous retinoic acid resulted in the loss of expression of all *dlx* genes in the branchial arches in a manner which strongly correlated with craniofacial cartilage dysmorphogenesis (Ellies et al., 1997).

One striking aspect of *Dlx* expression is its combinatorial nature. We have shown this for the zebrafish (Akimenko et al., 1994; Ellies et al., 1997), and there are extensive similarities between the combinatorial expression of zebrafish *dlx* genes and that of their

Table 1. Orthologous groups of vertebrate *Dlx* genes

Dlx1	*Dlx2*	*Dlx3*	*Dlx5*	*Dlx6*	*Dlx7*
human *Dlx1*	human *Dlx2*	human *Dlx3*	chicken *Dlx5*	human *Dlx6*	mouse *Dlx7*
mouse *Dlx1*	mouse *Dlx2*	mouse *Dlx3*	human *Dlx5*	mouse *Dlx6*	newt *NvHbox-5*
zebrafish *dlx1*	*Xenopus Xdll-4*	newt *NvHbox-4*	mouse *Dlx5*	*Xenopus Xdll*	zebrafish *dlx7*
	zebrafish *dlx2*	*Xenopus Xdll-2*	rat *rdlx*	zebrafish *dlx6*	zebrafish *dlx8*
	zebrafish *dlx5*	zebrafish *dlx3*	*Xenopus Xdll-3*		
	E. coqui dlx2	axolotl *Dlx-3*	zebrafish *dlx4*		
			E. coqui dlx4		

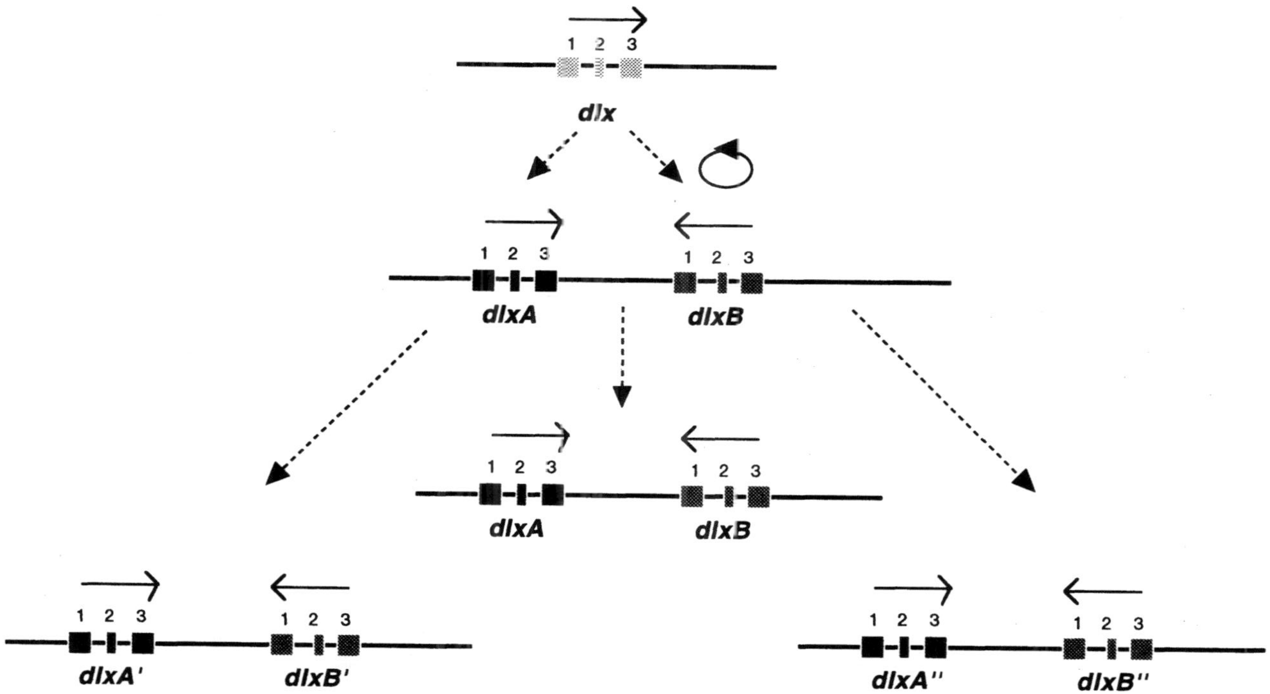

Figure 1. Proposed model for the evolution of *Dlx* genes. An ancestral *distal-less*-related gene duplicated and inverted to generate a first pair of inverted convergent genes. This probably occurred prior to vertebrate radiation. This primordial pair contained one gene from each of the two clades mentioned in Table 1. Duplication of chromosome regions containing the primordial pair resulted in the existence of the six *Dlx* genes found in several vertebrate species. This occurred early in vertebrate evolution, prior to the separation of lineages giving rise to mammals and teleost fish. For more details on *Dlx* evolution, see Stock et al., 1996.

orthologs in other vertebrate species such as the mouse (Dollé et al., 1992; Porteus et al., 1991; Price et al., 1991; Robinson and Mahon, 1994) and *Xenopus* (Dirksen et al., 1993; Papalopulu and Kintner, 1993). This suggests that combinatorial expression is an important aspect of *Dlx* function and is most likely the consequence of shared and conserved regulatory mechanisms that control multiple *Dlx* genes.

Partial or total genetic redundancy of genes that belong to a well conserved gene family has been proposed to explain the mild phenotypes or the absence of phenotype that have been observed in many selective gene inactivation experiments ("gene knockouts") in the mouse. This is especially to be expected when expression patterns of these genes overlap such as is the case for the *Dlx* genes. The phenotypes of *Dlx* null mutants in the mouse agree with this prediction. A null mutation in the mouse *Dlx-2* gene was shown to result in abnormal morphogenesis of the proximal derivatives of the first and second branchial arches and abnormal differentiation in the forebrain, whereas a mutation that affects both *Dlx-1* and *Dlx-2* causes stronger phenotypes both in the arches and in the forebrain (Anderson et al., 1997; Qiu et al., 1997; Qiu et al., 1995). However, other structures such as the limbs, where the *Dlx-2* and *Dlx-1* genes are also normally expressed, are unaffected (Qiu et al., 1995). This suggests a partial genetic redundancy that may involve other *Dlx* genes, such as *Dlx-3*, *Dlx-5*, and *Dlx-6* which share overlapping expression patterns with *Dlx-1* and *Dlx-2* (Qiu et al., 1995). The abnormalities observed in *Dlx* null mutants suggest a unique function for each *Dlx* gene only in a subset of the cells where it is expressed.

Little is known of the biochemical properties of Dlx proteins. We and others have shown that they act as transcriptional activators (Zerucha et al., 1997; Zhang et al., 1997). The high degree of similarity in the homeodomains of Dlx proteins argues for similar DNA binding specificities, although this has yet to be established. The remaining regions of the Dlx protein regions have yet to be studied. On the basis of the conservation of regions of the Dlx proteins outside the homeodomain in orthologs from distantly related vertebrates (Akimenko et al., 1994; Stock et al., 1996), we suggest that additional domains that confer specific biochemical properties to individual Dlx proteins likely exist.

4. GENOMIC ORGANIZATION AND THE COMBINATORIAL EXPRESSION OF *Dlx* GENES THROUGH SHARED REGULATORY MECHANISMS

Multigene families are generally formed as the result of gene duplication events or duplication of chromosomal regions. Furthermore, the genomic organization of linked genes has, in numerous instances, been shown to play a critical role in the concerted control of the linked genes through the action of shared *cis*-acting sequences. A well known example in developmental genetics is the genomic organization of the complexes of *Antennapedia* class homeobox genes (*Hox* genes in vertebrates) which is important for the coordinate spatial and temporal regulation of these genes (reviewed in (Duboule, 1994; McGinnis and Krumlauf, 1992).

The vertebrate *Dlx* genes are not part of large gene complexes but are generally organized in closely linked pairs in a conserved inverted convergent configuration (tail-to-tail arrangement). Linked pairs of *Dlx* genes always contain one member of each of the two clades (Table 1 and Fig. 1). Thus, the *Dlx-1* and *Dlx-2* genes from mouse or human are linked(McGuinness et al., 1996; Ozcelik et al., 1992; Simeone et al., 1994). We have determined that the zebrafish *dlx1* and *dlx2* show the same genomic organization (Ellies et

Table 2. Combinatorial expression of zebrafish *dlx* genes

	dlx1/dlx2 pair		*dlx3/dlx7* pair		*dlx4/dlx6* pair		*dlx5/dlx8* pair	
Ectodermal stripes in gastrula	—	—	*dlx3*	*dlx7*	—	—	—	—
Ventral forebrain	*dlx1*	*dlx2*	—	—	*dlx4*	*dlx6*	*dlx5*	—
Olfactory placodes	—	—	*dlx3*	*dlx7*	*dlx4*	*dlx6*	—	—
Migrating neural crest	—	*dlx2*	—	—	—	—	—	—
Visceral arches*	*dlx1*	*dlx2*	*dlx3*	*dlx7*	*dlx4*	*dlx6*	—	*dlx8*
Dorsal otic vesicle	—	—	*dlx3*	*dlx7*	*dlx4*	*dlx6*	—	—
Pectoral fin buds	*dlx1*	*dlx2*	*dlx3*	*dlx7*	*dlx4*	*dlx6*	*dlx5*	*dlx8*
Median fin fold	*dlx1*	*dlx2*	*dlx3*	*dlx7*	*dlx4*	*dlx6*	*dlx5*	*dlx8*

Linked *dlx* genes are indicated as pairs. *Visceral arch expression is complex both spatially and temporally and could not easily be summarized in this table.

al., 1997). The *Dlx-5* and *Dlx-6* genes of mammals and their orthologs in zebrafish, *dlx4* and *dlx6*, show a similar tail-to-tail configuration (Ellies et al., 1997; Simeone et al., 1994). Finally, using a panel of somatic zebrafish/mouse hybrids (Ekker et al., 1996, we have shown that the *dlx3* and *dlx7* genes constitute a third pair of inverted convergent *dlx* genes (Stock et al., 1996) as is the case for the mammalian genes (Nakamura et al., 1996). Therefore, the tail-to-tail arrangement of *dlx* genes is conserved amongst vertebrates. This organization of *distal-less* related genes appears to be ancient as it is also seen in the ascidian *Ciona intestinalis* (Di Gregorio et al., 1995).

A clear relationship emerges between the linkage of *Dlx* gene pairs and their overlapping expression. Thus, the *dlx1* and *dlx2* expression patterns overlap almost everywhere (Table 2), the same applies for the *dlx4/dlx6* pair and the expression patterns of the linked *dlx3* and *dlx7* are indistinguishable (Ellies et al., 1997). By analogy to other linked genes, it is to be expected that the driving force behind the combinatorial expression of pairs of *Dlx* genes is the sharing of crucial regulatory control elements. Regulatory sequences conferring tissue and stage-specific expression to developmental genes are often retained long after the duplication events resulting in gene family members that display partially overlapping embryonic expression pattern.

Preliminary evidence (not shown) suggest the conservation of sequences in the intergenic regions of pairs of *Dlx* genes between the zebrafish and the mouse. Presence of *cis*-acting regulatory sequences responsible for the coordinate expression of pairs of *Dlx* genes would provide selective pressure to maintain this close genomic organization if the coordinated expression of the two genes is beneficial to the animal.

Comparisons of the sequences and genomic organization of *Dlx* genes of the zebrafish and mouse should be helpful in the elucidation of the regulatory mechanisms that control the expression of these genes. The spatial and temporal manners in which the *Dlx* genes are expressed during development will likely influence aspects of the vertebrate body plan. Conservation of patterns of *Dlx* gene expression in distantly related vertebrates suggest that they participate in a conserved and thus fundamental genetic cascade. Therefore, analyses of fish developmental genes is useful not only for our understanding of fish biology but also that of vertebrate evolution.

ACKNOWLEDGMENTS

This research is supported by grants from the Medical Research Council of Canada and from the Natural Sciences and Engineering Research Council of Canada.

REFERENCES

Akimenko, M.-A., Ekker, M., Wegner, J., Lin, W. and Westerfield, M. 1994, *J. Neurosci.* **14**:3475–3486.

Anderson, S.A., Qiu, M., Bulfone, A., Eisenstat, D.D., Meneses, J., Pedersen, R. and Rubenstein, J.L.R. 1997, *Neuron* in press,

Asano, M., Emori, Y., Saigo, K. and Shiokawa, K. 1992,. *J. Biol. Chem.* **267**:5044–5047.

Baumgartner, S., Bopp, D., Burri, M. and Noll, M. 1987, *Genes Dev.* **1**:1247–1267.

Beauchemin, M. and Savard, P. 1992, *Dev. Biol.* **154**:55–65.

Côté, S., Preiss, A., Haller, J., Schuh, R., Kielin, A., Seifert, E. and Jäckl;e, H. 1987, *EMBO J.* **6**:2793–2802.

Di Gregorio, A., Spagnuolo, A., Ristoratore, F., Pischetola, M., Aniello, F., Branno, M., Cariello, L. and Di Lauro, R.1995, *Gene* **156**:253–257.

Dirksen, M.L., Mathers, P. and Jamrich, M. 1993, *Mech. Dev.* **41**:121–128.

Dollé, P., Price, M. and Duboule, D. 1992, *Differentiation* **49**:93–99.

Duboule, D. 1994, *Development suppl.*, 135–142.

Ekker, M., Akimenko, M.-A., Bremiller, R. and Westerfield, M. 1992,. *Neuron* **9**:27–35.

Ekker, M., Speevak, M.D., Martin, C.C., Joly, L., Giroux, G. and Chevrette, M. 1996, *Genomics* **33**:57–64.

Ellies, D.L., Langille, R.M., Martin, C.C., Akimenko, M.-A. and Ekker, M. 1997,. *Mech. Dev.* **61**:23–36.

Ellies, D.L., Stock, D.W., Hatch, G., Giroux, G., Weiss, K.M. and Ekker, M. 1997,. *Genomics* **45**:580–590.

Fang, H. and Elinson, R.P. 1996, *Eleutherodactylus coqui. Dev. Biol.* **179**:160–172.

Ferrari, D., Sumoy, L., Gannon, J., Sun, H., Brown, A.M.C., Upholt, W.B. and Kosher, R.A. 1995, *Mech. Devel.* **52**:257–264.

Gehring, W.J. 1987, *Science* **236**:1245–1252

Laughon, A. and Scott, M.P. 1984, *Nature* **310**:25–31.

McGinnis, W. and Krumlauf, R. 1992, *Cell* **68**:283–302.

McGinnis, W., Levine, M., Hafen, E., Kuroiwa, A. and Gehring, W.J. 1984, *Nature* **308**:478–483.

McGuinness, T., Porteus, M.H., Smiga, S., Bulfone, A., Kingsley, C., Qiu, M., Liu, J.K., Long, J.E., Xu, D. and Rubenstein, J.L.R. 1996, *Genomics* **35**:473–485.

Morasso, M.I., Mahon, K.A. and Sargent, T.D.1995, *Proc. Natl. Acad. Sci. USA* **92**:3968–3972.

Mullen, L.M., Bryant, S.V., Torok, M.A., Blumberg, B. and Gardiner, D.M. 1996, *Development* **122**:3487–3497.

Nakamura, S., Stock, D.W., Wynder, K.L., Bollekens, J.A., Takeshita, K., Nagai, B.M., Chiba, S., Kitamura, T., Freeland, T.M., Zhao, Z., Minowada, J., Lawrence, J.B., Weiss, K.M. and Ruddle, F.H. 1996, *Genomics* **38**:314–324.

Ozcelik, T., Porteus, M.H., Rubenstein, J.L.R. and Francke, U. 1992, *Genomics* **13**:1157–1161.

Pabo, C.O. and Sauer, R.T. 1984, *Annu. Rev. Biochem* **53**:293–321.

Papalopulu, N. and Kintner, C. 1993, *Development* **117**:961–975.

Porteus, M.H., Bulfone, A., Ciaranello, R.D. and Rubenstein, J.L.R. 1991, *Neuron* **7**:221–229.

Price, M., Lemaistre, M., Pischetola, M., Di Lauro, R. and Duboule, D. 1991, *Nature* **351**:748–751.

Qiu, M., Bulfone, A., Ghattas, I., Meneses, J.J., Christensen, L., Sharpe, P.T., Presley, R., Pedersen, R.A. and Rubenstein, J.L.R. 1997, *Dev. Biol.* **185**:165–184.

Qiu, M., Bulfone, A., Martinez, S., Meneses, J.J., Shimamura, K., Pedersen, R.A. and Rubenstein, J.L.R. 1995, *Genes Dev.* **9**:2523–2538.

Robinson, G.W. and Mahon, K.A. 1994, *Mech. Devel.* **48**:199–215.

Robinson, G.W., Wray, S. and Mahon, K.A. 1991, *New Biol.* **3**, 1183–1194.

Scherer, S.W., Heng, H.H.Q., Robinson, G.W., Mahon, K.A., Evans, J.P. and Tsui, L.-C. 1995, *Genome* **6**:310–311.

Scott, M.P., Tamkun, J.W. and Hartzell, G.W. 1989, *Biochim. Biophys. Acta* **989**:25–48.

Scott, M.P. and Weiner, A.J. 1984, *Proc. Natl. Acad. Sci. USA* **81**:4115–4119.

Simeone, A., Acampora, D., Pannese, M., Desposito, M., Stornaiuolo, A., Gulisano, M., Mallamaci, A., Kastury, K., Druck, T., Huebner, K. and Boncinelli, E. 1994, *Proc. Natl. Acad. Sci. USA* **91**:2250–2254.

Stock, D.W., Ellies, D.L., Zhao, Z., Ekker, M., Ruddle, F.H. and Weiss, K.M. 1996, *Proc. Natl. Acad. Sci. USA* **93**:10858–10863.

Tabata, T., Schwartz, C., Gustavson, E., Ali, Z. and Kornberg, T.B. 1995, *Development* **121**:3359–3369.

Zhang, H., Hu, G., Wang, H., Sciavolino, P., Iler, N., Shen, M.M. and Abate-Shen, C. 1997, *Mol. Cell. Biol.* **17**: 2920–2932.

Zhao, G.-Q., Zhao, S., Zhou, X., Eberspaecher, H., Solursh, M. and De Crombrugghe, B. 1994, *Dev. Biol.* **164**:37–51.

MEIOTIC CELL CYCLE CONTROL BY MOS IN ASCIDIAN OOCYTES

Gian Luigi Russo,[1,2,*] Keiichiro Kyozuka,[3] Marcella Marino,[1]
Elisabetta Tosti,[1] Martin Wilding,[1] Maria Laura de Simone, and Brian Dale[1]

[1]Stazione Zoologica 'Anton Dohrn'
80121 Napoli, Italy
[2]Institute of Food Science and Technology
National Research Council
83100 Avellino, Italy
[3]Asamushi Marine Biological Station
Asamushi, Aomori 039-34, Japan

1. ROLE OF MOS IN MEIOSIS PROGRESSION

Mitotic cell division cycle is regulated by several control mechanisms generally known as checkpoints, whose main function is to ensure that critical events in the cell division such as DNA replication and chromosome segregation occur with high fidelity and in the correct order and time. A recent view of cell cycle regulation, indicate that checkpoints work as signal transduction pathways with their initiating signals, sensors, transducers, and effectors (Elledge, 1996). Similarly, meiotic cell cycle regulation have specific checkpoints and two of them have been recently characterized at molecular level. The first ensures the completion of recombination before the formation of meiosis I spindle; the second blocks anaphase I until all the paired chromosomes are correctly attached to the spindle (Page and Orr-Weaver, 1997). However, the most intriguing difference between mitosis and meiosis respect to cell cycle regulation, is the ability of animal oocytes to arrest at specific stage during maturation, and maintain this block for extremely long time: from years (frog) to decades (human) (Sagata, 1996a,b). It is universally accepted that the meiotic block is due to the activity of a cytostatic factor (CSF) primarily identified by Masui (Masui and Markert, 1971).

* Author for communications: Dr. Gian Luigi Russo, Istituto Scienze Dell'Alimentazione - CNR, via Roma 52 A-C, 83100 Avellino, Italy. Phone: +39 825 781600, Fax: +39 825 781585; E-mail: *glrusso@isa.av.cnr.it*

New Developments in Marine Biotechnology, edited by Le Gal and Halvorson.
Plenum Press, New York, 1998.

The molecular components of CSF are only partially known. The first to be identified is the product of the proto-oncogene *mos*. Mos is a serine/threonine protein kinase specifically expressed and activated during oocyte maturation. The function of Mos, as CSF subunit, has been originally associated with the metaphase block in oocytes. Mos maintains high the activity of the maturation-promoting-factor (MPF or Cdc2/Cyclin B complex), and blocks the metaphase-anaphase transition (Sagata, 1996a,b).

In mammals and anurans, fully grown immature oocytes are arrested at prophase-I and resume the cell cycle in response to maternal hormones. The cell cycle is then arrested a second time at metaphase II and the oocyte remains in this state until activated by the fertilizing spermatozoon (Whitaker and Patel, 1990). In *Xenopus* oocytes, Mos synthesis is essential for oocyte maturation and for metaphase II arrest, through the activity of MAP kinase, actually considered as a component of CSF (Sagata, 1996a,b). More problematic is the function of Mos during meiosis I: a current idea is that Mos is not essential for oocyte maturation, or at least other pathways are present. In many deuterostomes, oocyte activation is accompanied by a large single calcium transient, generated from intracellular sources, that starts at the point of spermatozoon entry and traverses the oocyte in a wave to the antipode (Jaffe, 1985). This transient increase in calcium is thought to play a key role in the cells decision to exit the cell cycle (Jaffe, 1985; Whitaker and Patel, 1990), by decreasing the activities of CSF and MPF (Sagata et al., 1989); however, the interaction of calcium signals with MPF, CSF and the other events leading to the completion of meiosis have not yet been resolved.

2. MEIOSIS PROGRESSION IN ASCIDIAN AT FERTILIZATION

Meiotic progression in ascidian appears to be regulated differently to that in vertebrate oocytes: the signal for meiosis resumption is not known; however, the second meiotic block occurs at metaphase I (Dale, 1983). Consequently, activation by spermatozoon drives the cell through two nuclear cycles without any DNA replication. In ascidian oocytes, the large calcium wave is followed by a series of smaller repetitive waves (Russo et al., 1996). Recently, we have characterized biochemically MPF complex in the ascidian *Ciona intestinalis*, and demonstrated that the activity of the associated Cdc2/Cyclin B kinase is differently regulated by calcium in meiosis I compared to meiosis II (Russo et al., 1996). We proposed ascidian oocytes as an excellent model to study meiotic division both at physiological and molecular level. In fact, in this animal the meiotic block is at metaphase I; therefore, after sperm addition, it is possible to follow sperimentally meiosis I completion and the entire meiosis II, while in other systems (*e.g.* vertebrates) the block is in metaphase II, allowing solely the study of meiosis II progression. In addition, ascidian oocytes are relatively easy to microinject respect to other non-vertebrate oocytes blocked early in the meiotic cell cycle (*e.g.* clam oocytes, blocked in prophase I), and the total number of oocytes for each single animal is enough for biochemical analysis. Finally, the population of ascidian oocytes are perfectly synchronized in metaphase I, and meiosis is completed in relatively short time (one hour from fertilization to two-cell embryos).

The overall goal of our study is to clarify the different cell cycle regulation in meiosis I respect to meiosis II using a relatively simple and versatile model like ascidian oocytes. We expect that some conclusions of our study might be useful to understand meiosis regulation in vertebrate and humans. Therefore, after the characterization of ascidian MPF, we decided to investigate the role of CSF in meiosis progression of *C. intestinalis*. The preliminary data obtained are reported in the next paragraph.

3. CHARACTERIZATION OF CSF-LIKE FACTOR AND MOS IN *CIONA INTESTINALIS*

Recently, it has been proposed that members of the MAP kinase family are components of CSF factor, at least in vertebrate, and generally, the peak of MAP kinase activity corresponds to high level of *mos* expression and activity (reviewed in Sagata, 1996a,b). Since, there are no data on MAP kinase activity during fertilization in ascidian oocytes, we measured its activity at different times after sperm addition (Table 1). To avoid possible cross-reactivity between MAP kinase and Cdc2/Cyclin B (the two kinases share a similar consensus sequence for phosphorylation), we used a commercially available kit (Amersham) including as MAP kinase substrate a synthetic peptide not phosphorylated by Cdc2 kinase(s). Table 1 shows that MAP kinase activity is high in unfertilized oocytes of *Ciona intestinalis*, with a maximal value at five minutes from fertilization followed by a decrease at the time of extrusion of the first polar body (telophase I). The enzyme activity did not show any further increase during meiosis II. Using an antibody raised against mammalian p44 MAP kinase (Santa Cruz Biotechnology), we detected a band of 44 kDa in lysed oocytes of ascidian whose intensity did not change during fertilization (data not reported).

Based on this observation, we investigated the expression of Mos-like factor(s) in *C. intestinalis* using an antibody raised against a peptide mapping within the carboxy terminal domain of *Xenopus* Mos (Santa Cruz Biotechnology). As reported in Fig. 1, the antiserum detected a band at an apparent molecular weigh of 30 kDa. All the controls performed indicated that the signal might correspond to a Mos-like protein polypeptide in ascidian. The protein was expressed during the entire process of meiosis completion without any significant change in the level of expression. A similar observation has been reported in *Xenopus* oocytes and eggs, where the expression of Mos initiated after GVBD and remained constant until the end of meiosis II (Sagata, 1996b).

To further confirm the presence of a Mos-like protein in *C. intestinalis*, and to measure its activity, we set up a microinjection assay based on the classical CSF assay by Masui (Masui and Markert, 1971). Briefly, we prepared cell extract of *C. intestinalis* oocytes blocked in metaphase-I, and injected them into two-cell embryos of the same organism. The presence of a CSF activity was detected by a cleavage arrest at metaphase in one blastomer compared to its twin untreated, or injected with a control solution (Fig. 2). As reported in Table 2, a clear CSF-like activity was present in metaphase-I oocytes,

Table 1. MAP kinase activity in *Ciona intestinalis* at fertilization

Time after fertilization (min)	Meiotic phase	MAP kinase activity*
0	metaphase I	21.05 ± 1.3
5	anaphase I	26.25 ± 0.5
10	telophase I	15.60 ± 2.1
15	telophase I (prophase II)	8.57 ± 1.5
20	prophase II (telophase I)	9.00 ± 0.8
25	metaphase II	7.45 ± 0.6
30	anaphase II	5.38 ± 1.0
35	telophase II	4.00 ± 1.2

*The enzyme activity is expressed as pmoles of phosphate incorporated on a peptide substrate per min at 37°C ± S.D. Activity value are the mean of triplicate assays using few oocytes per assay.

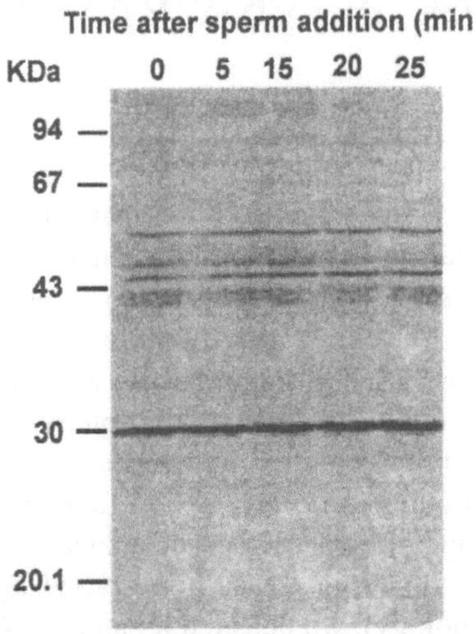

Figure 1. Expression of Mos-like protein after fertilization in *Ciona intestinalis*. Immunoblotting of oocyte lysed at different times (top) from sperm addition. Each lane conteins the same amount of total protein. Primary antibody was an anti-*Xenopus* Mos serum (see text), and band detection was performed using a secondary antibody, alkaline phosphatase conjugated, and a colorimetric assay for alkaline phosphatase.

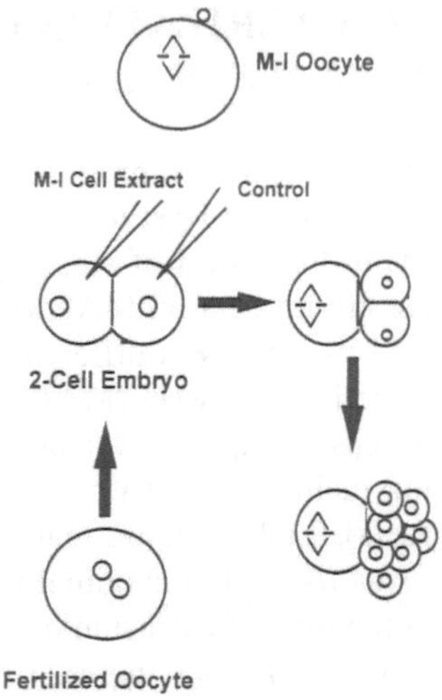

Figure 2. Scheme of CSF assay in *Ciona intestinalis* oocytes.

and this activity was blocked by incubation of anti-*Xenopus* Mos immunoglobulin with metaphase-I extracts before microinjection.

Although preliminary, these data indicate that a Mos-like protein is present during meiosis progression in ascidian oocytes, and it is essential for ascidian CSF activity.

4. CONCLUSION

In this preliminary report, we presented evidence that a CSF mechanism mediated by Mos is active in ascidian oocyte. Our data are partially in agreement with the proposed role of Mos in oocyte maturation in vertebrate. In fact, in *C. intestinalis* Mos protein seems to be expressed constitutively during the meiotic division as in *Xenopus*; however, in vertebrate, Mos is degraded by the proteasome complex at the metaphase-anaphase II transition, *i.e.* after the removal of metaphase block at fertilization. Same fate is due to MAP kinase activity. On the opposite, in *C. intestinalis* oocytes, Mos expression is still detected after metaphase I-anaphase I transition, and Cyclin B degradation (Russo et al.,

Table 2. CSF- and Mos-*like* activity in *Ciona intestinalis*

Injection into two-cell embryos	N. of cleavage
Metaphase-I extract	0/10
Control	14/15
Metaphase-I extract + XeMos IgG	6/6
Metaphase-I extract + control IgG	1/8

1996). One further contradiction is that MAP kinase is not active after metaphase I, while Mos is still present.

Our work confirms the discrepancy between the physiological role of Mos in meiosis I compared to meiosis II, as indicated by others (Sagata, 1996a,b), and proposes a new model to understand its regulation. Future studies will be devoted to clone *mos* homolog(s) in ascidian, an essential step in order to investigate the its role in meiosis progression.

REFERENCES

Dale, B. 1983, *Fertilization in animals*. London: Edward Arnold.

Elledge, S. J. 1996, *Science* **274:**1664–1672.

Jaffe, L. F. 1985, In *Biology of fertilization* vol. 3 (eds. C. B. Metz and A. Monroy), pp 127–165. San Diego: Academic Press.

Masui, Y. and Markert, C. L. 1971, *J. Exp. Zool.* **177:**129–146.

Page, A. W. and Orr-Weaver, T. L. 1997,. *Current Opinion in Genetics and Development* **7:**23–31.

Russo, G. L., Kyozuka, K., Antonazzo, L., Tosti, E. and Dale, B. 1996, *Development* **122:**1995–2003.

Sagata, N., Watanabe, N., Vande Woude, G. F. and Ikawa, Y. 1989,. *Nature* **34:**512–518.

Sagata, N. 1996a, *Trends Cell Biol.* **6:**22–28.

Sagata, N. 1996b, *BioEssays* **19:**1–9.

Whitaker, M. and Patel, R. 1990, *Development.* **108:**525–542.

26

ACTIVATION OF *CIONA INTESTINALIS* AT FERTILISATION IS CONTROLLED BY NICOTINAMIDE NUCLEOTIDE METABOLISM

M. Wilding,[1,*] G. L. Russo,[2] M. Marino,[1] L. Grumetto,[1] M. L. De Simone,[1] E. Tosti,[1] and B. Dale[1]

[1]Stazione Zoologica 'Anton Dohrn'
Villa Comunale, 1, 80121 Napoli, Italy
[2]Institute of Food Science and Technology
National Research Council
83100 Avellino, Italy

1. INTRODUCTION

Fertilisation is characterised by an ion current across the oocyte plasma membrane (DeFelice *et al.*, 1986), a transient increase in intracellular calcium (see Jaffe, 1985), and the release of the cell cycle block controlled at the level of MPF (Newport and Kirschner, 1984). The fertilisation calcium signal appears to trigger the resumption of the cell cycle at fertilisation in oocytes arrested at meiosis II or after the completion of meiosis (reviewed in Whitaker and Patel, 1990). However, the role of intracellular second messengers, including calcium, in the control of progression through the early meiotic cell cycle, is less well defined (see Whitaker and Patel, 1990; Whitaker, 1996).

We have used ascidian oocytes to study maturation and the metabolic processes that control this process. These oocytes are useful because the oocyte undergoes meiosis after fertilisation. Furthermore, the ease of exposure of the plasma membrane permits studies of plasma membrane ion channels using the whole cell voltage clamp technique. In these oocytes, the fertilisation current is triggered independently of calcium ions (Dale, 1987, 1988; Tosti and Dale, 1994). In this manuscript, we show that this current is primarily induced by the gating of a plasma membrane ion channel by ADPr (ADP ribose). Our data demonstrate the activation of an enzymatic pathway at fertilisation which leads to the breakdown of NN (Nicotinamide Nucleotides) to ADPr.

* To whom correspondence should be addressed.

New Developments in Marine Biotechnology, edited by Le Gal and Halvorson.
Plenum Press, New York, 1998.

2. RESULTS

2.1. Currents Triggered by NNs and NN Metabolites

Pressure injection of cADPr and other related compounds into *C. intestinalis* oocytes through a micropipette in the whole-cell voltage clamp configuration generated inward currents that varied in amplitude in a dose-dependent manner. The most potent compound was ADPr, which induced an inward current even at intracellular concentrations as low as 10 nM. Currents produced by the injection of NN's were blocked by nicotinamide, an inhibitor of NN metabolism, suggesting that NN's are metabolised to ADPr and that this molecule is the unique ligand for the channel.

2.2. The ADPr Channel Is the 'Fertilisation Channel'

The plasma membrane of ascidian oocytes is characterised by a large conductance, non-specific ion channel that is activated by ascidian sperm in a calcium-independent fashion (Tosti and Dale, 1994). Our data demonstrates that this channel is the ADPr channel described above (Table 1). Both whole-cell and single channel I/V curves for currents induced by both ADPr and cADPr point to a reversal potential of between +15 mV and +25 mV, suggesting a non-specific ion channel. The single channel data reveals that this channel has a unitary conductance of 140 pS. Our data further suggests that the ADPr channel is not calcium-activated because buffering of cytoplasmic Ca^{2+} did not affect the peak inward current gated by 5 μM cADPr or 10 nM ADPr.

We tested whether the ADPr channel was gated by NN metabolism by measuring the current induced by ascidian sperm in the presence of nicotinamide. When 50 mM nicotinamide was added into the bath, the peak current induced at fertilisation was eliminated. The fertilisation current observed when cytoplasmic calcium is buffered by the calcium chelator 1,2-bis[2-aminophenoxy]ethane-N,N,N′,N′-tetraacetic acid (BAPTA, Tosti and Dale, 1994) is also blocked by nicotinamide, showing that this current is triggered by a mechanism involving NN metabolism and is not calcium-dependent. These data suggest that ADPr is produced at fertilisation and that this is directly involved in gating the ascidian fertilisation current.

3. DISCUSSION

Nicotinamide nucleotides are fast emerging as precursors of second messengers that are important in many aspects of cell regulation including calcium release, DNA repair and cell differentiation (reviewed in Galione and White, 1994). In this context, the identification of cADPr as a potent Ca^{2+} mobilising enzyme has already increased our understanding of the role of NN metabolism in second messenger signalling (see Galione and

Table 1. Comparison between ADPr channel and 'fertilisation channel'*

	ADPr	'Fertilisation'*
Reversal potential	15–25 mV	0–15 mV
Conductance	140 pS	400 pS
Calcium-dependence?	Independent	Independent

*Data taken from De Felice and Dale, 1984.

White, 1994). Here, we describe multiple second messenger roles for ADPr during fertilisation and meiotic progression in ascidian oocytes.

We have shown that the calcium-independent inward current observed at fertilisation in ascidian oocytes is triggered by ADPr. Our data further suggest that ADPr is produced at fertilisation through the breakdown of NN's. These data suggest that the NN metabolic pathway is stimulated at fertilisation by ascidian sperm.

4. ABSTRACT

We have used the whole cell voltage clamp technique to study the calcium-independent plasma membrane ion channel activated at fertilisation in ascidian oocytes. We show that this channel is activated by the second messenger adenosine disphosphate ribose. This second messenger is made by sperm-induced stimulation of the metabolic pathway acting on nicotinamide nucleotides (NN's). Our data demonstrates that the NN metabolic pathway plays a major role as a cell signal during meiosis completion in ascidians, in addition to calcium signals.

REFERENCES

Dale, B. 1987, *Exp. Cell Res.* **172**:474–479.
Dale, B. 1988, *Exp. Cell Res.* **177**:205–211.
DeFelice, L., Dale, B. and Talevi, R. 1986, *Proc. R. Soc. Lond.* B **229**:209–214.
Galione, A. and White, A. 1994, *Trends Cell Biol.* **4**:431–436.
Jaffe, L. F. 1985, In *Biology of fertilization* vol. 3 (eds. C. B. Metz and A. Monroy), pp 127–165. San Diego: Academic Press.
Newport, J. W. and Kirschner, M. W. 1984, *Cell* **37**:731–742.
Tosti, E. and Dale, B. 1994, *Molec. Reprod. Dev.* **37**:473–476.
Whitaker, M. J. 1996, *Biol. Reprod.* **1**:127–135.
Whitaker, M. and Patel, R. 1990, *Development* **108**:525–542.

27

APOPTOSIS AS A NORMAL MECHANISM OF GROWTH CONTROL AND TARGET OF TOXICANT ACTIONS DURING SPERMATOGENESIS

Insights Using the Shark Testis Model

Gloria V. Callard, Leon M. McClusky, and Marlies Betka

Department of Biology
Boston University
5 Cummington St.
Boston, Massachusetts 02215

Production and development of mature male gametes is essential for normal reproductive efficiency and survival of species. Epidemiological research and observations of wildlife have reported an increase in male reproductive abnormalities, including declining sperm counts in human males, thus opening a debate as to the role of environmental pollutants and other factors. The complex organization of the testis of man and common laboratory mammals, and the lack of a good in vitro spermatogenesis system for direct study, has hampered routine chemical testing and frustrated attempts to define the extent of the problem. The challenge is further exacerbated by our incomplete understanding of normal testicular control mechanisms, despite intense research activity over the last 25 years and the acquisition of powerful new tools of cellular and molecular biology for accumulating data. Future progress in understanding human health effects, and the impact of environmental factors on survival of species and ecosystems, has been linked to the development of new animal models and test systems.

As part of a program of research to elucidate factors and mechanisms that regulate spermatogenesis, we have identified the spiny dogfish shark (*Squalus acanthias*) as an advantageous experimental model (Callard, 1991; Callard et al., 1997) (Fig. 1). The strategy of seeking simpler models is justified on the basis that spermatogenesis has been fundamentally conserved throughout the Vertebrata. Phyletic and species differences are seen mainly in the organization of the testis. Due to a unique combination of features, the testis of spiny dogfish sharks and other squaliformes are particularly well-suited for the stage-by-stage analysis of spermatogenesis. First, they have a cystic mode of spermatogenesis,

New Developments in Marine Biotechnology, edited by Le Gal and Halvorson.
Plenum Press, New York, 1998.

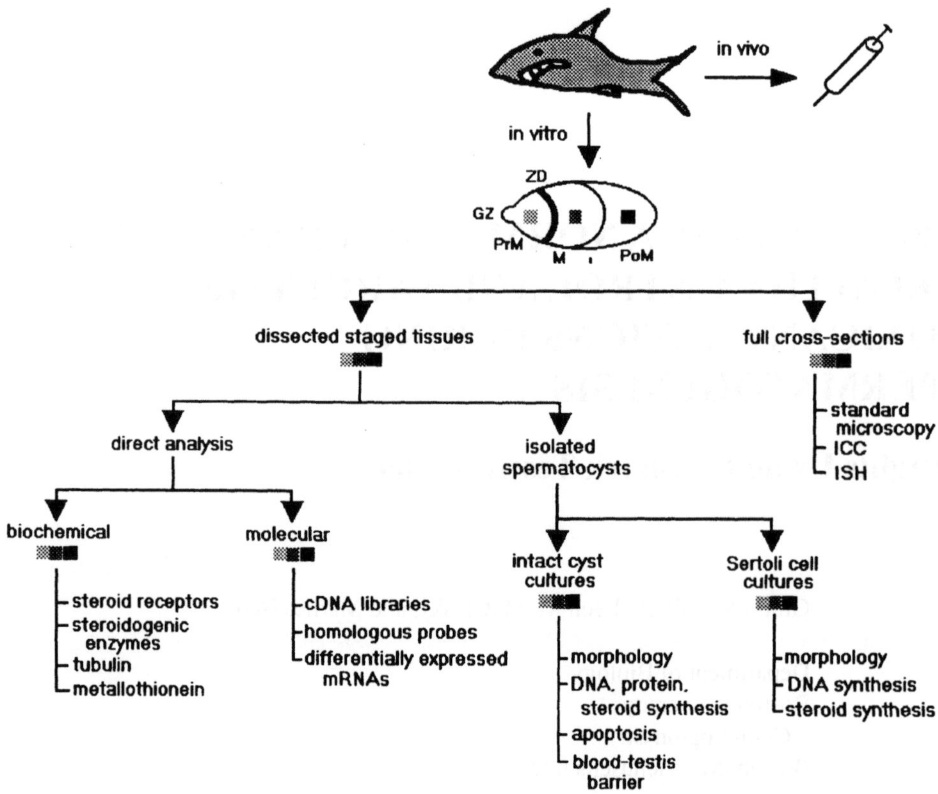

Figure 1. Strategies for investigating spermatogenesis using the shark testis model. GZ, germinal zone; PrM, premeiotic zone; M, meiotic zone; PoM, postmeiotic zone; ZD, zone of degeneration. For references to specific applications, see Callard, 1991; Piferrer & Callard,1995).

in which single germ cell clones and a second clonal population of stage-synchronized Sertoli cells form anatomically distinct, "follicle-like" units (spermatocysts). Second, spermatocysts in succeeding developmental stages are arranged in a simple linear order across the diameter of the testis, resulting in a readily visible zonation in testicular cross-sections. Finally, spermatogenesis relies mainly or exclusively on Sertoli cells for somatic cell support, Leydig cells being absent at this phyletic level. As shown in Fig. 1, staged tissues from untreated or treated sharks can be obtained for direct biochemical or molecular analysis, or processed further to obtain staged spermatocysts or Sertoli cells for investigating their functional properties under defined conditions in vitro. Because all developmental stages are present in a single cross-section, stage-related morphological and cytochemical differences are readily observed by standard tinctorial and cytochemical methods. Prior studies in this laboratory (Callard, 1991; Callard et al., 1997) have documented stage-dependent patterns in key steroidogenic enzymes, steroid receptors, and non-receptor steroid binding proteins. We have identified Sertoli cells as the source and target of steroids during spermatogenesis, and described an intratesticular vascular pathway of paracrine communication which operates between developmental stages (mature to immature). Using isolated spermatocysts, we have shown that stage-related growth and differentiated traits are maintained up to 7 d in vitro (e.g., DNA synthesis, protein synthesis/secretion; steroid synthesis/secretion) and are sensitive to regulation by steroids, growth factors and known or suspected spermatotoxicants (Callard, 1991; Callard et al., 1995; Piferrer & Callard, 1995).

Our initial study of apoptosis in the zonal testis of the spiny dogfish shark (Callard et al., 1995) was prompted by a report in the late '60s describing a zone of degenerating spermatocysts (ZD) that first appears in May at the transition of spermatogonial (premeiotic, PrM) to spermatocyte (meiotic, M) stages, and is progressively displaced peripherally by the proliferation of new cysts during the ensuing period of spermatogenic activity (Fig. 1). ZD is thought to represent the previous winter period of spermatogenic inactivity and may resemble the situation in embryonic and immature sharks, in which development of new cysts is outpaced by their degeneration prior to meiosis. In adults, ZD can be induced out of season by hypophysectomy, indicating it is due to a gonadotropin deficit. We obtained evidence of nucleosomal fragmentation products on agarose gels (DNA "ladders", an early marker of apoptosis) in extracts of PrM and ZD-stage tissues, but not in extracts from M or postmeiotic (PoM) stages (Callard et al., 1995), indicating that the apoptotic or regulated form of degeneration (programmed cell death) is limited to spermatogonial generations (n = 13 in *Squalus*). Cytological features of apoptosis were seen in germ cells but not Sertoli cells of a subset of PrM cysts, and degeneration was all-or-none in affected clones. However, Sertoli cells contain phagocytosed germ cell corpses in advanced stages of cyst degeneration (ZD).

In recent studies, we developed a procedure for visualizing the condensed nuclei characteristic of apoptosis in living spermatocysts using acridine orange (AO) as a vital stain (McClusky et al., 1996). AO is taken up by metabolically competent cells and is visualized as intense green or yellow fluorescent spots in germ cells with condensed chromatin, or as orange fluorescent masses in Sertoli cells with phagocytosed germ cell corpses. Accordingly, AO labels apoptotic but not necrotic cells (Fig. 2A). The AO procedure has been verified by electron microscopy of AO-labeled cysts and by the terminal deoxynucleotidyl transferase (TdT)-mediated dUTP nick end-labeling (TUNEL) assay (McClusky & Callard, 1997) (Fig. 2B). TdT labels blunt ends of double-stranded DNA

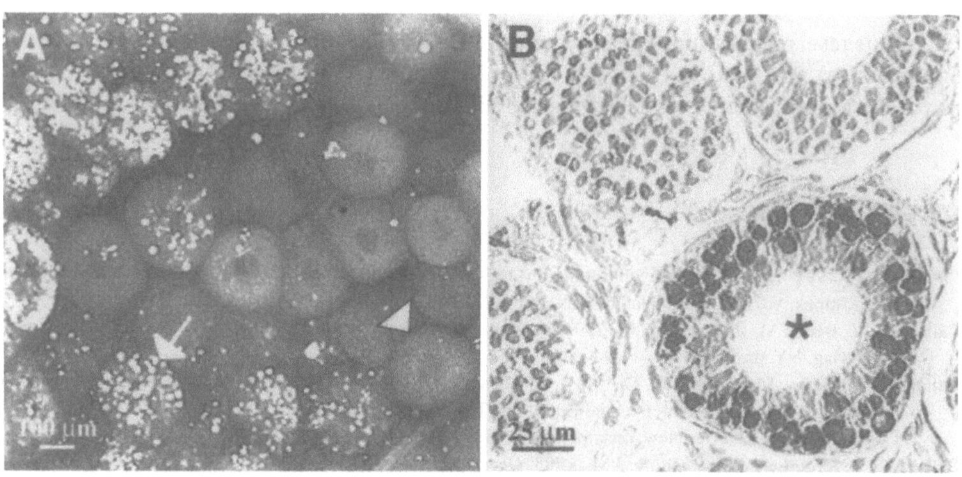

Figure 2. (A) Fluorescent microscopic image showing acridine orange (AO) vital staining of PrM spermatocysts. Condensed chromatin of apoptotic cells is visualized as bright green spots in some (arrow) but not all (arrowhead) cysts. (B) Light micrograph of TUNEL-positive (*) PrM cyst. In affected cysts, nuclei of germ cells (peripheral) but not Sertoli cells (adluminal) are labeled.

breaks, the predominant type of DNA fragmentation during apoptosis and thus is able to distinguish nucleosomal fragmentation products from the randomly fragmented DNA generated during necrosis. Whereas AO has utility for testing direct effects of regulators and toxicants on apoptosis in cultured cysts in vitro, TUNEL allows analysis of the process in testis in situ after experimental manipulation of animals in vivo.

As measured by AO vital staining and TUNEL, the apoptotic form of programmed cell death was confirmed as the mechanism of degeneration occurring in PrM and ZD cysts (Fig. 2). When apoptotic rate was quantified as a function of spermatogonial substage [i.e., distance from the germinal zone (GZ) in cross-sections, or cyst diameter in dispersed cysts], we observed that mature spermatogonial generations were most frequently affected. That is, apoptosis was rare in GZ but increased progressively in succeeding stages to a maximum of 30% of the cysts prior to entry into meiosis. A single injection of cadmium ($CdCl_2$, 5 mg/kg), a prototypical spermatotoxicant, increased apoptosis in PrM cysts within 1 day of treatment; however, mid- and late-stage spermatogonial generations were more susceptible than immature stages. In the same experiment, Cd had no effect on DNA or protein synthesis. In basal medium in vitro, the percentage of apoptotic cysts increased progressively over 6 d, indicating deprivation of one or more survival factors present in vivo. Although Cd added in vitro had no effect on apoptosis, apoptosis decreased after addition of IBMX, estradiol, and EGIF (Epigonal Growth Inhibitory Factor), agents known to decrease DNA synthesis in PrM cysts (Callard et al., 1994; Callard et al., 1997; Piferrer & Callard, 1995). These results suggest (a) that effects of Cd on germ cell survival are mediated by one or more natural survival factors; and (b) that entry into the apoptotic pathway is via the cell division cycle. We conclude that apoptosis is a normal mechanism of growth control during proliferative stages of spermatogenesis and a possible target of hormone and spermatotoxicant action. Advantages of testicular organization in the shark testis model recommends its wider use to examine problems of male reproduction and toxicology.

ACKNOWLEDGMENTS

This research was supported by grants to LM (Foundation for Research Development, South Africa) and GVC (NIEHS P45 ES-07381 and EPA G6 M01763) and carried out in part at the Mount Desert Island Biological Laboratory, Salsbury Cove, ME 04672, USA.

REFERENCES

Callard, G.V., 1991, In: Pang P. and Schreibman M.(eds.) Vertebrate Endocrinology: Fundamentals and Medical Implications, Vol 4A, Academic Press, NY. pp. 303–341.

Callard, G.V., Betka, M., Jorgensen, J.C., 1994, In: Bartke A. (ed). Function of Somatic Cells in the Testis. Springer Verlag, NY. pp. 27–54.

Callard, G.V., Betka, M., and Miller, D., 1997,. In: (Sipes GI, McQueen CA, Gandolfi AJ, eds.) Comprehensive Toxicology, Vol 10 Reproductive and Endocrine Toxicology (Boekelheide K, Chapin RE, Hoyer PB, Harris C, volume eds.), Elsevier New York, pp. 235–247.

Callard, G.V., Jorgensen, J.C., Redding, J.M., 1995. *Develop Genetics* **16**:140–147.

McClusky, L.M., Betka, M., Miller, D., and Callard, G.V., 1996, *The Bulletin, Mount Desert Island Biological Laboratory* **35**:96–97.

McClusky, L.M., Callard, G.V. 1997, *The Bulletin, Mount Desert Island Biological Laboratory* **36**:106–107.

Piferrer, F., Callard, G.V., 1995, *Biol Reprod* **53**:390–398.

28

MEDAKAFISH EMBRYONIC STEM CELLS AS A MODEL FOR GENETIC IMPROVEMENT OF AQUACULTURE LIVESTOCKS

Yunhan Hong,[1] Songlin Chen,[1] Christoph Winkler,[2] and Manfred Schartl[1]

[1]Physiologische Chemie I
Biozentrum der Universität Würzburg
97074 Würzburg, Germany
[2]Present address: Department of Pharmacology
University of Washington
Seattle, Washington 98185-7750

1. INTRODUCTION

Transgenesis, the transfer into the germ line of genes controlling particular traits of interest, is attractive for genetic improvement of aquaculture production. For example, salmon transgenic for a homologous growth hormone (GH) gene driven by a piscine regulatory sequence averaged more than 11-fold heavier than the non-transgenic controls (Devlin *et al.*, 1994). Up to now, transgenic fish have been produced by introduction of a gene expression cassette directly into freshly fertilized eggs or oocytes. Unfortunately after gene delivery, the fate of transgenes is not controllable or selectable. In most cases the founder fish which developed from manipulated eggs were highly mosaic for the introduced gene. Desired integration of a single copy of the transgene and its proper expression is rarely achieved (for review see Iyengar *et al.*, 1996). This problem of transgene integration and expression may be overcome by using embryonic stem (ES) cell lines as an alternative approach for gene transfer.

ES cell lines are undifferentiated cell cultures which retain their pluripotency after long-term cultivation *in vitro* (Evans and Kaufman, 1981; Martin, 1981). Upon reintroduction into early embryos they can enter various cell lineages and most importantly, colonize the germ line. This ability is maintained after gene transfer and selection for the desired, but rare events of transgene integration and expression. Therefore, ES cell lines provide a bridge between *in vitro* manipulations of the vertebrate genome and germline transmission of the transgene. Furthermore, ES cells offer a unique opportunity to identify novel genes of particular interest to study their physiological functions by gene targeting. They also may be utilized in strategies for preservation of biodiversity (Hong and Schartl, 1996a).

New Developments in Marine Biotechnology, edited by Le Gal and Halvorson.
Plenum Press, New York, 1998.

The key for ES cell-based germline transgenesis is the availability of pluripotent ES cell lines. However, ES cell lines have so far been limited to the mouse. Despite of numerous attempts in different mammalian species, in the chicken and zebrafish, embryo-derived cells could be maintained only for limited periods or their pluripotency was lost after a few subcultures.

Here we summarize our work on the establishment of fish ES cell lines and of related technologies by using a small aquarium fish, the Japanese medaka (*Oryzias latipes*). Our results show that it is possible to derive ES cell lines in teleost fish and to use them for transgenesis.

2. STABLE PLURIPOTENT ES CELL LINES OF MEDAKA

The major challenge for the establishment of ES cell lines is to inhibit their spontaneous differentiation. In the mouse this is achieved by cultivating the blastocyst-derived cells on a layer of feeder cells (Evans and Kaufman, 1981), in conditioned media (Martin, 1981) or in the presence of leukemia inhibitory factor (LIF; Nichols *et al.*, 1990) and related cytokines. Application of all these approaches originally developed in the mouse to non-murine vertebrates achieved only partial success. If the possibility to obtain ES cell lines is not a unique feature of the mouse, then the culture conditions for ES cells appear to be very specific for each species. In order to derive medaka ES cell lines, we have developed feeder layer cell-free culture conditions which support long-term proliferation of midblastula embryo (MBE)-derived cells in an apparently undifferentiated state.

The start population of cells we used for initiating medaka ES cell lines were dissociated MBE cells. At the midblastula stage, all blastomeres are strongly positive for alkaline phosphatase staining (Fig. 1), a marker for undifferentiated mouse ES cells. By transplantation of such midblastomeres (Wakamatsu *et al.*, 1993) medaka germline chimeras have been produced.

Since no suitable feeder cells have been established for maintaining fish ES cells in an undifferentiated state, we developed feeder cell-free culture conditions under which MBE cells can be grown on gelatin-coated surface (Hong and Schartl, 1996b). A medium has been formulated for the initiation and maintenance of MBE cells. Growth response assays revealed that several supplements besides fetal calf serum are essential for the stable growth. These include fish serum, 2-mercaptoethanol, basic fibroblast growth factor and most importantly, extract from fish embryos. Interestingly, recombinant human LIF has no effect on the proliferation and differentiation of MBE cells (Hong and Schartl, 1996b).

From those MBE cultures we obtained several stable medaka ES-like (MES) cell lines. One of them, MES1, has been extensively characterized *in vitro* and *in vivo*. *In vitro* this line shows all features known from mouse ES cells: stable growth, the typical morphology of ES cells (small size with relatively large nuclei and prominent nucleoli, a round or polygonal shape), a normal diploid karyotype, high alkaline phosphatase activity, the ability to form compacted colonies when seeded at low density and to form embryoid body-like structures in suspension culture. MES1 cells have the potential to differentiate under defined conditions into various cell types including melanin-synthesizing pigment cells, contracting muscle cells, nerve cells, fibroblasts and others. Evidence that MES1 cells are pluripotent came from subclonal culture experiments: all descendants from a single colony of cells uniformly displaying the ES cell morphology were able to give rise not only to ES-like cells but also to all differentiated derivatives (Hong *et al.*, 1996). This property is retained after long-term cultivation (>100 passages during two years of culture) and is not abolished by cryostorage.

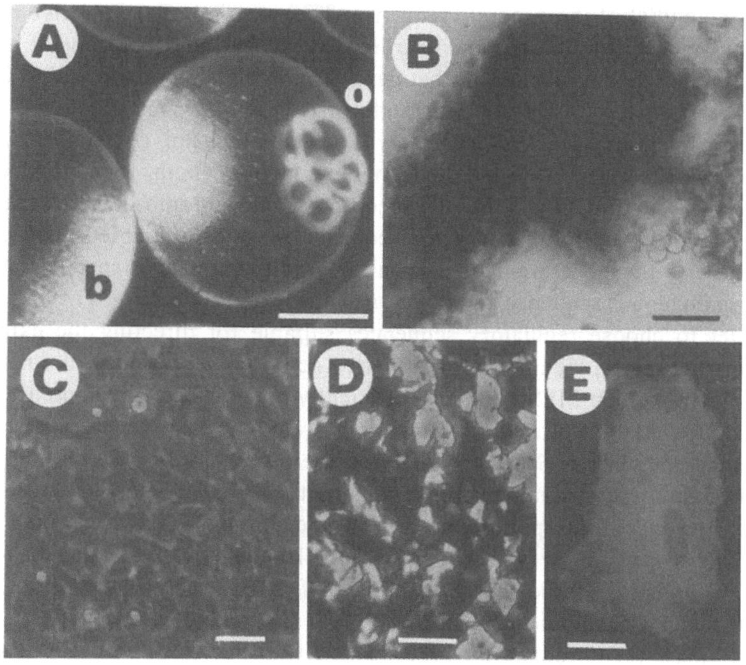

Figure 1. Medaka embryos and ES cells. (A) Embryos at the midblastula stage, each consisting of approximately 1000 cells in the blastodisc (b) opposite to oil droplets (o). (B) Blastomeres after dissociation from the midblastulae and staining for alkaline phosphatase activity. (C) MES1 cells (at passage 80) displaying a small size and compacted morphology with relatively large nuclei. (D) MES1 cells showing high alkaline phosphatase activity. (E) Fluorescent micrograph of MES1 cells expressing the GFP transgene. Bars represent 400 μm in (A), 100 μm in (B), 50 μm in (C and D) and 10 μm in (E).

3. *IN VITRO* GENETIC MANIPULATIONS OF MES1 CELLS

One major application of ES cells is gene targeting (GT), the replacement of an endogenous genomic gene sequence by a genetically engineered version via homologous recombination (HR). This is the most powerful tool for introducing defined genetic modifications into specific sites of the mouse genome, allowing the elucidation of the physiological functions of genes under study. However, this approach is so far limited to the mouse. This approach is made possible by pluripotent ES cell cultures, because following transfection ES cells can be screened for the desired modifications of the genome such as HR events prior to germline transmission.

HR events are very rare compared to random integrations. Thus, in GT experiments suitable selectable markers are required in order to distingush HR from random integration and to enrich for targeted ES cell colonies. A positive selectable marker permits the selection for cells containing an integrated HR sequence, while a negative selectable marker allows for killing cells harboring one or more copies of randomly integrated HR sequence. This dual selection system is called the positive-negative selection (PNS) procedure which was firstly devised in mice (Mansour *et al.*, 1988), but is also possible in several cell lines of other mammals. Usually, the bacterial neomycin gene *neo* encoding resistance to G418 is inserted within one exon of a gene in question, while the Herpes simplex virus thymidine kinase (*tk*) gene encoding sensitivity to gancyclovir (Gc) is attached to the end(s) of the genomic sequence. The insertion of the positive marker *neo* will result in frameshift mutations and/or truncated translation of the gene of interest and therefore, will abolish the function of that gene. The expression of *neo* and *tk* will confer on the cells resistance to G418 selection and sensitivity to Gc, respectively. Thus, a cell with randomly integrated HR vector DNA will, on the one hand, contain both *neo* and *tk* gene. This cell will survive G418 but die from Gc selection. On the other hand, during HR only the *neo* gene is cointegrated into the specific site on a chromosome, but the *tk* gene is lost. Consequently, a cell with a targeted sequence survives a double selection of G418 and Gc, ensuring the generation of a specific modification at a specific site via HR between an endogenous sequence and its *in-vitro* altered form (see Fig. 2).

In order to choose regulatory sequences suitable for directing the expression of selectable marker genes, we assayed the relative activities of various promoter/enhancer sequences. We found that MES1 cells can efficiently be transfected by a modified calcium phosphate coprecipitation procedure (Hong *et al.*, 1993). Sensitivity assays showed that MES1 cells cannot survive G418 selection. On the other hand, Gc had no effect on cell growth. Based on these data, a bicistronic construct was made which contains expression cassettes for *neo* and *tk*. Our preliminary selection experiments show that the PNS procedure using this pair of genes is functional in MES1 cells as well as other fish cell lines. Therefore, we constructed the first generation of fish HR vectors using both selectable gene cassettes and a genomic fragment of *Xmrk-Y*, the melanoma oncogene of *Xiphophorus* receptor kinase (Wittbrodt *et al.*, 1989). This HR vector was introduced into a *Xiphophorus* melanoma cell line. Following double selection with G418 and Gc, the number of cell colonies was greatly reduced compared to selection with G418 alone.

4. CHIMERAS DERIVED FROM MES1 CELLS

Chimera formation is a stringent criteria to test the pluripotency of ES cells. At first, we modified the procedure of Wakamatsu *et al.* (1993) to transplant dispersed blastomeres of a wild-type pigmented strain to albino blastula recipients. This led to the efficient production of chimeras (>90%) displaying donor-derived wild-type melanocytes as early as two days of embryonic development (Fig. 3). To determine if MES1 cells maintain their pluripotency *in vivo*, we established a procedure to transplant cultured cells. Upon microinjection of MES1 cells of 20–40 passages into albino blastulae 90% of the fry were chimeric as revealed by genotype-specific PCR-assays. Moreover, we obtained pigmented chimeras in a series of four independent experiments. The morphology and distribution of the pigmented melanocytes were similar to those of chimeras from blastomere injection.

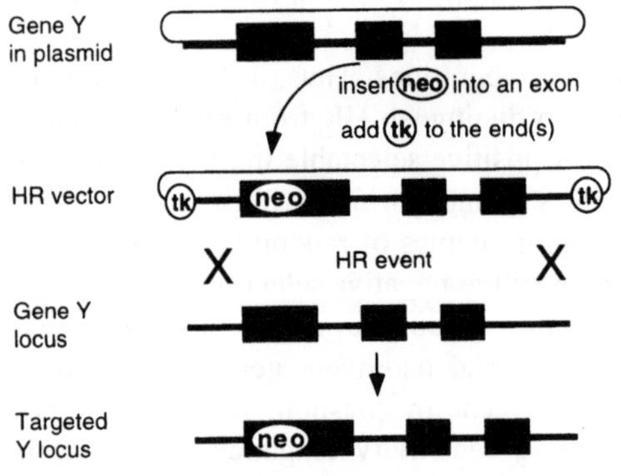

Figure 2. Scheme for gene targeting by homologous recombination. Selectable gene cassettes are shown in circle, exons in box, introns in thick line and the plasmid backbone in thin line. Note that HR between the HR vector and the chromosomal gene results in a disrupted loss-of-function locus. Following HR, the *neo* gene cassette is cointegrated, whereas the *tk* gene cassette is lost, conferring on the cells the resistance to selection with both G418 and Gc.

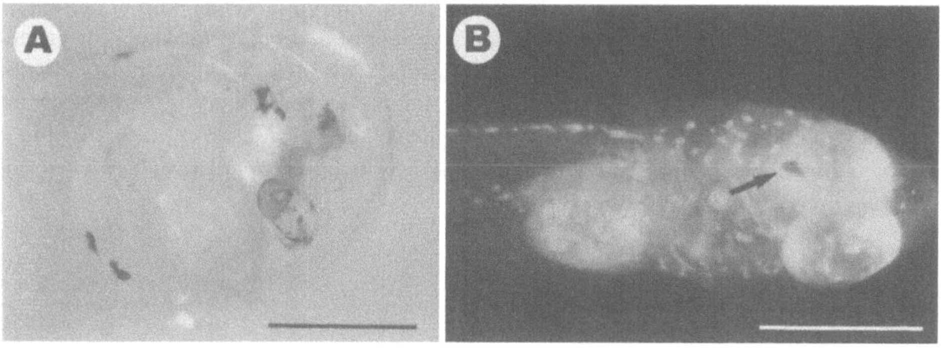

Figure 3. Chimeras derived from albino blastula recipients transplanted with wild-type blastomeres (A) and long-term cultured MES1 cells (B). Note that a wild-type pigmented melanocyte (arrow) is clearly seen on the head of the MES1-derived chimeric fry. Bars represent 500 μm.

To investigate whether MES1 cells were able to contribute not only to the pigment cell lineage but also to other cell lineages, they were labelled by transfection with a construct containing the green fluorescent protein (GFP) cDNA driven by a strong constitutive, ubiquitously active promoter. Two days after transfection 7% of the cells were GFP-positive. Such transfected culture was used for transplantation to host blastulae and the resulting embryos were examined up to the hatching stage. More than 90% of these blastulae developed into GFP-positive fry. These chimeras contained from one to more than 50 GFP-expressing MES1 cells which were found in one to several different areas. These GFP-positive cells contributed to all major organ systems of all three germ layers, for example, epithelial cells in undulating fins, contracting muscles in hearts and extraembryonic cells on the yolk sac.

These results demonstrate that the MES1 line retains its pluripotency *in vivo* even after gene transfer. Together with its normal karyotype, this line is a fish equivalent to mouse ES cell lines and thus promising to be used as a cellular vehicle for transgenesis.

5. SUMMARY

Embryonic stem (ES) cells are undifferentiated cells derived from early embryos which retain their totipotency after long-term culture. They can be genetically manipulated *ex vivo* and reintroduced into normally developing embryos. ES cells represent a powerful system to study the totipotency and differentiation of cells. They provide a bridge linking *in vitro* and *in vivo* manipulations of the genome. Small aquarium fish, like the medaka and zebrafish, offer an excellent opportunity to combine embryological, genetic and molecular analyses of vertebrate development. For this research the availability of the ES-cell technology in fish is expected to provide a similar useful tool as in the mouse for the analysis of gene functions. ES cells also have enormous potential for rapid genetic improvement of aquaculture livestocks. We have developed in the medaka feeder cell-free culture conditions supporting the long-term growth of blastula-derived embryonic cells. This led to the establishment of several medaka ES-like cell lines (MES). The MES cells exhibit stable growth over 18 months of culture with more than 100 passages using defined culture conditions. They have a normal karyotype and form colonies of densely packed, alkaline phosphatase-positive cells resembling undifferentiated mouse ES cells. In suspension culture they form embryoid bodies, and under appropriate conditions are able to differentiate into a variety of cell types. Methods for genetic manipulations and selection conditions are currently developed. After transplantation into recipient blastulae, the MES cells can participate in host embryogenesis. Furthermore, we could show that medaka ES cells after extended periods of culture are able to produce chimeras with high efficiency. Thus MES1 cells show all tested *in-vitro* and *in-vivo* characteristics of true ES cells.

ACKNOWLEDGMENTS

We thank Dr. Y. Hyodo-Taguchi for founder fish of our medaka stocks, and J. Deng for breeding of fish. This work was supported by grants to M.S. from the Deutsche Forchungsgemeinschaft, the Commission of the European Community and Fonds der Chemischen Industrie.

REFERENCES

Evans, M.J., and Kaufman, M.H., 1981, *Nature* **292**:154–156.
Delvin, R. H. *et al.*, (1994, *Nature* **371**:209–210.
Hong, Y., Winkler, C., Brem, G., and Schartl, M.,1993, *Aquaculture* **111**:215–226.
Hong, Y., and Schartl, M,1996a, In: Refrigeration Science and Technology. Proceedings, pp. 475–488.
Hong, Y., and Schartl, M., 1996b, *Mol. Mar. Biol. Biotechnol.* **5**:93–104.
Hong, Y., Winkler, C., and Schartl, M.,1996, *Mech. Dev.* **60**:33–44.
Iyengar, A., Müller, F. and Maclean, N., 1996, *Transgenic Res.* **5**:147–166.
Mansour, S. L., Thomas, K. R., and Capecchi, M. R., 1988, *Nature* **336**:348–352.
Martin, G.R., 1981, *Proc. Natl. Acad. Sci. USA* **78**:7634–7638.
Nichols, J., Evans, E.P., and Smith, A.G., 1990, *Development* **110**,1341–1348.
Wakamatsu, Y. *et al.* ,1993, *Mol. Mar. Biol. Biotechnol.* **2**:325–332.
Wakamatsu, Y., Ozato, K. and Sasado, T.,1994, *Mol. Mar. Biol. Biotechnol.* **3**:185–191.
Wittbrodt, J. *et al.*,1989, *Nature* **341**:415–421.

29

THE TROPICAL ABALONE *HALIOTIS ASININA* AS A MODEL SPECIES TO INVESTIGATE THE MOLECULAR AND CELLULAR MECHANISMS CONTROLLING GROWTH IN ABALONE

Regina T. Counihan,[1] Nigel P. Preston,[2] and Bernard M. Degnan[1,*]

[1]Department of Zoology
University of Queensland
Brisbane, Queensland 4072, Australia
[2]CSIRO Division of Marine Research
Cleveland, Queensland 4116, Australia

1. INTRODUCTION

Many marine invertebrates, including bivalves (clams, oysters, mussels, cockles), gastropods (abalone, *Trochus*) and crustaceans (prawns, crabs, lobster), are being cultivated for food around the world. While success and expansion of these shellfish aquaculture industries depend upon the economical production of a marketable product, in many cases a species' life history may impede progress in obtaining this goal. An understanding of the molecular and cellular bases of commercially-problematic life history features, such as prolonged or complex larval stages and slow growth rates, provides an essential foundation for future alleviation of this problem. This background information, along with the ability to produce transgenic animals, has the potential to improve growth and reproduction rates, disease resistance and other features of cultivated organisms (Degnan, 1997).

Despite huge interest in, and expenditure on abalone mariculture, the relatively slow growth rates of most species have hampered industry advances and have retarded efforts to select desirable traits by traditional selective breeding (Morse, 1984; Powers et al., 1995). We are investigating the factors regulating growth in the tropical abalone *Haliotis asinina* with the aim of improving growth rates, not only in this species, but in all abalone. *H. asinina* has a number of life history features that make it the appropriate model species for this task (e.g. rapid growth and early onset of sexual maturity) (Singhagraiwan & Sasaki,

* Corresponding author.

New Developments in Marine Biotechnology, edited by Le Gal and Halvorson.
Plenum Press, New York, 1998.

1991a,b; Capinpin, 1995; Capinpin & Corre, 1996). The characterisation and functional analysis of putative growth-promoting genes is facilitated in *H. asinina* because it has a short generation time and growth rates that can yield results in a relatively short period.

2. MATERIALS AND METHODS

2.1. Animals, Embryos, and Larvae

Haliotis asinina adults were collected from the reef flat of Heron Reef, Great Barrier Reef, Australia (23°27′S; 151°55′E) prior to predicted spawning time (Jebreen et al., submitted) and maintained in flowing, ambient temperature (24°C) sea water. Males and females were separated prior to spawning. Eggs and sperm were combined in 0.22 μm filtered sea water (FSW) for 10 min, zygotes then were transferred into fresh FSW using a 60 μm filter cup, and embryos and larvae were cultured at 24°C. Development was monitored by microscopy.

2.2. PCR Amplification, Cloning, and Sequencing

cDNA was prepared from total RNA from *H. asinina* adult foot muscle and amplified by PCR with degenerate oligonucleotide primers that annealed to conserved regions of the tropomyosin gene—primer T1 (5′ atggaygcnathaaraaraaratgcargcn 3′) annealed to codons 1–10, primer T2 (5′ YTCNGCNCKNGTYTCNGCYTCYTTNARYTT 3′) to codons 240–231 (Degnan et al., 1997). The resultant RT-PCR product was blunt-ended, gel purified and cloned into pBSK+ (Stratagene), and recombinant plasmid DNAs were sequenced as described in Degnan and Morse, 1993. The 5′ end of the foot muscle tropomyosin mRNA was determined by using the RACE amplification method described in Degnan et al., 1997, with a gene-specific oligonucleotide primer (TGS1: 5′ CTTCTGCTCTTCGGTATCCCTT 3′) and an anchor primer (5′ GGCCACGCGTCGACTAGT-ACGGGIIGGGIIGGGIIG 3′). The major 5′ RACE products was purified, cloned and three randomly-selected recombinant colonies cycle sequenced using TGS1.

3. RESULTS AND DISCUSSION

3.1. Rapid Development and Growth of *Haliotis asinina*

The rate of embryonic and larval development of *Haliotis asinina* is significantly more rapid than in other abalone (see Hann, 1989 for a compilation of abalone development and growth; Table 1). As in most marine invertebrates, the rate of *H. asinina* development is temperature dependant, with the trochophore hatching 5 and 8 h after fertilisation at 29 and 24°C, respectively (Capinpin and Corre, 1996; this study). Of the investigated abalone, only the development rate of *H. diversicolor supertexta* is reasonably similar, being about 27% slower at 24°C; most temperate abalone normally develop more than twice as slow (Table 1). Since the highest rate of mortality of many marine invertebrates occurs during early life history stages, the rapid development of *H. asinina* is an advantage in culture.

H. asinina postlarval-juvenile growth *in situ* and *in vitro* is estimated to be ~55 and 43 mm of shell/year, respectively (Table 1). As with early development, these are the fastest rates documented in any species of abalone (cf. Hahn, 1989). *H. asinina* has a similarly

Table 1. Comparison of embryonic developmental, juvenile growth rates and spawning seasons of some commercially-important abalone

Species	Hatching time hours (temp.)	Juv. growth rate mm shell/year	Spawning season
H. asinina	5–8 (24–29°C)	43–55	October–April
H. diversicolor supertexta	11 (24°C)	25–43	July–November
H. iris	13 (18°C)	20	March–April
H. tuberculata	13 (20°C)	18	June–November
H. discus hannai	13.3 (20°C)	20–36	September–October
H. corrugata	17 (16°C)	25	May–September
H. kamtschatkana	24 (10–13°C)	13–20	June–August
H. rufescens	25 (17°C)	24–25	January–December

Most of the data in this table is from Hahn 1989, except *H. asinina* development (Singhagraiwan and Sasaki, 1991a; this study), growth (Singhagraiwan & Sasaki, 1991b; Mc Namara & Johnson, 1995) and reproduction (Jebreen et al., submitted); and *H. diversicolor supertexta* development (Chen, 1989).

rapid onset of sexually maturity, with 30 mm (est. 9 months old) animals spawning viable gametes (unpublished data). With the shortest generation time of any abalone (e.g. temperate Australian abalone, *H. roei* and *H. laevigata*, typically develop mature gonads at 3 and 5 years of a age, respectively; Hahn, 1989), traditional selective breeding is a realistic prospect in *H. asinina*. This combination of rapid growth and short generation time facilitates the experimental analysis of gene function in that the length of time required to monitor (i) the outcome of a transient assay (e.g. the impact of a particular gene construct on growth rates) and (ii) if introduced genes are stably incorporated into the genome and inherited are significantly shortened.

The spawning seasons are known for many abalone, although in many species the exact time of natural spawning events cannot be predicted (Table 1). However, for many cultivated abalone, gonad maturation and spawning can be induced by a range of methods (reviewed in Hahn, 1989), hence alleviating the problems associated with natural unpredictable spawning cycles. Unlike other abalone which spawn a few times a year, *H. asinina* inhabiting the Southern Great Barrier Reef naturally spawn predictably and sychronously every two weeks for about six months (Jebreen et al., submitted). The predictability of spawning and the length of the reproductive season allows the correlation of gene expression patterns and neuroendocrine peptide production with vitellogenesis, ovulation and spawning in a very short period compared to other abalone. A typical *H. asinina* female will spawn ~5×10^5 viable eggs every two weeks and contain cohorts of oocytes within the gonad for future spawns (Jebreen et al. submitted). The length of time *H. asinina* gametes remain viable after spawning is also greater than that reported for other abalone. Fertilisation rates greater than 90% can be achieved at 5.5 h post spawning (Counihan et al. submitted) compared to 4 h for *H. discus hannai* and 3 h for *H. rufescens* (Kikuchi & Uki, 1974; Ebert & Hamilton, 1983). This extended period allows for more extensive manipulation of gametes, including the delivery of exogenous DNA

3.2. *Haliotis asinina* and *H. rufescens* Tropomyosin Sequences Are Highly Conserved

The nucleotide sequence of a *Haliotis asinina* muscle-specific tropomyosin cDNA was obtained by a combination of RT-PCR with degenerate primers and 5′ RACE with a gene-specific primer in a manner identical to that described for *H. rufescens* (Degnan et

```
Ha TCTTGTGCATTCCTCCTGTTTTGGCCCACTGTCAAACCATTCGCCAGAAACCTCGCTACAGTTCTGAGATCTCTCTTTCGGAGTCTAAGTCCAAGC.ACAGACGCTCTCAGTTCAGTTGG 119
Hr ------------------------------------T----------------T---------------G----------.---------------A----------------------.

Ha TGAGGCACTCTCGTCCGCAGTTGCGATCATTCCCTTCTGCTTGGAATTCCCTTTCTTTTATACTTTTGTATTTGTTAGCGCTCCACACCACAAACTACAGCCATGGATGCCATCAAGAAG 239
Hr -TG------------T-------------.------------------------------------------------------GTA-------G-------------------------
                                                                                                         M  D  A  I  K  K    6

Ha AAGATGTTGGCCATGAAGATGGAAAAGGAGAATGCTGTTGACAGGGCCGAACAGAACGAGCAGAAACTAAGGGATACCGAAGAGCAGAAGGCCAAGATTGAAGAAGATCTGAACAACCTG 359
Hr -----------------------G-----A--------G-----------------------A---------------------------------------------------------
   K  M  L  A  M  K  M  E  K  E  N  A  V  D  R  A  E  Q  N  E  Q  K  L  R  D  T  E  E  Q  K  A  K  I  E  E  D  L  N  N  L    46

Ha CAAAAGAAATGCGCAAACTTGGAAAACGACTTCGACAATGTAAACGAGCAACTGCAGGAAGCCATGGCCAAGTTGGAAACTTCCGAGAAGAGAGTTGCAGAGATGGAACAAGAGGTGAGC 479
Hr -----------------T-----G-------------G-------------G------TT-----------------------------------GA----------------------T
   Q  K  K  C  A  N  L  E  N  D  F  D  Ns V  N  E  Q  Lv Q  Ev A  M  A  K  L  E  T  S  E  K  R  V  At E  M  E  Q  E  V  S    86

Ha GGCACCACCCGAAAGATCACACTGCTGGAGGAAGACCTGGAGAGGAATGAGGAGAGGCTTCAAACTGCAACAGAGAGACTCGAGGAGGCATCCAAAGCTGCAGATGAGAGTGAGAGAGGG 599
Hr --A------A-------------------------------------C--------------G-----C---------------------------CTG---------------------
   G  T  T  R  K  I  T  L  L  E  E  D  L  E  R  N  E  E  R  L  Q  T  A  T  E  R  L  E  E  A  S  K  Al A  D  E  S  E  R  G   126

Ha CGCAACCTGCTTGAGAGCAGGAGTCTTGCTGATGATGAAAGGATAGATCAGTTGGAAGCTCAACTTAAAGAGGCTAAATACATTGCAGAAGATGCTGAGCGTAAATACGATGAGGCTGCC 719
Hr -----GG---------------------------------------------------------------------------------------------C-----T------------
   R  Nq Lv L  E  S  R  S  L  A  D  D  E  R  I  D  Q  L  E  A  Q  L  K  E  A  K  Y  I  A  E  D  A  E  R  K  Y  D  E  A  A   166

Ha CGTAAACTCGCGATCACAGAGGTTGACCTTGAGCGTGCAGAGGCTCGTCTTGAAGCTGCAGAAGCGAAGATTCTCGAGCTTGAGGAGGAGTTGAAGGTTGTGGGCAACAATATGAAATCA 839
Hr --------------------------------------------------------C---------------------------------------------------------------
   R  K  L  A  I  T  E  V  D  L  E  R  A  E  A  R  L  E  A  A  E  A  K  I  L  E  L  E  E  E  L  K  V  V  G  N  N  M  K  S   206

Ha CTGGAAATCAGTGAACAAGAGGCATCTCAGAGAGAGGACAGCTATGAGGAGACTATCCGTGACCTCACACAG 911
Hr --------------------------------G------G----C--------C-----G-----G------
   L  E  I  S  E  Q  E  A  S  Q  R  E  D  Sg Y  E  E  T  I  R  D  L  T  Q   232
```

Figure 1. Partial *Haliotis asinina* (*Ha*) tropomyosin nucleotide and predicted amino acid sequence. The *H. rufescens* (*Hr*) nucleotide sequence (Degnan et al., 1997) is aligned to the *H. asinina* sequence, with dashes representing identity and dots equalling gaps. Differences in amino acid sequence between the two abalone tropomyosins is denoted by an uppercase (*H. asinina*) followed by a lowercase (*H. rufescens*) letter code. Nucleotide and amino acid numbering is to the right.

al., 1997). Exact sequence overlap between RT-PCR and RACE products confirmed that the cDNA sequence corresponded to a single tropomyosin mRNA. In addition, identical genomic DNA sequence was obained for the 5′ end of the open reading frame (ORF) and 5′ untranslated region (5′ UTR) by suppression PCR (Siebert et al., 1988) using the TGS1 primer (Counihan and Degnan, unpublished data). These analyses yielded 911 bp of sequence—221 bp of 5′ UTR and 690 bp of the 5′ end of the ORF (Fig. 1).

As in *H. rufescens*, only a single type of tropomyosin cDNA was amplified from the muscle of *H. asinina*, supporting the notion that multiple tropomyosin genes or alternatively spliced transcripts may not be present in abalone or other molluscs (Degnan et al., 1997). In situ hybridisation analysis of tropomyosin gene expression in ~1 month old abalone confirmed that this gene is expressed in foot, columellar and a range of smaller muscles associated with other organs (Counihan and Degnan, data not shown). Comparison of the *H. asinina* tropomyosin sequence with the *H. rufescens* sequence revealed a very high level of conservation, with nucleotide and amino acid sequences being 95.0 and 96.6% identical, respectively (Fig. 1). The 5′ UTRs and ORFs were 93.7 and 95.4% identical. 21 of the 32 bp differences in the ORF were silent changes that have no impact on the amino acid sequence. 4 of the 8 amino acid differences between *H. asinina* and *H. rufescens* were conservative and probably have little effect on the overall structure of tropomyosin (Fig. 1).

This high level of sequence similarity between a southern hemisphere tropical abalone (*H. asinina*) and a northern hemisphere temperate abalone (*H. rufescens*) suggests either (i) nucleotide sequence divergences in nuclear structural genes may be low amongst the haliotids or (ii) that functional constraints are restricting significant evolutionary change of the abalone tropomyosin gene. Analysis of the evolution and systematics of Haliotidae by analysis of the sperm lysin ORF and 3′ UTR sequences reveals that the temperate abalone inhabiting the Northeast Pacific are the most distantly related haliotids to the Indo-West Pacific tropical abalone (Lee & Vacquier, 1995). While *H. asinina* was not included in this survey, subsequent phylogenetic analyses of Australian and South African abalone using sequences from the mitochondrial cytochrome oxidase II gene has shown *H. asinina* is most closely related to the other Indo-Pacific tropical abalone, *H. varia* and *H. ovina* (S. Degnan, unpublished data).

4. CONCLUSIONS

Given the extreme conservation of regulatory genes involved in similar developmental and physiological processes in disparate taxa (i.e. even organisms in different phyla) (see Degnan, 1997) and the fact there little difference between the sequences of divergent abalone, it is likely that homologues of genes involved in controlling growth or reproduction in *H. asinina* will have the same function in all haliotids. With rapid development, growth and attainment of sexual maturity, and a highly predictable spawning cycle, *H. asinina* is a ideal model species to analyse the molecular and cellular processes regulating commercially-important characteristics of the abalone.

5. ABSTRACT

Efforts to select desirable traits by the traditional selective breeding of abalone have been hindered by slow growth rates and relatively long generation times. We are currently using the abalone *Haliotis asinina* as model to investigate the molecular basis of growth in abalone and other commercially-important molluscs. *H. asinina* is a rapidly-growing tropical species distributed widely in coastal reefs of the Indo-Pacific that attains a shell length of ~55 mm in one year and sexual maturity in less than a year. *H. asinina* also has an extremely predictable reproductive cycle, with gravid animals spawning every two weeks for six months of the year. Molecular analysis and comparison of the muscle-specific tropomyosin gene indicate that there is a remarkably high level of sequence conservation between disparate haliotid taxa; *H. asinina* and *H. rufescens* nucleotide and amino acid sequences are 95% and 96.6% identical. These life history and evolutionary features facilitate the rapid analysis of commercially-important characters in *H. asinina* and suggests that insights into molecular and cellular processes gained by studying the tropical abalone will be applicable to the slower growing temperate abalone species.

ACKNOWLEDGMENTS

We thank K. Townsend for collecting and handling abalone and the staff of the Heron Island Research Station for their assistance in maintenance of the animals. This work was supported by an Australian Research Council grant to BMD and a CSIRO Fisheries Scholarship to RC.

REFERENCES

Capinpin, E.C. 1995. *Philippine J. Sci.* **124**, 215–232.
Capinpin, E.C., Corre, K.G. 1996. *Aquaculture* **144**, 81–89.
Chen, H. 1989. In: CRC Handbook of Culture of Abalone and Other Marine Gastropods (Ed. K. Hahn). CRC Press, Florida. pp. 265–283.
Degnan, B.M. 1997. *In:* Molecular Approaches to the Study of the Ocean (Ed. K. Cooksey). Chapman and Hall, London, pp. 343–358.
Degnan, B.M., Degnan, S.M., Morse, D.E. 1997. *Dev. Genes Evol.* **206**, 464–471.
Degnan, B.M., Morse, D.E. 1993. *Mol. Mar. Biol. Biotech.* **2**, 1–9.
Ebert, E.E., Hamilton, R.M. 1983. *Calif. Fish Game.* **69**, 115–120.
Hahn, K.O. 1989. CRC Handbook of Culture of Abalone and Other Marine Gastropods. CRC Press, Florida.
Kikuchi, S., Uki, N. 1974. *Bull. Tohoku Reg. Fish. Res. Lab.* **34**, 73.
Lee, Y.-H., Vacquier, V.D. 1995. *Mar. Biol.* **124**, 267–278.
McNamara, D.C., Johnson, C.R. 1995. *Mar. Freshwater Res.* **46**, 571–574.
Morse, D.E. 1984. *Aquaculture.* **39**, 263–282.
Powers, D.A., Kirby, V.L., Cole, T., Hereford, L. 1995. *Molc. Mar. Biol. Biotech.* **4**, 369–376.
Siebert et al. 1995. *Nucleic Acids Res.* **23**, 1087–1090.
Singhagraiwan, T., Sasaki, M. 1991a. *Thai Mar. Fish. Res. Bull.* **2**, 83–94.
Singhagraiwan, T. Sasaki, M. 1991b. *Thai Mar. Fish. Res. Bull.* **2**, 95–100.

30

NORTH AMERICAN *PORPHYRA* CULTIVATION

From Molecules to Markets

I. A. Levine[1] and D. Cheney[2]

[1]Coastal Plantations International, Inc.
Poland, Maine 04274
[2]Marine Science Center
Northeastern University
Nahunt, Massachusetts 01908

1. INTRODUCTION

The red alga *Porphyra* (nori) is utilized by humans as a food source throughout the world. In 1992, approximately 15 billion sheets (approximately 45,000 dry metric tons) of *Porphyra* were produced with an annual value of $1.8 billion (Jensen, 1993). The consumption of nori, long prized by the Japanese, Chinese, Koreans and other people of Asia and the western Pacific as a complement to rice, sushi, soups and salads. Additionally, nori is a staple in macrobiotic diets (Mumford and Miura, 1988), is a source of taurine which controls blood cholesterol levels and a valuable source of phycoerythrin, a red pigment used in the medical diagnostic industry. The consumption of nori in the United States was wholly dependent upon importation from Japan, Korea and China prior to Coastal Plantations International entering the market.

Commercial Cultivation of the red alga *Porphyra yezoensis* (nori) was initiated in Cobscook and South Bays, Maine in 1991 by Coastal Plantations International, Inc. (CPI) (Levine and Crawford, 1997; Yarish et al., 1996a,b, 1997). CPI's attempt to cultivate nori in the country's northeastern coastal waters follows Washington State's efforts of the 1980s. The transfer and modification of Japanese, Korean, Chinese and Washington State's nori cultivation and processing technologies to Maine has resulted in the world's first internationally certified, organically cultivated and processed nori in addition to the establishment of a nori:finfish integrated polyculture system.

The development of CPI has resulted in the formation of two divisions. The Maine Nori Company represents CPI's effort to produce and market organic nori and develop new products and marketing strategies for nori in North America and beyond. PhycoGen, a biotechnology research and product development effort whose goals are to produce new

New Developments in Marine Biotechnology, edited by Le Gal and Halvorson.
Plenum Press, New York, 1998.

industrial products and drugs through the development of transgenics, natural product discovery, site directed mutagenesis, phycoremediation, and genetic transformations via protoplast fusions. The following applications are representative selections of pending and future aquatic biotechnology projects: natural sunscreen, nori pigments, phycoremediation, edible vaccines, and a sugar substitute.

CPI imported the Japanese cultivar *Porphyra yezoensis* to the waters of Cobscook Bay, Maine due to the non-commercial qualities of the indigenous species. The introduced *P. yezoensis* could be improved by incorporating certain characteristics of Maine's local varieties. To that effect, a cultivar improvement program was initiated utilizing proprietary protoplast fusion and mutagenesis techniques to be followed by strain selection.

While visiting our Maine facilities, a Japanese nori industrialist commented "CPI and its management have exhibited *frontier spirit* for what it has accomplished in North America". He further stated he had no reservations concerning the prediction of our future success, because CPI possesses the "knowledge and passion" for nori cultivation, research and marketing.

2. RESULTS AND DISCUSSION

- Convened International Seminar "Potential *Porphyra* (Nori) Cultivation and Processing Industry in Maine"
- Awarded Maine Aquaculture Innovation Center, National Coastal Research Institute, National Marine Fishery Service, and National Sea Grant College Program Research Grants totally 1.5 million dollars.
- Purchased and assembled nori processing plant and cultivation system
- Successfully seeded and cultivated *Porphyra yezoensis* (Japanese cultivar)
- Produced market quality nori sheets under the Maine Nori Company logo
- Successfully established direct sales of organic nori sheets
- Awarded permits, licenses and approvals from; local (Eastport), state (DEP and DMR), federal (Army Corps of Engineers, US Coast Guard, and EPA), Canadian (Federal Dep. Fisheries and Oceans, Provincial Departments of Fisheries and Aquaculture and the Environment) and global (UN's International Conference for Exploration of the Sea (ICES)
- Established Japanese nori farmer exchange program
- Sponsored Marine Agronomy Fellowship Program
- Developed bulk nori sales to biomedical industry
- Awarded organic certification from the OCIA International, Inc.
- Domesticated local wild nori and Genetically improved several strains of nori
- Initiated salmon:nori integrated polyculture

The development of a United States based marine agronomy and integrated polyculture industry, is complete. The farm has reached a maximum of 700 saku with two nets/saku/season. Production values have been inconsistent, ranging from 0 to 8000 sheets per net equivalents. The farm and processing units have developed the experience to produce and market a crop of organically produced nori sheets distributed to health food stores, direct mail order, and as a source of fluorescent labelers marketed directly to biomedical concerns.

CPI operates a cultivation and processing facility in Eastport, Maine, maintains three aquaculture lease sites in Cobscook and South Bay, Maine totaling 128 acres, while the corporate offices and laboratory are located in Poland, Maine. Additionally, the company has established satellite research efforts in Massachusetts, North Carolina, Florida, and California. A second farm and processing facility are envisioned for the waters of Long Island Sound.

Cultivation expansion has been achieved through the recruitment, training and support of independent nori farmers. A commercial fisherman in Grand Manan, New Brunswick, Canada and a nori growing cooperative in central Maine are the first two contract nori operations. CPI supports the new nori concerns by supplying seeded nets and purchasing the harvested nori. Additional contract farms are anticipated from Nova Scotia, Canada south to Connecticut, USA.

Presently, 38 putative protoplast fusion products that have either been grown to whole plants or are in the process; 20 of these plants or cultures are currently under further investigation. Utilizing isoenzymes as genetic markers to distinguish species, several plants exhibit unusual combination of features that differentiate them from either parental species. Unusual characteristics include: unusual blade morphology or the blade morphology of one species and the isoenzyme phosphoglucose isomerase (PGI) banding of the other species (Cheney, 1997; Watson et al., 1996, 1997). Several of these latter plants have recently had their DNA content studied by Donald Kapraun (methods as described in Dutcher and Kapraun, 1994; Kapraun et al. 1991). The results indicate a plant produced by protoplast fusion FE#9–13 has a *Porphyra yezoensis* blade shape and PGI isoenzyme pattern with a chromosome number of n=6 instead of n=3, the normal chromosome number for *P. yezoensis* blades (see Cole, 1990) and therefore may be a polyploid or hybrid. Initial growth rates measured from FE#9–13 are 14% per day in laboratory culture, a particularly rapid rated suggesting the potential commercial nature of this fusion product.

3. SUMMARY

The transfer and adaptation of the Asian *Porphyra* (nori) cultivation and processing technologies to the coastal waters of Maine, USA and the Maritime Provinces of Canada by Coastal Plantations International, Inc. (CPI) has completed five years of effort. The introduction of the Japanese winter *P. yezoensis* to Maine during the summer and autumn months represents the first of many biological challenges. CPI's farming effort entails 700 saku (net space) along with the addition of independent family, cooperative and corporate nori farmers. The commercialization of *Porphyra* has resulted in new sources of employment and the formation of PhycoGen, CPI's internal biotechnology R&D division. In addition to the traditional uses of nori, CPI is providing the raw materials for the production of bioluminescent markers.

The enhancement of domesticated local species and/or the modification of the introduced species is one of our goals. Utilizing newly developed protoplast fusion techniques, we are producing putative hybrids that will exhibit enhanced growth in Maine waters and produce novel pigment compositions. Analysis of a hybrid between *P. umbilicalis* and *P. yezoensis* is underway.

ACKNOWLEDGMENTS

Support from the National Coastal Research Institute and the National Sea Grant Program is gratefully acknowledged.

REFERENCES

Cheney, D. 1997, Genetic manipulation and strain improvement of commercially valuable seaweeds. In: *Recent Advances in Marine Biotechnologies* (in press).

Cole, K. 1990, Chromosomes. In: (Eds.) K. Cole and R. Sheath, *The biology of the red algae*. Cambridge Univ. Press, NY, pp 73–101.

Dutcher, J. And D. Kapraun. 1994, Random amplified polymorphic DNA (RAPD) identification of genetic variation in three species of *Porphyra* (Bangiales, Rhodophyta). *J. Applied Phycology* **6**: 267–273.

Jensen, A. 1993, Present and future needs for algae and algal products. *Hydrobiologia* **260/261**:15–23.

Kapraun, D., T. Hinson and A. Lemus. 1991, Karyology and cytophotometric estimation of inter- and intraspecific nuclear DNA variation in four species of *Porphyra* (Rhodophyta). *Phycologia* **30**: 458–466.

Levine, I.A. and S. Crawford. 1997, North American *Porphyra* cultivation - Marine Agronomy, Integrated *Polyculture, and Biotechnology. Phycologia* **36**(4): 60.

Mumford, T. And A. Miura. 1988, *Porphyra* as food: cultivation and economics. In: (eds.) C. Lembi and J. Waaland). *Algae and Human Affairs*. Cambridge Univ. Press. NY. Pp 87–117.

Watson, K., K. Roberts, and D. Cheney. 1996, Protoplast isolation and fusion of North Atlantic *Porphyra* species and *P.yezoensis*. *J. Phycology* **32**(4): 50.

Watson, K., K. Roberts, D. Cheney, and I. Levine, 1997, *Porphyra* Hybridization and Strain Improvement for Cultivation in North America. *Phycologia* **36**(4): 121.

Yarish, C., A.E. Sperr, X.G. Fei, A.C. Mathieson, and I.A. Levine. 1996a, Field and Culture Studies for the Development of a Commercially Viable Nori Aquaculture Industry in New England. *J. Phycol Suppl.* **32**(3):45.

Yarish, C., A.E. Sperr, X.G. Fei, A.C. Mathieson, and I.A. Levine. 1996b, Field and culture studies of several New England *Porphyra* species in order to characterize the distributional ecology and develop strains. *J. Phycol. Suppl.* **32**(3):52.

Yarish, C., A.C. Mathieson, E. Hehre, A.S. Klein, C.D. Neefus, G.G. Mitman, X.G. Fei, and I. Levine. 1997., Domestication of *Porphyra* for Northeast America. *Phycologia* **36**(4): 128.

31

OXYGEN TRANSPORT BY HEMOCYANINS

Kensal E. van Holde

Department of Biochemistry and Biophysics
Oregon State University
Corvallis, Oregon 97331-7305

Although much study of oxygen transport and utilization in animals has centered on the hemoglobins of vertebrates, a very large number of invertebrate marine species utilize very different proteins, the hemocyanins. Both molluscs and arthropods possess these proteins; in each phylum they utilize binuclear copper centers as oxygen binding sites, rather than an iron heme. The molecular structure, and mechanisms of regulation of oxygen binding appear to be quite different in the two phyla. We have concentrated on the hemocyanin of the cephalopod, *Octopus dofleini*. The functional molecule in the blood is immense, with molecular weight of about 3.6×10^6; this is comprised of ten identical subunits, each containing 2898 amino acid residues. We have recently completed the c-DNA sequencing of this chain. It is divided into seven neatly equivalent domains, each containing one oxygen binding site. The structure of one such domain has recently been determined in collaboration with the laboratory of Dr. W. Hendrickson. The seventy oxygen binding sites in this molecule make it one of the largest allosteric assemblies known. Exactly how such unusual proteins serve in facilitating oxygen transport to the tissues of the major invertebrates remains a matter of continued study. For a comprehensive review, see van Holde and Miller, 1995.

Comparison of the sequences and three-dimensional structures of the arthropod and molluscan hemocyanins is now possible. The two types exhibit almost entirely different molecular architectures and amino acid sequences, with the exception of a small region containing the histidine residues chelating one of the copper atoms (the *copper-B-site*). The other (*copper-A-site*) is entirely different in arthropod and molluscan hemocyanins. The molluscan A-site resembles in sequence that found in many tyrosinases. On the basis of these differences and similarities, we can propose a general hypothesis for the evolution of hemocyanins and related proteins from primitive ancestral copper proteins.

Hemocyanins have long been known to be extremely potent antigens. Recently, this behavior has been exploited in the use of these proteins as immunostimulants in cancer therapy (Swerdlow et al., 1995). Continued study of the structure of these proteins is essential to the further development of such applications.

New Developments in Marine Biotechnology, edited by Le Gal and Halvorson.
Plenum Press, New York, 1998.

REFERENCES

van Holde, K.E. and Miller, K.I. 1995, *Advances Protein Chem.* **47**, 1–81.
Swerdlow, R., Miller, K.I., Lee, P., Bonaventura, C. and Ebert, R.F. 1995, *Comp. Biochem. Physiol.* **113B**, 537–548.

32

THE INK GLAND OF *SEPIA OFFICINALIS* AS BIOLOGICAL MODEL FOR INVESTIGATIONS OF MELANOGENESIS

Anna Palumbo,[1] Ida Gesualdo,[1] Anna Di Cosmo,[2] and Luigi De Martino[1]

[1]Zoological Station
Naples, Italy
[2]Department of Zoology
University of Naples Federico II
Naples, Italy

Melanins and melanogenesis are currently the subject of keen interest because of their increasing biotechnological applications in a variety of processes of biomedical, industrial and cosmetic relevance. Because of the notorious difficulties in the direct investigation of melanogenesis in mammals, considerable attention has traditionally been directed to elucidate the process of pigment formation in the ink gland of *Sepia officinalis*, which provides a most convenient and active melanin producing system. Ultrastructural studies have shown that within ink gland cells melanin formation takes place in specific subcellular organelles termed melanosomes by oxidation of tyrosine catalyzed by tyrosinase (Ortonne et al., 1981). Recent biochemical studies carried out in our laboratory have shown that, in addition to tyrosinase (Prota et al., 1981), other enzymatic activities are involved to a various extent and with different modalities in the pigment pathway. In particular, the ink gland contains a novel enzyme called dopachrome rearranging enzyme (Palumbo et al., 1994), which catalyzes the rearrangement of dopachrome to 5,6-dihydroxyindole (DHI), and a peroxidase (Palumbo & Jackson, 1995; Gesualdo et al., 1997) which catalyzes the polymerization of DHI and 5,6-dihydroxyindole-2-carboxylic acid (DHICA) to melanin (Fig. 1). However, whether and to what extent these enzymes co-localize in the melanogenic compartments and interact is an open question. To address this aspect, polyclonal antibodies have been raised against *Sepia* melanogenic enzymes and their specificity has been analyzed by immunoprecipitation and western blot analysis. Immunoelectron microscopy revealed that peroxidase has a different subcellular localization pattern from tyrosinase and dopachrome-rearranging enzyme. Whereas peroxidase is located in the rough endoplasmic reticulum (RER) and in the matrix of premelanosomes and melanosomes, tyrosinase and dopachrome-rearranging enzyme are present in the

New Developments in Marine Biotechnology, edited by Le Gal and Halvorson.
Plenum Press, New York, 1998.

Figure 1. Schematic outline of melanogenesis in *Sepia officinalis*.

RER-Golgi transport system, at the level of trans-Golgi cisternae (TGC), trans-Golgi network (TGN) and coated vesicles, and in melanosomes on pigmented granules (Palumbo et al., 1997). Taken together, our results allow the elaboration of a model of melanin formation in the ink gland which highlights the dynamic interactions of the melanogenic enzymes at molecular and ultrastructural levels. As reported for the formation of other particulate melanosomes, the biogenesis of *Sepia* melanosomes involves two separate cellular compartments (Fig. 2). The process begins as vesicles bleb from the RER with a concomitant removal of the ribosomes (step 1). These vesicles, which represent the premelanosomes and the matrix of the mature melanosomes, contain active peroxidase as revealed by classical enzymatic histochemistry using diaminobenzidine and hydrogen

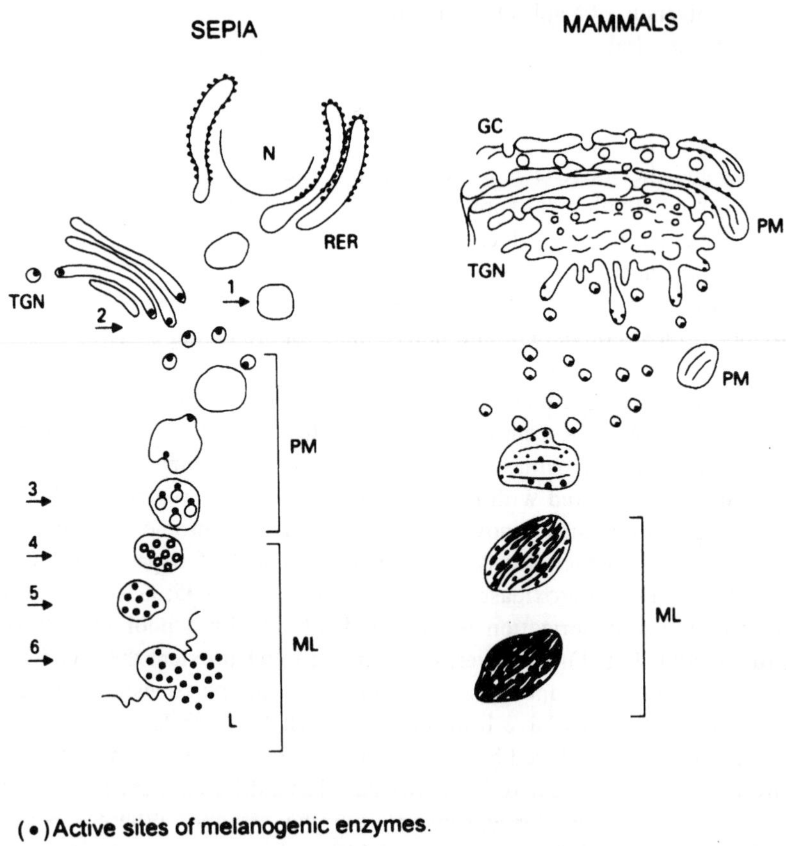

Figure 2. A comparison between melanin formation in the ink gland of *Sepia officinalis* and mammals. N, nucleus; RER, rough endoplasmic reticulum, TGN, trans-Golgi network; PM, premelanosome, ML, melanosome; L, lumen; GC, Golgi complex.

peroxide (Palumbo et al., 1997). At the same time, small vesicles, called coated vesicles, which we have found to contain tyrosinase and dopachrome rearranging enzyme are released from the TGN into the cytoplasm (step 2). We can assume that in *Sepia*, tyrosinase and dopachrome rearranging enzyme are oriented in the membrane as in mammals with the active site into the lumen of the coated vesicle. Coated vesicles then fuse with the membrane of premelanosomes and are subsequently incorporated into it (step 3). It has been reported in the literature that this incorporation involves inversion and reformation of the coated vesicles inside the premelanosomes. So we can hypothesize that after incorporation and inversion of the coated vesicles, the enzymes will expose their catalytic sites to the melanosomal matrix. On this basis, we can speculate that tyrosine, present within the melanosomal matrix, is converted by the enzymes present on the external side of the coated vesicles membrane to the products dopachrome and the indoles DHI and DHICA according to the scheme reported in Fig. 1. The indoles are then oxidized and polymerized by peroxidase, already within the melanosomal matrix. Following the action of peroxidase, melanin synthesis begins at around the periphery of the incorporated coated vesicles (step 4) and subsequently proceeds to completely cover their surfaces (step 5). So the incorporated coated vesicles not only provide enzymes to premelanosomes, but also function as a structural foundation for melanin synthesis. The size and the number of these vesicles remain as such even during melanin synthesis which results in the particulate melanosomes. When the maturation of the epithelial cell is complete, the membranes of the melanosomes fuse with those of the apical pole of the cell and all the cellular constituents, including the melanin granules, are excreted into the lumen of the ink gland (step 6). From this scheme of melanin formation in *Sepia* it is clearly evident that *Sepia* melanosomes possess a complex organization surprising for an invertebrate which could represent along the phylogenetic scale an intermediate step in the evolution of the melanin producing system. Although it is difficult to speculate to what extent the information obtained for *Sepia* can be extended to the mammalian system, it is evident that the major difference between these two melanin producing systems lies in the structure of the melanosomes. Whereas *Sepia* possesses particulate melanosomes, typical of primitive vertebrates, in mammals melanosomes are fibrillar and are characterized by a lamellar matrix completely filled by melanin. In the course of evolution there was the transition from particulate to fibrillar melanosomes. The peculiar organization of melanin in *Sepia* which occurs without the complete filling of the matrix as in mammalian melanosomes, provided an opportunity to investigate at subcellular level the mechanism of interaction of the melanogenic enzymes. We think that in spite of the transition from particulate to fibrillar melanosomes, it is reasonable to hypothesize that the molecular mechanisms leading to melanin formation have remained almost unchanged in the course of evolution.

REFERENCES

Gesualdo, I., Aniello, F., Branno, M. and Palumbo, A. 1997, *Biochim. Biophys. Acta* **1353**:111–117.

Ortonne, J. P., Voulot, C., Chatchadourian, C., Palumbo, A. and Prota, G. 1981, In Pigment Cell 1981: Phenotypic Expression in Pigment Cells. M. Seiji, editor. University of Tokyo Press, Tokyo. 49–57.

Palumbo, A., and Jackson, I. J. 1995, *Biochim. Biophys. Acta* **1247**:173–178.

Palumbo, A., d'Ischia, M., Misuraca, G., De Martino, L. and Prota, G. 1994, *Biochem. J.* **299**:839–844.

Palumbo, A., Di Cosmo, A., Gesualdo, I. and Hearing, V.J. 1997,. *Biochem. J.* **323**:749–756.

Prota, G., J. P. Ortonne, J. P., Voulot, C., Chatchadourian, C., Nardi, G. and Palumbo, A. 1981, *Comp. Biochem. Physiol.* **68B**:415–419.

33

RECOMBINANT FACTOR C FROM *CARCINOSCORPIUS ROTUNDICAUDA* BINDS ENDOTOXIN

A. W. M. Pui,[1] S. D. Roopashree,[1] B. Ho,[2] and J. L. Ding[1,*]

[1]School of Biological Sciences
[2]Department of Microbiology
National University of Singapore
Kent Ridge, Singapore 119260

1. INTRODUCTION

The coagulation cascade in the circulating amoebocytes of the Singapore horseshoe crab, *Carcinoscorpius rotundicauda*, plays a vital role in the crab's defence against invading Gram negative bacteria present in its habitat. Factor C, a serine protease zymogen, is activated by Gram negative bacterial endotoxin or lipopolysaccharide (LPS) to initiate this coagulation cascade. The *Limulus* gelation assay for detection of endotoxin is based on the affinity of the endotoxin-binding domain of Factor C for LPS [Ho, 1983; Levin, 1985]. Consequently, *C. rotundicauda* Factor C cDNA has been cloned and expressed in the methylotrophic yeast, *Pichia pastoris*, to produce recombinant Factor C (rFC) [Roopashree et al., 1996] which will serve as an alternative source of the enzyme to the conventional amoebocyte lysate. This paper describes the biological activity of full-length rFC viz its immunoreactivity by Western detection and particularly, the functionality of its endotoxin-binding domain to bind specifically to LPS and lipid A.

2. MATERIALS AND METHODS

Two recombinant Factor C constructs, namely pHIL-D2/CrFC21 and pPIC-9/CrFC26SN were used for the study. The pHIL-D2/CrFC21 construct contains the full-length CrFC21 cDNA of 3.4 kb (GenBank Database Accession No. S77063) with its native translational start and signal sequence cloned as a non-fusion fragment into

* Author for correspondence.

New Developments in Marine Biotechnology, edited by Le Gal and Halvorson.
Plenum Press, New York, 1998.

pHIL-D2 vector [Roopashree et al., 1996; Ding et al., 1995]. The pPIC-9/CrFC26SN construct is a 2.4 kb SalI-NotI 3′ fragment of CrFC26 (GenBank Database Accession No. S77064) [Ding et al., 1995] lacking the endotoxin-binding region and cloned in fusion downstream of the pPIC-9 vector start site and secretion signal. Both clones were transformed into the methylotrophic yeast, *Pichia pastoris*.

The recombinant *Pichia* clones were grown to mid-log phase in minimal glycerol growth medium and subsequently in induction medium containing 0.5% methanol for 8 h. Induced cells were harvested and lysed mechanically with chilled acid-washed glass beads (0.45 mm diameter, Sigma). Total protein content in the yeast cell lysate was estimated by Bradford assay [Bradford, 1976].

Immunoblotting was employed to confirm the expression of the recombinant Factor C, both full-length rFC as well as the truncated product, rFCSN, from pHIL-D2/CrFC21 and pPIC-9/CrFC26SN, respectively. Protein samples were electrophoresed on denaturing SDS/10% polyacrylamide gel [Laemmli, 1970] and electroblotted onto Immobilon PVDF membrane (Millipore). Factor C-related proteins were detected with anti-Factor C antibody and visualised with secondary horseradish peroxidase-conjugated antibody using 4-chloro-1-naphthol (Sigma) and H_2O_2 as substrates.

A modified Western blot of LPS (*E. coli* 055:B5; Sigma) and lipid A (*E. coli* K12, D31m4, List Biologicals, USA) [Ding et al., 1997] was developed to investigate the ability of rFC to bind LPS or lipid A. Briefly, Immobilon PVDF membrane-blotted LPS or lipid A, cut into strips, were reacted with yeast lysate containing rFC. Detection using anti-Factor C antibody was followed by visualisation with secondary alkaline phosphatase-conjugated goat anti-rabbit antibody.

3. RESULTS AND DISCUSSION

As detected on the immunoblot (Fig. 1), the full-length rFC has a molecular size of ~132 kDa which corresponds to the apparent molecular weight of the native Factor C

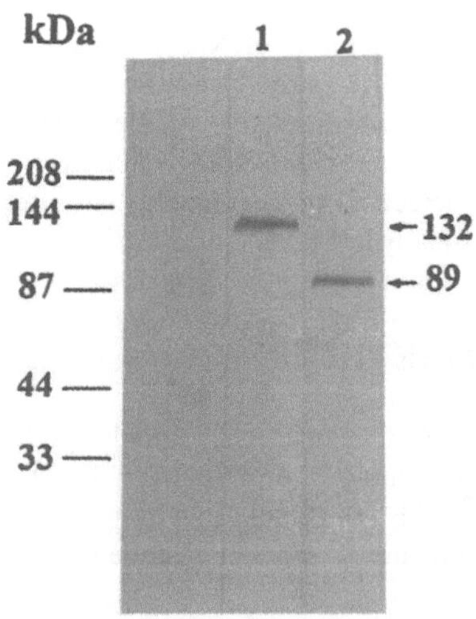

Figure 1. Immunoblot showing expression of rFC (pHIL-D2/CrFC21; lane 1) and rFCSN (pPIC-9/CrFC26SN; lane 2).

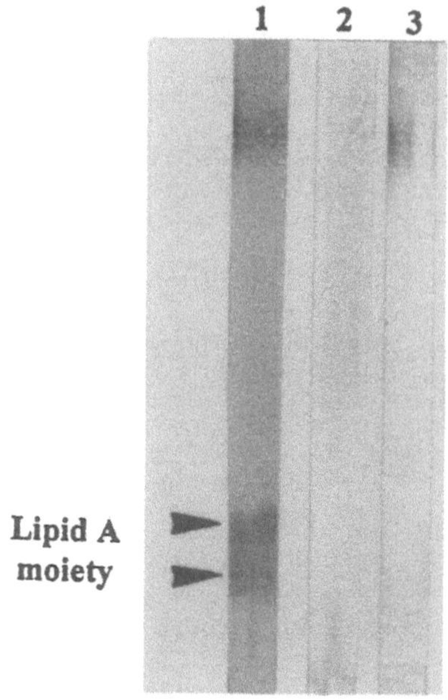

Figure 2. Modified Western showing the binding affinities of **1**) rFC, **2**) pHIL-D2 and **3**) rFCSN to electroblotted lipid A. The lipid A bands (7–20 kDa) were indicated.

[Ding et al., 1993]. Since the rFC has a molecular size approximately 11 kDa higher than the calculated size based on the cDNA insert, it could be envisaged that the rFC was post-translationally modified by glycosylation as there are six potential N-glycosylation sites [Ding et al., 1995]. Thus, the expression of rFC indicates a successful recognition of the CrFC21 translational start and stop signals by *P. pastoris*. Similarly, the truncated rFCSN was expressed as an 89 kDa protein.

Besides its immunoreactivity, rFC was able to bind specifically to the 7–20 kDa lipid A moiety of LPS showing distinctive bands on the modified Western blot (Fig. 2). The lipid A binding is deemed to be specific to the endotoxin-binding domain of rFC because protein samples containing the parent vector pHIL-D2 alone or rFCSN lacking the N-terminus of Factor C did not show any binding to the LPS/lipid A strips. It has been reported that the first 250 amino acid residues of the N-terminus of Factor C are involved in endotoxin binding [Miura et al., 1992]. Therefore, it could be envisaged that rFC possesses a functional endotoxin-binding domain which further suggests correct folding of the protein in *P. pastoris*.

The lipid A moiety of LPS is responsible for the pathophysiological reactions of Gram negative bacterial infections [Rietschel & Brade, 1992]. This study suggests the potentials of using rFC, after further purification, for the detection and/or removal of LPS or lipid A in pharmaceutical products, other biologicals and chemicals, and clinical specimens.

ACKNOWLEDGMENTS

We thank Mr. H. K. Yip for photography. This work was supported by National University of Singapore, Grant RP950303.

REFERENCES

Bradford, M.M. 1976, *Anal Biochem* **72**:248–254.
Ding, J.L., Chai, C., Pui, A.W.M., Ho, B. 1997, *J Endotoxin Res* **4**:33–43.
Ding, J.L., NavasIII M.A.A, Ho, B. 1993, *Biochim Biophys Acta* **1202**:149–156.
Ding, J.L., NavasIII M.A.A., Ho, B. 1995, *Mol Mar Biol Biotech* **4**:90–103.
Ho, B. 1983, . *Microbios Lett* **24**:81–84.
Laemmli, U.K. 1970, *Nature* **227**:680–684.
Levin, J. 1985, In: Cate J.W., Buller H.R., Sturk A. and Levin J. (eds), Progress in Clinical and Biological Research Vol. **189**. pp 3–28. Alan Liss, Inc., New York.
Miura, Y., Tokunaga, F., Miyata, T., Moriasu, M., Yoshikawa, K., Iwanaga, S., 1992, *J Biochem* **112**:476–481.
Rietschel E.T., Brade, H. 1992, Bacterial endotoxins. *Sci Am* **Aug**:26–33.
Roopashree, S.D., Ho, B., Ding, J.L. 1996, *Mol Mar Biol Biotech* **5**:334–343.

34

MOLECULAR AND IMMUNOLOGICAL CHARACTERIZATION OF SHELLFISH ALLERGENS

Patrick S. C. Leung[1,*] and Ka-Hou Chu[2]

[1]Division of Rheumatolology/Allergy and Clinical Immunology
School of Medicine
University of California, Davis
Davis, California 95616
[2]Department of Biology
The Chinese University of Hong Kong
Shatin, Hong Kong, China

1. INTRODUCTION

Seafood is a popular component of human diet particularly in the United States, Japan and Europe. In 1995, US consumers spent more than $38 billion in seafood with an average consumption of 15 lbs of seafood, including 2.5 pounds of shrimp per capita. During the last decade with the rapid growth of shrimp farming, the amount of shrimp consumed has doubled to 1 billion pounds per year in the United States, followed by Japan and the European Union. Shrimp is a heart-healthy diet because it has almost no fat, and most importantly no saturated fat. Thus, it is not surprising that shrimp and other shellfish are becoming more popular in our diet. Despite the high nutritious value and the increasing demand of shellfish, some individuals develop hypersensitive reactions upon ingestion of shellfish and cannot enjoy these delicious seafood. It is estimated that 0.3 to 7.5% of the U.S. population have some forms of food allergy (Taylor, 1985).Shellfish allergies are among the more commonly encountered food allergies in adults (Taylor and Bush, 1988). More than 250,000 people are having a high risk of developing allergic reactions to the ingestion of seafood products. (Henriksen, 1994; Lehrer et al., 1992). Commonly consumed shellfish includes shrimp, lobster, crab and crayfish (class Crustacea) and a variety of mollusks of class Gastropoda, Bivalvia and Cephalopoda (Table 1). Here, we will describe the common allergic reactions to shellfish and current discoveries on the biochemical, molecular and immunological characterization of shellfish allergens.

* To whom correspondence should be addressed.

New Developments in Marine Biotechnology, edited by Le Gal and Halvorson.
Plenum Press, New York, 1998.

Table 1. Taxonomic classification of commonly consumed shellfish

Phylum Arthropoda	
Class Crustacean	Shrimp, Lobster, Crab, Crayfish
Phylum Mollusca	
Class Gastropoda	Abalone, Whelk, Snail, Limpet
Class Bivalvia	Mussel, Scallop, Oyster, Clam
Class Cephalopoda	Cuttlefish, Squid, Octopus

2. ALLERGIC REACTIONS TO SHELLFISH

Allergic reactions to shellfish is highly exemplified by the hypersensitive reactions to ingestion of crustaceans. The reaction is a IgE mediated type I hypersensitive reaction, manifested by flushing, urticaria, angioedema. Other reactions may include gastrointestinal (nausea, diarrhoea, abdominal cramping), respiratory (laryngospasm, wheezing) and ocular (conjunctivitis). The initial development of a specific IgE response is mediated through the presentation of the allergen by antigen presenting cells in the context of MHC class II to T lymphocytes which in turn stimulate B lymphocytes to mature to plasma cells that secrete the specific IgE antibodies. Upon reexposure to the allergen, the sensitized subject will develop an allergic reaction through interaction of IgE bound on IgE receptors on surfaces of mast cells and basophils. The recognition of IgE epitopes of these allergens will trigger the release of chemical mediators such as histamines, leukotrienes, slow reacting substance of anaphylaxis (SRA-A), eosinophil chemotactic factor of anaphylaxis (ECF-A), resulting in increase in vascular permeability, constriction of bronchial smooth muscles, mucus secretion and migration of inflammatory cells to the reaction site (Figure 1). The reactions are usually immediate, beginning minutes to an hour after the food is taken in. Life threatening reactions of systemic anaphylaxis may occur after ingestion of shellfish among sensitized individuals (Sheffer, 1985). Such symptoms include urticaria, angiodema, generalized pruritus, palate itching, difficulty in swallowing upper airway obstruction, hypotension with a sensation of faintness and loss of consciousness in serious cases.

In addition to hypersensitivity reaction to shellfish by ingestion, occupational asthma to shellfish have been reported in fisherman, processing workers, shell grinders, cooks as well as restaurant workers (Cartier et al., 1986; Droszca et al., 1981; Gaddir et al., 1980; Hudson et al., 1985). Lobster, shrimp and crab have been documented to cause occupational dermatitis, rhinitis, conjunctivitis (Goransson, 1981; Lehrer and O'Neil, 1992; Nagano et al., 1984; Orford and Wilson, 1987). Thus allergic reactions to shellfish is a significant concern in shellfish industry and environmental health.

3. CHARACTERIZATION AND IDENTIFICATION OF MAJOR SHRIMP ALLERGEN

Although hypersensitive reactions caused by shellfish has long been known, biochemical and immunological studies on shellfish allergies had only begun in the 1980s. Using crossed immunoradioelectrophoresis, Lehrer et al.(Lehrer et al., 1985) identified seven allergens in shrimp. Hoffman et al. (Hoffman et al., 1987) reported the purification of two allergens, termed antigen I and antigen II from raw and cooked shrimp respectively. Antigen II is a heat stable IgE reactive protein of 38 kD and 341 amino acid

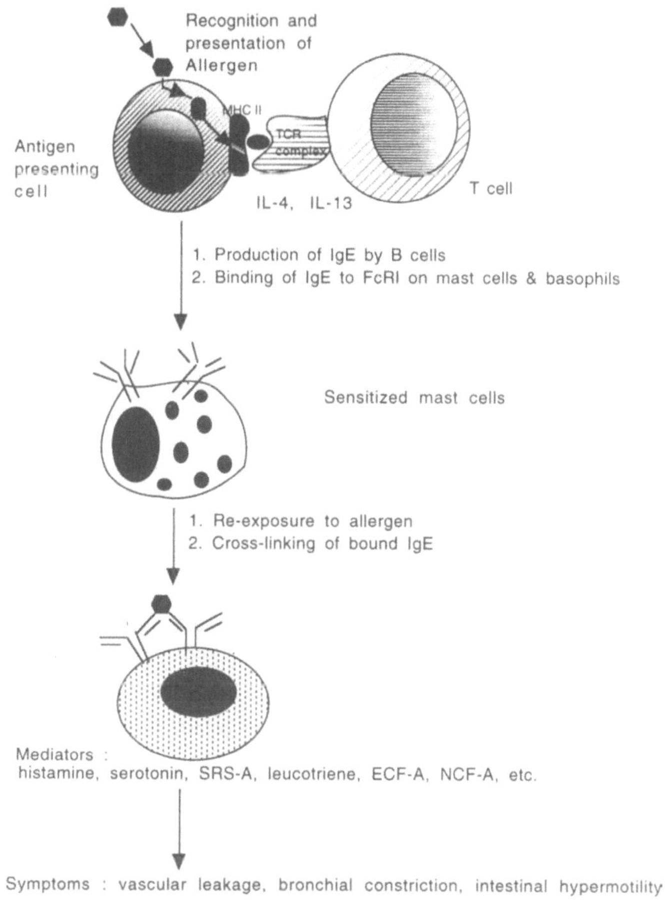

Figure 1. Cascade of hypersensitive reactions upon sensitization by an allergen. First, the allergen is presented by antigen presenting cells in the presence of MHC to T cells which in turn stimulate B cells to produce allergen specific IgE. These IgE are bound on surfaces of mast cells and basophils. Upon re-exposure to the same allergen and cross-linking of bound IgE to the allergens, the mast cells and basophils release a variety of mediators resulting in increase in vascular permeability, constriction of bronchial smooth muscles, mucus secretion, and migration of inflammatory cells to the reaction site.

residues with an estimated pI of 5.4–5.8. By SDS-PAGE, Daul et al. (1991) showed that a 36kD protein from *Penaeus aztecus* was able to inhibit 75% of the RAST activity to the whole shrimp. This protein, *Pen a I* is rich in aspartic acid and glutamic acid and is similar to antigen II described by Hoffman et al. A similar protein from *Penaeus indicus* designated as Sa-II was described (Nagpal et al., 1989). In 1993 and 1994, three groups independently reported the identification of the major shrimp allergen as tropomyosin (Daul et al., 1994; Leung et al., 1994; Shanti et al., 1993).

To characterize the shrimp allergen at the molecular level, we isolated an IgE reactive clone, coined *Met e I* from a cDNA library from *Metapenaeus ensis*. *M. ensis* (family Penaeidae) is a commercially important shrimp in South China Sea and is extensively cultured in Southeast Asia. Serum antibodies from patients with shrimp allergy demonstrated specific positive IgE reactivity to recombinant *Met e I* by immunoblotting whereas sera from non-allergic control do not react (Figure 2). *Met e I* has an open reading frame of 281

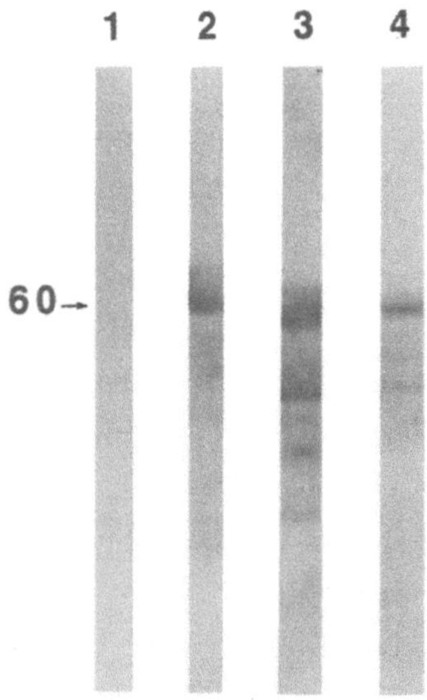

Figure 2. IgE reactivity of sera from subject with shellfish allergy with recombinant *Met el* (lane 2), *Pan s I* (lane 3), *Hom a I* (lane 4) by immunoblotting. Purified recombinant proteins of *Met e I*, *Pan s I* and *Hom a I* were resolved on SDS-PAGE and analysed for IgE reactivity with sera from subjects with shellfish allergy. Please note that the sera reacted to recombinant fusion proteins of 60 kD in each of the recombinant proteins (lanes 2–4) but not to the control (lane 1). Sera from normal controls did not react to the recombinant proteins. The minor bands below 60 kD are probably due to breakdown products of the recombinant proteins.

amino acids, coding for a protein of 34kD (Leung et al., 1994). Comparison of amino acid composition of *Met e I* with that of other shrimp allergens such as *Pen a I*, Sa-II and antigen II showed that they are highly similar, suggesting that the major shrimp allergen is likely to include multiple isoforms of shrimp tropomyosin. (Table 2). Furthermore, the deduced amino acid sequence of *Met e I* is highly homologous to that of *Drosophila melanogaster*. Thus now, it is well established that the shrimp tropomyosin is the major heat stable shrimp allergen.

In addition to tropomyosin, Nagpal et al. (1987) also reported that shrimp tRNA as a shrimp allergen. However, the data was obtained only with sera from two patients and the nucleic acid preparation was not totally protein free. Since then, this data has not been reproduced.

4. IDENTIFICATION OF ALLERGENS IN OTHER CRUSTACEANS

Hefle et al. (1995) reported the presence of multiple IgE reactive proteins of 25–45 kD and 5 IgE proteins of 14 kD or less in snow crab meat and cooking water extracts. Moreover, they reported that 5/18 serum samples studied showed IgE binding at 38–41 kD which may represent the snow crab tropomyosin. When boiled muscle extracts from shrimp, spiny lobster and mud crab were subjected to immunoblotting against a panel of sera from subjects with a clinical history of shrimp allergy, nine sera, which have been previously shown to react to recombinant shrimp tropomyosin (Leung et al., 1994) reacted to muscle proteins at approximately 38 kD in the crustacean samples (Table 3) (Leung et al., 1996). Sera preabsorbed with recombinant shrimp tropomyosin lost their IgE reactivity totally against the 38 kD protein band in shrimp, spiny lobster and mud crab, suggesting the presence of similar, if not identical IgE epitopes among these crustaceans.

To further investigate the actual molecular identity of the lobster allergen, we have constructed a lgt11 expression cDNA library from the spiny lobster *Panulirus stimpsoni* (Family Palinuridae). This library was screened for IgE reactive clones using sera from subjects with crustacean allergy. A 2 kb cDNA, coined *Pan s I* was identified. Expression

Table 2. Amino acid composition of tropomyosin from shrimp, lobster, crab, fruitfly, and chicken

	Shrimp				Lobster			Crab		
Amino acid	*Met e I*[#]	*Pen a I*[+]	Sa-II^	Antigen II*	*Pan s I*[Ψ]	*Hom a I*[Ψ]	Ha sTm	*Chaf I*[Ψ]	Fruitfly[τ]	Chicken[π]
Alanine (A)	31	33	21	31	31	33	33	33	25	34
Cysteine (C)	1	ND	3	2	0	0	0	0	3	1
Aspartic acid (D)	17	40	39	58	17	18	16	16	22	25
Glutamic acid (E)	55	80	75	61	55	55	57	56	51	54
Phenylalanine (F)	4	4	6	9	4	3	3	3	5	1
Glycine (G)	5	10	6	20	4	4	4	4	3	3
Histidine (H)	1	1	3	4	1	1	1	0	1	1
Isoleucine (I)	4	6	6	12	4	6	8	5	7	9
Lysine (K)	25	26	27	27	25	28	29	25	35	40
Leucine (L)	33	35	30	30	33	33	31	31	28	38
Methionine (M)	6	8	6	9	6	8	8	6	8	4
Asparagine (N)	18	ND	ND	ND	16	16	15	15	13	6
Proline (P)	0	2	3	6	0	0	0	0	0	0
Glutamine (Q)	17	ND	ND	ND	16	17	17	17	19	12
Arginine (R)	22	26	30	19	21	21	21	22	17	13
Serine (S)	15	14	12	15	14	15	13	14	14	17
Threonine(T)	9	12	9	12	10	9	11	11	14	10
Valine (V)	14	13	15	19	13	13	13	12	17	9
Tryptophan (W)	0	ND	4	ND	0	0	0	0	0	0
Tyrosine (Y)	4	4	6	7	4	4	4	4	2	7
Total	281	312	301	341	274	284	284	274	284	284

[#]Leung et al., 1994; [+]Musmand et al., 1993; ^Nagpal et al. 1989; *Hoffman et al., 1987; [Ψ]Chu et al., 1997; [τ]Basi and Storti, 1986; [π]Goodling et al., 1987. Ha sTm: *Homarus americanus* slow muscle tropomyosin (Myk es et al., 1997). ND: not detected.

Table 3. IgE reactivities of shellfish allergic sera against animal muscle extracts

Organisms	Reactive band (kD)	No. reactive sera/ total sera tested*
Phylum Arthropoda		
Class Crustacea		
shrimp (*Metapenaeus ensis*)	38	9/9
spiny lobster (*Panulirus homarus*)	38	9/9
mud crab (*Scylla serrata*)	38	9/9
Class Insecta		
grasshopper (*Mecopoda elongata*)	38	9/9
cockroach (*Periplaneta americana*)	38	9/9
fruitfly (*Drosophila melanogaster*)	38	9/9
Phylum Mollusca		
Class Gastropoda		
abalone (*Haliotis diveriscolor*)	38	9/9
whelk (*Hemifusus ternatana*)	38	9/9
	49	8/9
Class Bivalvia		
mussel (*Perna viridis*)	38	9/9
fan shell (*Pinna atropupurea*)	38	9/9
	49	4/9
	68	1/9
scallop	38	9/9
	72	1/9
oyster (*Crassostrea gigas*)	38	9/9
	49	2/9
	69	1/9
clam (*Lutraria philipinarum*)	38	9/9
Class Cephalopoda		
cuttlefish (*Sepia madokai*)	38	9/9
squid (*Loligo edulis*)	38	9/9
	49	7/9
	72	1/9
octopus(*Octopus luteus*)	28	1/9
	38	9/9
	72	1/5
Phylum Chordata		
chicken	—	—
mouse	—	—

*Sera were tested at 1:10 dilution for their IgE reactivity against muscle proteins from various organisms by immunoblotting. Normal sera did not react to any of the preparations in this study.

of *Pan s I* in plasmid vector pGEX produced a 60 kD recombinant fusion protein reactive to the IgE antibodies from crustacean allergies (Figure 2). This protein consists of a 26 kD glutathione transferase fused to protein with an opening reading frame of 274 amino acids. *Pan s I* is the first identified major allergen of the spiny lobster and showed significant homology to tropomyosin of shrimp and fruitfly *Drosophila melanogaster* (Basi and Storti, 1986; Chu et al., 1997). In addition, we have also expressed recombinant tropomyosin from the American lobster, *Homarus americanus* (family Nephropidae) (Mykles et al., 1997) and tested for the IgE recognition of this protein, coined *Hom a I* by immunoblotting. *Hom a I* was also recognized by IgE from patients with crustacean allergies (Figure 2). Antibodies from subjects with crustacean allergies, when preabsorbed with recombi-

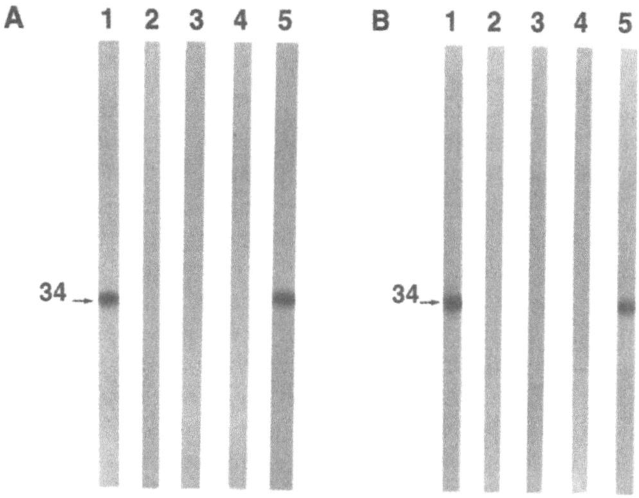

Figure 3. Specific inhibition of IgE reactivity against (A) *Panulirus homarus* and (B) *Homarus americanus* proteins by *Pan s I*, *Hom a I* and *Met e I*. Muscle extracts of *Panulirus homarus* and *Homarus americanus* were resolved on SDS-PAGE, transferred to nitrocellulose filters and probed with sera. Note the presence of a reactive band with unabsorbed sera at 34 kD (lane 1) and the loss of reactivity when the sera were absorbed by recombinant proteins of *Pan s I* (lane 2) *Hom a I* (lane 3) and *Met e I* (lane 4). Sera absorbed by an irrelevant control recombinant protein PDC-E2 did not inhibit the reactivity (lane 5).

nant proteins *Pan s I* or *Hom a I* lost their IgE reactivity to muscle extract of spiny lobster *P. homarus* and *H. americanus*. Likewise, preincubation of crustacean allergy sera with the shrimp recombinant tropomyosin *Met e I*, removed their IgE reactivity to *Homarus americanus* and *Panulirus homarus* (Figure 3). These data demonstrate that subjects with crustacean allergies have common IgE epitopes in shrimp and lobster and lobster tropomyosin is a major lobster allergen. Similarly, using a cDNA library from *Charybdis feriatus* (family Portunidae) a common crab species consumed as seafood throughout its geographical range extending from Japan to the east coast of Africa, we have obtained an IgE reactive clone coined *Cha f I* and demonstrated that the major allergen in crab is also tropomyosin.

5. CROSS-REACTIVITY BETWEEN CRUSTACEAN AND MOLLUSK ALLERGENS

Although crustaceans are well known allergens in susceptible individuals, similar adverse reactions have been described in patients consuming mollusks such as cuttlefish (Shibasaki et al., 1989), abalone (Morikawa et al., 1990), limpet (Carrillo et al., 1991; Carrillo et al., 1994; Morikawa et al., 1990) and squid (Carrillo et al., 1992). Significant RAST reactivity to oyster extracts has been found in sera from crustacean-sensitive subjects and a number of patients having sensitivity reactions to both crustaceans and oysters have been reported to have elevated IgE antibodies to extracts of both raw and boiled oysters(Lehrer and McCants, 1987). Moreover, inhibition experiments also demonstrated that there is significant correlation of IgE reactivity to oyster with that to crustacean allergens suggesting that oyster and crustacean may share common antigenic epitopes (Lehrer and McCants, 1987). Thus, the molecular basis of cross-reactivity between allergy to crustaceans and mollusks is of particular interest .

Recently, we have analyzed the pattern of IgE reactivities of a panel of sera from subjects with a clinical history of hypersensitive reactions to shrimp using a variety of edible mollusks including class Gastropoda, class Bivalvia and class Cephalopoda. IgE reactivity of the nine sera obtained from patients with a documented clinical history of shrimp hypersensitivity was tested against ten common mollusks in three different classes

(Table 3) (Leung et al., 1996). All nine sera demonstrated IgE reactivity to the 38 kD protein present in the boiled muscle extracts of all mollusks. Normal sera did not react with any of the muscle samples tested.

In addition to the major cross-reactive 38 kD allergen, other minor allergens were found in the immunoblotting experiments. In particular, a 49 kD antigen was detected in eight out of the nine sera tested in whelk; in seven out of the nine sera tested in squid; in four out of the nine sera tested in fan shell; and in two out of the nine sera tested in oyster (Table 3). It was also found that a particular serum detected other antigens in some mollusk samples tested; including a 68 kD protein in oyster and fan shell extracts; and a 72 kD protein in scallop, octopus and squid extracts. In addition to these antigens, a 28 kD antigen in octopus was recognized by this serum. Furthermore, preabsorption of the tested sera with recombinant shrimp tropomyosin resulted in complete inhibition of their IgE reactivity to the 38 kD protein present in all the tested mollusks. A representative inhibition immunoblot is shown in Figure 4. Moreover, preabsorption with the shrimp allergen also completely blocked their IgE reactivity to the other allergens detected by the tested sera. Unabsorbed sera and sera absorbed with an irrelevant recombinant control, a GST fusion protein of human PDC-E2 did not lose their reactivities against the 38 kD band (data not shown). The results suggest that the 38 kD heat-stable muscle protein is the major cross-reactive allergen present in crustaceans and mollusks. This finding is in accordance with previous reports identifying a 38 kD muscle protein as the common crustacean allergen (Halmenpuro et al., 1987; Lehrer, 1986; Lehrer et al., 1985; Musmand et al., 1993) as well as a major cross-reactive 35–39 kD allergen in shrimp and scallop (Goetz and Whisman, 1990).

Our data showed that sera from shrimp allergy subjects but not normal controls react with a protein band of approximately 38kD in both crustaceans and mollusks. Furthermore, the reactivity to this 38kD protein in mollusks can be totally absorbed by preincu-

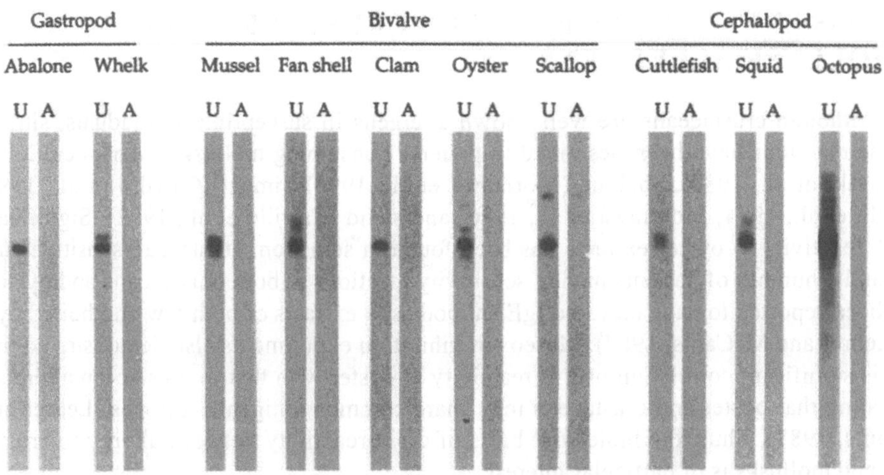

Figure 4. Representative inhibition immunoblot of the IgE reactivity against mollusk muscle extracts. Briefly, 1ml of a 1:10 diluted sera were incubated with 50 mg of purified recombinant shrimp tropomyosin at 4°C overnight. The preabsorbed sera were tested for their IgE reactivities against mollusk extracts by immunoblotting. Note the presence of reactive band in the unabsorbed lanes (U) and the loss of reactive bands in the absorbed lanes (A). Unabsorbed sera and sera absorbed with PDC-E2 were run in parallel as controls. Sera absorbed with PDC-E2 did not absorb out IgE reactivity against the mollusk preparations.

bating the sera with recombinant shrimp tropomyosin. In accordance with our findings, a 38 kD squid (*Todarodes pacificus*) protein, termed *Tod p I* which is reactive to sera from patients with shellfish allergy as well as shrimp and squid allergen specific monoclonal antibodies has been reported. Amino acid sequence analysis of *Tod p I* showed that it has marked homology with the snail tropomyosin (Miyazawa et al., 1996). The presence of this common allergen explains the cross-reactivity between crustaceans and mollusks and enhances our understanding of the molecular basis of the hypersensitive response against a highly conserved molecule.

6. CONCLUSION

Knowledge on molecular and immunological characterization of shellfish allergens has increased significantly during the last decade. Not only were we able to isolate the allergens and determine their biochemical properties, recently we were able to clone and define these allergens at the molecular level. The availability of recombinant clones and proteins will allow further study on the molecular and physiological basis of shellfish allergy as well as providing valuable reagents for us to ask questions that were not able to address in the past: most importantly the reason why some individuals become sensitized and develop allergies to shellfish. Furthermore, future development of clinical reagents for diagnosis and therapeutic applications will be made more feasible with knowledge accumulated from further studies using these recombinant clones.

7. SUMMARY

IgE mediated type I hypersensitivity to shellfish (crustaceans and mollusks) are amongst the most common and serious forms of food allergy. Despite years of immunological and biochemical studies, the major shrimp allergen has just recently been identified. Our laboratory has cloned, sequenced and identified the cDNA encoding the major heat stable allergen from the shrimp *Metapenaeus ensis* as the muscle protein tropomyosin. Subsequent studies on lobster (*Panulirus stimpsoni* and *Homarus americanus*) and crab (*Charybdis feriatus*) demonstrated that tropomyosin is the major common allergen in crustaceans. Studies on cross-reactivity using immunoblotting suggested that tropomyosin is also the predominant allergen in mollusks. By specific absorption studies using recombinant shrimp tropomyosin, the protein was shown to share immunodominant IgE epitopes among different species of arthropods and mollusks, but not vertebrates. The knowledge on molecular and immunological characterization of shellfish allergens will be useful in understanding the molecular immunobiology of shellfish allergy as well as for diagnostic and therapeutic applications.

ACKNOWLEDGMENTS

The authors would like to thank Miss Yen-chen Chen, Mr. Wing-Kuen Chow, Mr. Shun-Hang Wong and Mr. Chi Pang Li for their technical help in preparing this manuscript. The work is partially supported by a research grant from the Croucher Foundation Hong Kong and an earmarked grant from the Research Grants Council (CUHK4072/97M), Hong Kong.

REFERENCES

Basi, G. S., and Storti, R. V. 1986, *J. Biol. Chem.* **261:**817–827.

Carrillo, T., Castillo, R., Caminero, J., Cuevas, M., Rodriguez, J. C., Acosta, O., and de Castrio, F. R. 1992, *Ann. Allergy* **68:**483–487.

Carrillo, T., de Castillo, R., Cuevas, M., Caminero, J., and Cabrera, P. 1991,. *Allergy*,**46**:515–519.

Carrillo, T., de Castrio, F. R., Blanco, C., Castillo, R., Quiralte, J., and Cuevas, M. 1994,. *Ann. Allergy*,**73**:504–508.

Cartier, A., Malo, J. L., Ghezzo, H., McCants, M., and Lehrer, S. B. 1986, *J. Allergy Clin. Immunol.*, **78**:344–348.

Chu, K. H., Chow, W. K., Chen, Y. C., Gershwin, M. E., Kwan, H. S., and Leung, P. S. C. 1997, Paper presented in the Fourth International Marine Biotechnology Conference, Italy.

Daul, C. B., Slattery, M., Reese, G., and Lehrer, S. B. 1994,. *Int. Arch. Allergy Appl. Immunol.* **105**:49–55.

Droszca, W., Kowalski, J., Piotrowska, B., Pawlowicz, A., and Pietruszewska, E. 1981, *Int. Arch. Occup. Environ. Health*, **49**:13–19.

Gaddir, J., Legge, J. S., Friend, J. A. R., and Reid, T. M. S. 1980, *Lancet*, **2**:1350–1353.

Goetz, D. W., and Whisman, B. A. 1990, *J. Allergy Clin. Immunol.*, **85**:531(abstract).

Goodling, C., Reinach, F. C., and Macleod, A. R. 1987, *Nucl. Acid. Res.*, **15**:8105.

Goransson, K. 1981, *Contact Dermatitis.*, **7**:282–283.

Halmenpuro, L., Salvaggio, J. E., and Lehrer, S. B. 1987, *Int. Arch. Allergy Appl. Immmunol.*, **84**:165–172.

Hefle, S. L., Bush, R. K., Lehrer, S. B., Malo, J.-L., and Cartier, A. 1995, *J. Allergy Clin. Immunol,* . **95**:322 (abstract).

Henriksen, O. 1994, Net major shrimp allergen. *National Center for Research Resources Reporter.* **XVIII**:10–11.

Hoffman, D. R., Day, E. D., and Miller, J. S. 1987, *Ann. Allergy,* **47**:17–22.

Hudson, P., Cartier, A., Pineau, L., Lafrance, M., St-Aubin, J. J., Dubois, J. Y., and Malo, J. L. 1985, *J. Allergy Clin. Immunol.*, **76**:682–688.

Lehrer, S. B. 1986, *Ann. Allergy*, **57**:267–272.

Lehrer, S. B., Helbing, A., and Dual, C. B., 1992, *J. Food Safety.*, **13**:61–76.

Lehrer, S. B., and McCants, M. L. 1987,. *J Allergy Clin. Immunol.*, **80**:133–139.

Lehrer, S. B., McCants, M. L., and Salvaggio, J. E. 1985, *Int. Arch. Allergy Appl. Immunol.*, **77**:192–194.

Lehrer, S. B., and O'Neil, C. E. 1992, *Food Technol.*, **46**:153–156.

Leung, P. S., Chu, K. H., Chow, W. K., Ansari, A., Bandea, C. I., Kwan, H. S., Nagy, S. M., and Gershwin, M. E. 1994, *J Allergy Clin. Immunol.*,**94**:882–890.

Leung, P. S. C., Chow, W. K., Duffey, S., Kwan, H. S., Gershwin, M. E., and Chu, K. H. 1996, *J. Allergy Clin. Immunol.*, **98**:954–961.

Miyazawa, H., Fukamachi, H., Inagaki, Y., Reese, G., Daul, C. B., Lehrer, S. B., Inouye, S., and Sakaguchi, M. 1996, *J. Allergy Clin. Immunol.*, **98**:948–953.

Morikawa, A., Kato, M., Tokuyama, K., Kuroume, T., Minoshima, M., and Iwata, S. 1990, *Ann. Allergy*, **65**:415–417.

Musmand, J. J., Daul, C. B., and Lehrer, S. B. 1993, *Clin. Exp. Allergy*, **23**:722–732.

Mykles, D. L., Cotton, J. L. S., Taniguchi, H., Sano, K. I., and Maeda, Y. 1997, *J. Muscle Res. Cell Motil.* In Press.

Nagano, T., Kanao, K., and Sugai, T. 1984, *J. Allergy Clin. Immunol.* **74**:489–493.

Nagpal, S., Rajappa, L., Metcalfe, D. D., and Subba Rao, P. V. 1989, *J. Allergy Clin Immunol.*, **83**:26–36.

Orford, R. R., and Wilson, J. T. 1987. *Am. J. Ind. Med.*, 7:155–169.

Shanti, K. N., Martin, B. M., Nagpal, S., Metcalfe, D. D., and Subba Rao, P. V. 1993, *J. Immunol.*, **151**:5354–5363.

Sheffer, A. L. 1985, Anaphylaxis. *J. Allergy Clin. Immunol.*, **75**:227–233.

Shibasaki, M., Ehara, T., and Takita, H. 1989,. *Ann. Allergy*, **64**:421–422.

Taylor, S. L. 1985,. *Food Technol.*, **39**:98–105.

Taylor, S. L., and Bush, R. K. 1988,.Allergy by ingestion of seafoods., Volume 3, A. T. Tu, ed. (New York: Dekker).

35

CELL CULTURES FROM THE ABALONE *HALIOTIS TUBERCULATA*

A New Tool for *in Vitro* Study of Biomineralization

D. Sud, S. Auzoux-Bordenave, M. Martin, and D. Doumenc

Laboratoire de Biologie des Invertébrés Marins et Malacologie
URA CNRS 699
Muséum National d'Histoire Naturelle
57, Rue Cuvier, 75 005 Paris, France

1. INTRODUCTION

The internal part of most molluscan shells is made of mother-of-pearl which is secreted by the mantle. Because of its economically importance, particularly in perliculture, the structure and formation of nacre has long been studied by several authors (Gregoire, 1972; Wilbur et Saleuddin, 1983; Cuif, 1992). Lot of works lead on the specificity of an organic matrix secreted by the cells of the mantle which initiates the crystal nucleation (Erben and Watabe, 1974; Weiner and Traub, 1984; Belcher *et al.*, 1996; Bowen and Tang, 1996). However, the biological mechanisms involved in biomineralization remain relatively unknown because of the difficulty to analyse the deposit of crystal.

Cell cultures of the mantle of abalone, a nacreous Gastropod, is a new approach to study *in vitro* the mechanisms involved in biomineralization, and in particular the secretion of the extracellular matrix on which nacre settles. The purpose of this work was to establish a suitable cell culture system from the abalone *Haliotis tuberculata* for further investigations on the mechanisms of deposition and on the initiation of biomineralization. Successful long-term primary cell cultures of abalone mantle are reported here, with a particular emphasis on viability of cultured cells.

2. MATERIALS AND METHODS

2.1. Source and Maintenance of Animals

Abalones, *Haliotis tuberculata*, were purchased from an aquacole farm on the Atlantic coast. The animals were maintained in a closed-circuit aquarium containing aerated artificial seawater under controlled conditions (salinity: 30‰; temperature: 20–22°C).

New Developments in Marine Biotechnology, edited by Le Gal and Halvorson.
Plenum Press, New York, 1998.

2.2. Histology of the Mantle

An one centimeter long abalone was fixed in a Bouin solution. After embedding in paraffin, cross-sections of 5 µm have were stained with hemalun-picro indigo carmin (Martoja and Martoja, 1967).

2.3. Cell Culture

Mantles were taken from 6–7 centimeters long healthy abalones *Haliotis tuberculata*. In order to prevent contamination of cell cultures, mantles were cut into 5 mm^3 pieces in sterile conditions and then, samples were subjected three times to a sterile "wash solution" consisting of artificial seawater supplemented with broad spectra antifungal and antibacterial components: Chloramphenicol 200 µg/ml, Neomycin 2 Units/ml, Bacitracin 2 Units/ml, Penicillin 300 Units/ml, Streptomycin 300 Units/ml, Flagyl 125 µg/ml, Ampicillin 20 µg/ml, Polymyxin B 400 Units/ml. Then samples were stored 2–4 days at +4°C, and washed every day with the sterile "wash solution".

A cellular suspension was obtained from tissue samples by enzymatic dissociation with 0.1% collagenase in a saline solution containing 0.1% glucose and 1% Penicillin-Streptomycin during 45 minutes. The cellular suspension was twice filtered (380 µm, 70 µm), centrifugated and washed twice in a 50% antibiotic solution 50% Leibovitz-L15. Then cells were resuspended in a culture medium (Leibovitz-L15 commercial formula modified: Domart-Coulon *et al.*, 1994) for *in vitro* cultures in 96 well microplates (Nunclon). Plates were incubated at 20°C.

2.4. MTT Reduction Assay

The MTT reduction assay is an enzymatic test based on the determination of the activity of mitochondrial deshydrogenase enzymes. This test has been adapted to evaluate the viability of marine invertebrate cells (Domart-Coulon *et al.*, 1994; Naganuma *et al.*, 1994). For assays, 10 µl of stock MTT solution (5 mg MTT/ml artificial seawater filtered through a 0.22 µm filter) were added to each well to be tested. Plates were incubated at 20°C for 6 hours. Reaction was stopped by adding 100 µl of isopropanol containing 0.04N HCl and by shaking the plates for 10 minutes at 20°C. Absorbances at 580 nm were measured with a reference wavelength of 630 nm (O.D. 580 nm/ref. 630 nm).

3. RESULTS

3.1. Histology of the Abalone Mantle

An histological study of the abalone mantle allowed us to describe the different cell types involved before the establishment of cell cultures. The cross section of abalone (Fig. 1) reveals the inner epithelium and the body tissues. The shell was decalcified during treatment. The inner epithelium is made by a serie of parallepipedic cells in which occur some mucous cells responsible for the secretion of a muccopolysaccharide film. This film protects the epithelia. Inside, muscular and collagen fibers are crossed by hemolymphatic lacuna.

3.2. Evolution of Cell Viability

Primary cultures obtained are heterogeneous and contain the principal cell types identified in the mantle *in vivo*. These main cell types could thus be distinguished as

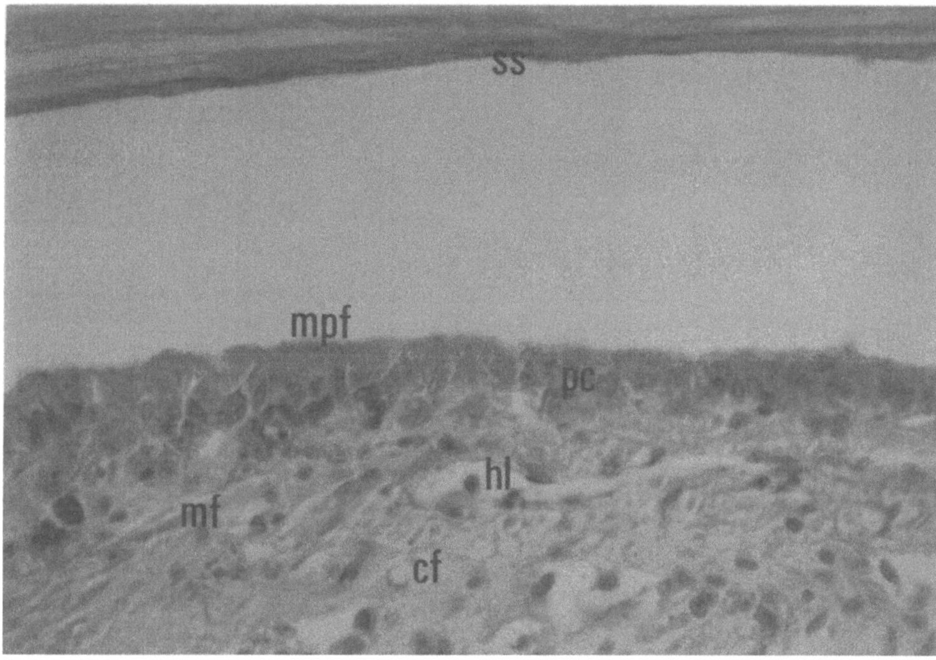

Figure 1. Cross-section of 5 µm of a one centimeter long abalone *Haliotis tuberculata*; hemalun-picro-indigo-carmin coloration (× 40). cf: collagen fiber; hl: hemolymphatic lacuna; mpf: muccopolysaccharide film; mf: muscular fiber; pc: parallepipedic cells; ss: shell surface.

epitheliocytes, hemocytes and muscular cells. Epitheliocytes constitute the major cell type (Fig. 2); most of them remain at the bottom of the well weakly adherent. Hemocytes are epithelial-like cells. Muscular cells are poorly represented in the cultures.

MTT assay reduction test was used to assess the viability of primary cultures 24 days long for increasing concentrations of cells. Cellular viability (corresponding to the global MTT response) increased during the first two weeks (Fig. 3), reaching a maximum between the 11th and the 13th day. Then cellular response decreased slightly but remained higher than the initial viability after 24 days. For each assay, we observed this augmentation the first two weeks.

4. DISCUSSION

The aim of this work was to establish a suitable *in vitro* model for studying the mechanisms of biomineralization. Primary cell cultures of mantle of *Haliotis tuberculata* are co-cultures, in which principal cells type involved in mineralization *in vivo* are present: epithelial cells and hemocytes. Indeed epithelial cells are the cells type directly involved in mineralization: the internal epithelium of the mantle covers the body surface and admits ions from the outer medium to the hemolymph wich carries ions to the external epithelium, responsible for the transfer of these ions to the extrapallial fluid, for the secretion of an organix matrix and for the deposition of crystals (Wilbur and Saleuddin, 1983). Microscopic examinations as well as viability measurement demonstrated the *in vitro* maintenance of mantle cells with a high cellular diversity for up to 4 weeks. Moreover, in

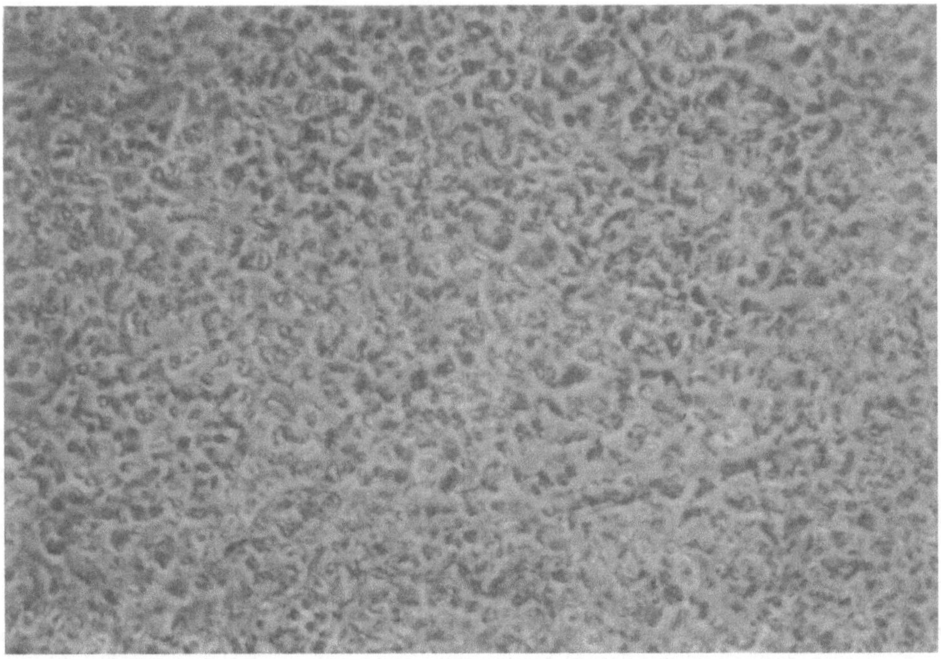

Figure 2. 25-day-old primary cultures of abalone *mantle cells* showing epithelial-like cells and hemocytes; initial concentration: 2×10^5 viable cells/well. (× 160, contrast phase).

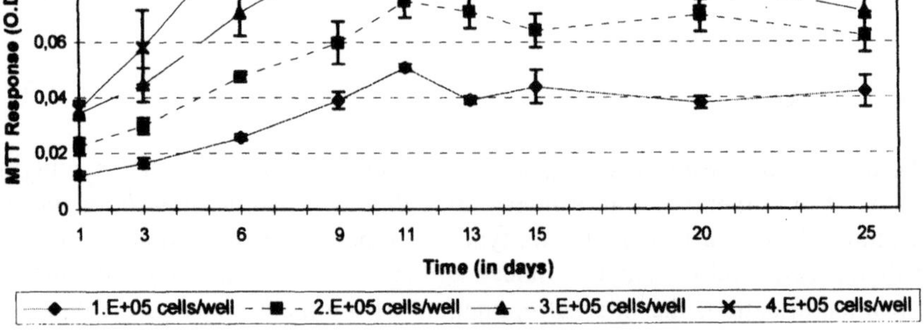

Figure 3. MTT response of abalone mantle cells for different cellular density during a 24 day period of culture. Each data point (O.D. 580 nm/ref. 630 nm) represents the mean±standard deviation of 6 replicate cultures, seeded at four concentrations in modified L15-Leibovitz.

order to obtain suitable cultured cells, the conjonctive tissue must be poorly dissociated, so that muscular fibers are eliminated while straining.

For preliminary studies, we used a serum free medium without any growth factor. Cell viability first increase during more than 10 days for each assay. Such an augmentation of cell viability is not frequent in invertebrate cell culture (Auzoux-Bordenave, 1995; Lebel *et al.*, 1997). This augmentation of total MTT response could be due either to an increase of metabolic activity following the *in vitro* establishment of cells or to a multiplication of mantle cells in culture. Further experiments, namely cell proliferation tests based on DNA content, are underway to confirm these hypothesis. It would also be interesting to separate cell types in order to notice part of each cell type in MTT response. Cells are weakly attached to the surface of the plates. By addition of Concanavaline A in the culture medium, Lebel *et al.* (1997) observed fibroblast-like hemocytes. In mantle cell cultures, hemocytes do not adhere to the plates and appeared to be epithelial-like cells. However, survival is not always correlated to adhesion (Toullec *et al.*, 1996).

Primary cultures contain the principal cell types involved in mineralization *in vivo*. The MTT reduction assay allow rapid quantifying of cellular viability and provide a global measure of cellular metabolic acticity.

Primary cell cultures of mantle are thus suitable for *in vitro* study of biomineralization. We presently focus on the research of an enzymatic biomarker to study mineralization *in vitro*. Alkaline phosphatase is a biomarker of the initiation of ossification in vertebrate species. Recently, a group from the laboratory of physiology (Muséum National d'Histoire Naturelle, Paris, France) discover the properties of nacre to induce mineralization by human osteoblasts (Silve *et al.*, 1992). Similarity of processes in mineralization of vertebrates and invertebrates leads us to think that alkaline phosphatase should be implicated in molluscan shell biomineralization. Besides alkaline phosphatase is detected in mantle and nacre of *Haliotis tuberculata* (Duvail, 1997). Measurement of alkaline phosphatase have been already realized in mantle cell culture. Dosable alkaline phosphatase presents a linear correlation with cell number, but its level is very variable depending on the abalone sample (Sud, unpublished data). Further experiments are underway to confirm the choice of this enzyme as a marker of biomineralization. Then, compounds known to induce mineralization in mammalian cell cultures will be tested for their ability to stimulate the activity of mantle cells. We also plan to test mantles of different aged abalone in relation with their growth rate.

5. ABSTRACT

In this work primary cell cultures from the abalone *Haliotis tuberculata* were developped from the mantle tissue in order to study the *in vitro* formation of nacre in marine Molluscs. Cells were obtained after an enzymatic dissociation of mantle, and were cultured in a L15-Leibovitz modified medium. Physiological cellular response was quantified using the tetrazolium dye reduction test (MTT) previously adapted to molluscan cells. Preliminary results demonstrated the *in vitro* maintenance of mantle cells with a high cellular diversity for up to 4 weeks.

REFERENCES

Auzoux-Bordenave, S., 1995, La branchie de palourde *Ruditapes decussatus* (Mollusque, Bivalve). Un modèle expérimental en pathologie et en écotoxicologie marine. Thèse de Doctorat de 3ème cycle. Muséum National d'Histoire Naturelle, 200 p.

Belcher, A.M., X.H., Wu, R.J., Christensen, P.K., Hansma, G.D., Stucky and D.E., Morse, 1996, *Nature*, **381**:56–58.
Bowen, C.E. and H. Tang, 1996, *Comp. Biochem. Physiol.*, **115**A(4):269–275.
Cuif, 1992, In Doumengue F. and A. Toulemont, Mother-of-pearls and pearls, ed. by Monaco-Musée Océanographique, pp. 77–87.
Duvail, L., 1997, Caractérisation de molécules apparentées aux peptides dérivés du gène de la calcitonine et rôle du CGRP au cours de la croissance chez l'ormeau *Haliotis tuberculata*. Diplôme pratique de l'Ecole Pratique des Hautes Etudes, 93 p.
Erben, H.K. and N., Watabe, 1974, *Nature*, **248**:128–130.
Gregoire, C., 1972, In Florkin M. and B.T. Scheer, Chemical Zoology, **II**, ed. by New York Academic Press, pp. 45–102.
Lebel, J.M., W., Giard, P., Favrel, and E., Boucaud-Camou,1996, *Biol. Cell.*, **86**:67–72.
Martoja, R. and M., Martoja, 1967, Initiation aux techniques de l'histologie animale. Ed. by Masson et Cie Paris, 345 p.
Naganuma, T., B.M., Degnan, K., Horikoshi and D.E., Morse, 1994, *Mol. Mar. Biol. Biotechnol.*, **3**:131–140.
Silve, C., E., Lopez, B., Vidal, D.C., Smith, S., Camprasse, G., Camprasse and G., Couly, 1992, *Calcif. Tissue Int.*, **51**, pp. 363–369.
Toullec, J.Y., Y., Crozat, J., Patrois, and P., Porcheron, 1996, *Journal of Crustacean Biology*, **16**(4):643–649.
Weiner, S. and W., Traub, 1984, *Phil. Trans. R. Soc. Lond.*, B**304**:425–434.
Wilbur, K.M. and A.S.M., Saleuddin ,1983, Shell formation. The Mollusca, **4**, ed. by Academic Press, pp. 235–287.

36

THE ARCHITECTURE OF DEGRADATIVE COMPLEX POLYSACCHARIDE ENZYME ARRAYS IN A MARINE BACTERIUM HAS IMPLICATIONS FOR BIOREMEDIATION

Ronald Weiner, Devi Chakravorty, and Lynne Whitehead

Department of Microbiology
University of Maryland (UMPC)

1. INTRODUCTION

As human population pressures build and more food is required, agricultural (Bayer & Lamed, 1992), aquacultural (Gomez-Pinchetti & Garcia-Reina, 1993), and algalcultural (Powell & Weiner, 1993) wastes threaten to become an increasingly serious problem. The wastes are mostly recalcitrant insoluble complex polysaccharides (ICP), namely cellulose, chitin, and often agar, respectively. These ICP are composed of monosaccharides, many of which can provide valuable feedstock when hydrolyzed. Indeed, in developing nations, this sort of reclamation is becoming progressively important. Unidentified autochthonous microbiota degrade and recycle shrimp, algal and other mariculture wastes in Asia (Ajisaka & Chiang, 1993) and South America (Punja & Zhang, 1993). These technologies have historically moved from the utilization of ìwild populationsî to carefully controlled, tamed specialists. In brewing and other endeavors, the gain in reliability, efficiency and control have been dramatic. Aqua-, algal- and agri-cultural practices are now in this transition (Doi et al., 1994).

Bioremediation, using prokaryotes, may be grouped into two basic approaches: those that use cell products, such as enzyme extracts, and those that use the entire metabolizing organism to grow on and degrade the biopolymer. The protocols and guidelines for such remediation are analogous to those proposed for biotreatment of hazardous waste (Gudz et al., 1986). However, it has been discovered that the complexity of enzyme system synthesis, cellular organization and regulation is commensurate with the daunting task of degrading chemically shielded ICP. To achieve full bioremediation potential, it is essen-

New Developments in Marine Biotechnology, edited by Le Gal and Halvorson.
Plenum Press, New York, 1998.

tial to know where the degradative enzymes act, how they may act in concert, and most importantly, the environmental conditions that induce their synthesis. Prokaryotes may also be used to control biofouling slimes (Matheeson & McCleary, 1985), which is an additional compelling future application since biofouling represents a major economic loss, more than 300 billion dollars annually in prevention and remediation (Stosz, 1994). New approaches in bioremediation can alleviate these problems, degrading the waste (Bertram et al., 1993) and recycling them into feedstock (Durairatnam, 1989) and other reusable products (Levin & Gealt, 1993).

2. RESEARCH APPROACH AND BACKGROUND

Some marine microorganisms, often faced with oliogotrophic conditions in the pelagic zone, have evolved symbiosis with photosynthesizing biota, becoming epiphytes. Others have evolved powerful enzyme systems to take advantage of the ubiquitous marine snow (Alber & Valiela, 1994), which are potential oases in the nutritionally poor open waters. As a consequence, selected marine species have developed very efficient mechanisms to utilize ICP. Marine bacterium 2-40 has been identified as a potentially important bioremediation species, since it synthesizes an unusually large number of degradative enzymes. A microbial strategy for the degradation of insoluble polymers is to site direct enzyme complexes to structures that make direct contact and synergistically act in organized concert on the polymer. We are testing that strategy in 2-40, particularly focusing on agarases, chitinases and alginases. If, as preliminary evidence indicates, the site directed, enzyme-complex is the mode of action, then the entire organism or enzyme complexes may be more efficacious in bioremediation than isolated enzymes.

2.1. Characteristics of 2-40

The marine/estuarine bacterium, 2-40, is a periphytic organism isolated from a salt marsh growing on Spartina alterniflora. It is Gram negative, pleomorphic, rod-shaped, and motile. The organism has yet to be classified, however it probably represents a new species (Montgomery & Kirchman, 1993). This aerobe requires sea salts and carbohydrates for growth. It produces numerous proteases, lipases and carbohydrases that allow 2-40 to degrade a variety of ICP of plant, fungal, and animal origin. Several of these carbohydrates, identified by Marx (Louis-Blaiseau & Lafay, 1992), include alginic acid, araban, carrageenan, carboxymethylcellulose, chitin, glycogen, laminarin, pullulan, starch, xylan, and agar. These ICP are produced by a wide variety of biota, including marsh grasses, algae, fungi, insects, zooplankton, and crustacea. Because these ICP are so abundant in nature, their degradation is essential for recycling carbon and nitrogen in marine ecosystems.

2.2. Enzyme Targeting and Activity

For the degradation of recalcitrant exopolymers, enzyme localization relative to other enzymes and biomolecules is often as important for enzyme efficiency as is the chemistry of its active site. In Gram negative bacteria the location may be cytoplasmic, periplasmic, on the outer membrane, or transported outside of the cell membrane. The enzyme may function alone or in conjunction with other cellular components. If the latter, its placement is crucial.

Relatively recently, a novel structure relating to insoluble substrate degradation was discovered in a Gram positive bacterium. It was a cellulose-binding and multicellulase-containing, cell-surface protuberance produced by Clostridium thermocellum (Bayer & Lamed, 1986) and C. cellulovorans. Coined ìcellulosomesî they were found to attach directly to the insoluble substrates, via special cellulose binding proteins (Bayer & Lamed, 1986). Thus, they bring cellulases into contact with cellulose, targeting the enzyme-substrate complex. Cellulosomes are comprised of at least 14 different proteins (Fukusa et al., 1992). Cip A refers to the largest of the cellulosome proteins, approximately 250 kDa. It serves as the scaffolding protein, binding and anchoring the enzymatic components and securing the entire cellulosome on the cell surface. This protein has a 166 amino acid sequence that is repeated 9 times and is a receptor for the enzymatic cellulosome components, such as CelD (Béguin & Aubert, 1994). CelD is a 68 kDa endoglucanase isolated from the cellulosome whose carboxy-terminus has a docking sequence that binds to the CipA receptors (Béguin & Aubert, 1994).

Such structures (we propose the general term degradosome for them) may be involved in the depolymerization of other ICP in addition to cellulose. Degradosome components could also consist of spatially arrayed enzymes, adhesions and scaffold protein. Not only do degradosomes maintain the released monomer product close to the cell for metabolic utilization, but also the degradosome may place a cascade of hydrolytic enzymes in proper juxtaposition for optimal enzyme activity (Bayer & Lamed, 1986). It would also be an attachment organelle, incorporating specific polymer binding proteins (MacGuire, 1982). Because of whole cell/degradosome efficiency and the potential for continued enzyme synthesis, the use of living bacteria as bioreactors in the degradation of not only cellulose, but also potentially chitin (aquaculture) algae slimes (algal culture) and biofouled surfaces may be quite advantageous.

In fact we found that 2-40 expresses cell surface protuberances on its outer membrane and does so coincidentally with insoluble biopolymer degradation (Andrykovitch et al., 1994; Sugano et al., 1993) (Fig. 1). Furthermore, preliminary results suggest that ICP degradation is indeed most efficient in 2-40 when the carbohydrase systems are induced and degradosome structures are expressed on the outer membrane of this Gram negative rod (Sugano et al., 1993). Thus for 2-40, it is proposed that a system of enzymes in the degradosome, acting in concert, degrade a portion of the carbohydrate to monomers, converting waste into usable nutrients. To our knowledge, degradosomes have not been reported in any other marine bacterium.

2.3. Agarose Degradation

Agar is a sulfated galactan, produced by the agarophytes of Rhodophytae (red seaweed). Agarose is the linear, non-ionic, gelatinous portion of agar made up of repeating (1-4 linked 3-6-anhydro-L-galactose and 1-3-linked (-D-galactose (Marx, 1986). Genera, shown to synthesize agarases and/or agarase systems include Vibrio (Schurr et al., 1995), Alteromonas (Schurr et al., 1995), Flavobacterium (Gonzales, 1995), Streptomyces (Brown et al., 1992), and Pseudomonas (Wall Street Journal, 1991). Fungal phages have also carry agarases (Gal et al., 1997). In addition to the fact that it degrades more different complex carbohydrates than any other reported bacteria (making it potentially versatile for applications such as slime remediation), we have determined that 2-40 produces four agarases with activities that are analogous to those of P. atlantica. However, the 2-40 agarases have different molecular weights, higher specific activity and are generally more resistant to denaturation than those of other species (Santelices, 1993).

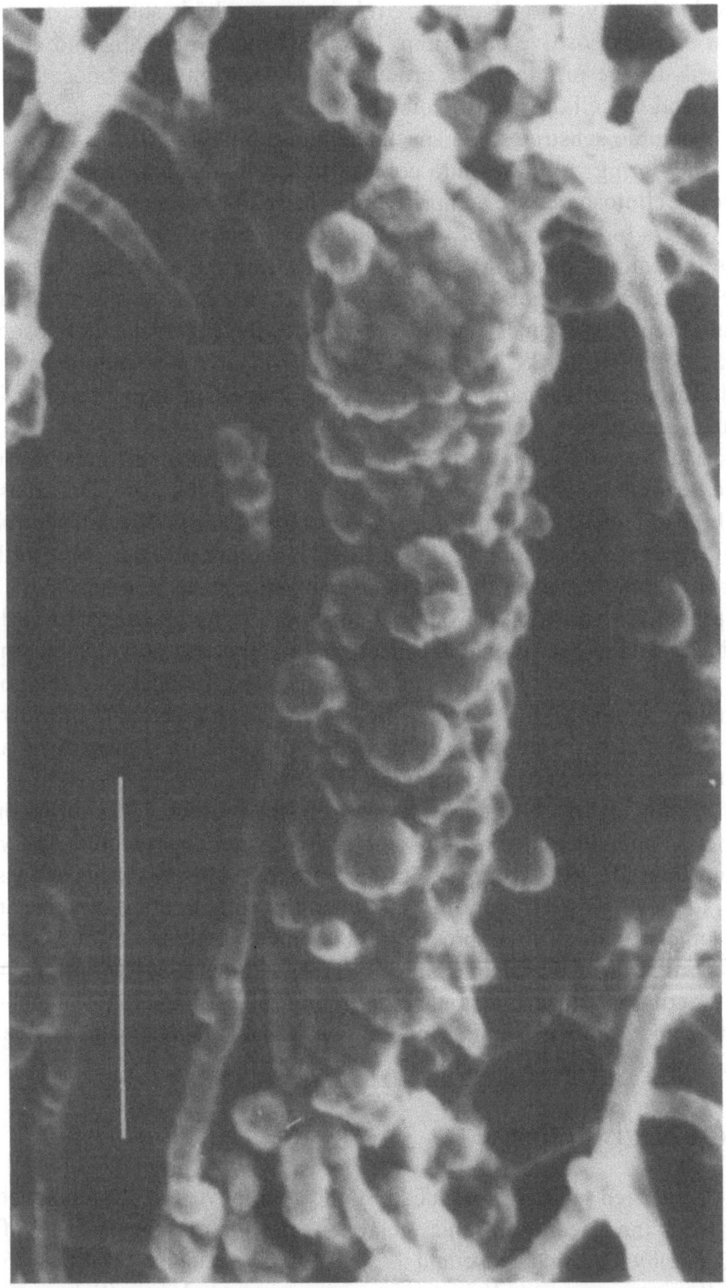

Figure 1. 2-40 grown in chitin minimal medium showing protuberances on the outer membrane of the cell. Bar = 500 nm.

2.4. Chitin Degradation

Chitin, comprised of repeating (-1,4 linked N-acetyl-D-glucosamine (NAG) residues, is the second most abundant ICP in nature. Among the bacteria, they have been identified >10 different genera. In addition, they are produced by invertebrates, protozoa, fungi (Liu & Suflita, 1993) and even plants as anti-invasive factors In 2-40, unlike Vibrio, we propose that chitinases are packaged in chitinosomes.

2.5. Alginic Acid

Alginic acid is a linear polymer composed of (-D mannuronic acid (M) and its C-5 epimer (-L guluronic acid (G) residues in (1-4 glycosidic linkages. The uronic acid residues are arranged as homopolymeric MM and GG or heteropolymeric GM tracts. Alginate is commonly produced by both algae (Bold & Wynne, 1985) such as Macrocystis pyrifera and prokaryotes (Santelices & Doty, 1989) such as Azotobacter vinelandii, and is consequently a major component of many biofilms. Thus far, none of the purified alginases are commercially available.

3. SUMMARY

2-40 is a unique organism, having >12 ICP degradative enzyme complexes. The systems studied to date are packaged in degradosomes and, under certain conditions, these ordered arrays in living cells might be more efficient in bioremediation than purified enzymes.

REFERENCES

Ajisaka, T. and Y.M. Chiang. 1993, Fourteenth International Seaweed Symp. Brest, France.
Alber, M. and I. Valiela. 1994, *Limnol. Oceanogr.* **39**:717–723.
Andrykovitch, G., V. Smith, M. Vergara and V. Chandhoke. 1994, Abstr. of Amer. Soc. for Microbiol.
Bayer, E.A. and R. Lamed. 1992, *Biodegradation.* **3**.171–188.
Bayer, E.A. and R. Lamed. 1986, *J. Bacteriol.* **167**:828–836.
BÈguin, Pierre and Jean-Paul Aubert. 1994, *FEMS Microbiol. Letters.* **13**:25–58.
Bertram, P.A., C.S. Buller, G.C. Stewart and J.M. Akagi. 1993, *J. Appl. Bacteriol.* **74**:460–466.
Bold, H.C. and M.J. Wynne. 1985, Introduction to the algae structure and reproduction. Prentice-Hall, Inc., NJ.
Brown, K.L., S. Wood and M.J. Buttner. 1992, *Mol. Microbiol.* **6**: 1133–1139.
Doi, R.H., M. Goldstein, S. Hashida, J.-S. Park, and M. Takagi. 1994, *Crit. Rev. Microbiol.* **20**:87–93.
Durairatnam, M. 1989, *J. Mar. Biol. Asso. India.* **31**:117–121.
Fukusa, K. Y. Teramoto, and S. Hayashida. 1992, *Biosci. Biotechnol. Biochem.* **56**:8–12.
Gal, L., et al.. 1997, *Appl. Environ. Microbiol.* **63**:903–909.
Gomez-Pinchetti, J.L. and G. Garcia-Reina. 1993, *Mar. Biol.* **116**:553–558.
Gonzalez, M.L., M.C. Perez, D.A. Lopez and C.A. Pino. 1993, *Aquaculture* **115**:87–95.
Gudz, S.P., S.G. Matviyko and M.F. Dumych. 1986, *Mikrobiologiya.* **55**:311–315.
Levin, M.A. and M.A. Gealt. 1993, In Biotreatment practices and promises. McGraw-Hill, Inc., New York., pp. 1–18.
Liu, S. and J.M. Suflita. 1993, *Trends in Biotechnology.* **11**:344–352.
Louis-Blaiseau, P. and J.F. Lafay. 1992, *Gene* **120**:243–248.
Marx, I. 1986, Ph.D. Dissertation , George Mason U. Library: 82pp.
Matheson, N.K. and B.V McCleary. 1985, In G.O. Aspinall (ed), The Polysaccharides. Academic Press, Inc., N.Y., p. 105.
McGuire, J. 1982, In C.F. Kulpa, R.L. Irvine and S.A. Sojka (eds.), Impact of applied genetics in pollution control, pp. 26.
Montgomery, M.T. and D.L. Kirchman. 1993, *Appl. Environ. Microbiol.* **59:**373–379.
Powell, D.M. and R.M. Weiner. 1993, ASM Conference on halophilic Bacteria. Williamsburg, VA.,Punja, Z.K. and Y.Y. Zhang. 1993, *J. Nematol.* **25**:526–540.
Santelices, B. and M.S. Doty. 1989, *Aquaculture* **78**:95–133.
Santelices, B., R. Westermeier and M. Bobadilla. 1993, *J. Appl. Phycol.* **5**:517–524.

Schurr, M.J., H. Yu, J.C. Boucher, N.S. Hibler, and V. Deretic. 1995, *J. Bacteriol.* **177**:5670–5679.
Stosz, S.K. 1994, Doctoral Dissertation. University of Maryland, College Park, MD.
Sugano, Y., T. Matsumoto, H. Kodama and M. Noma. 1993, *Appl. Environ. Microbiol.* **59**:3750–3756.
Wall Street Journal, April, 19, 1991.
Whitehead, L., et al. 1995, Abstr. of Amer. Soc. for Microbiol. 95th ASM Annual Meeting.
Yamaura, I., T. Matsumoto, M. Funatsu, H. Shigeiri and T. Shibata. 1991, *Agric. Biol. Chem.* **55**:2531–2536.

37

MANGANESE OXIDATION BY SPORES OF THE MARINE *BACILLUS* sp. STRAIN SG-1

Application for the Bioremediation of Metal Pollution

Bradley M. Tebo, Lorraine G. van Waasbergen, Chris A. Francis, Liming M. He, Deeanne B. Edwards, and Karen Casciotti

Marine Biology Research Division and Center for Marine Biotechnology and Biomedicine
Scripps Institution of Oceanography
University of California, San Diego
La Jolla, California 92093-0202

Microbial oxidation of soluble Mn(II) and the subsequent precipitation of insoluble Mn(III) and (IV) oxides and oxyhydroxides are key controls on Mn cycling in natural waters and are primary sources of these reactive solid phases in natural environments. Mn oxides often have high surface-area tunneled or layered structures, high sorptive capacities for metal ions, and are strong oxidants capable of non-specifically oxidizing a number of toxic aromatic and other organic compounds, as well as inorganic compounds such as reduced metals, hydrogen peroxide, and hydrogen sulfide. Thus, Mn(II)-oxidizing bacteria indirectly catalyze a variety of different reactions and impact the biogeochemical cycling of many elements. These bacteria may have useful biotechnological applications for metal recovery or the bioremediation of metal pollution.

The spore-forming marine Mn(II)-oxidizing *Bacillus* sp., strain SG-1, is one of the better characterized Mn(II)-oxidizing bacteria [Tebo et al., 1997]. The fact that it is the spores and not the vegetative cells that oxidize metals makes this organism particularly promising for use in biotechnological applications. Bacterial spores are naturally resistant to a variety of physical and chemical stresses which may be encountered in a waste stream or polluted environment and because they are dormant resting stages, growth does not have to be sustained. SG-1, which was isolated from a shallow marine sediment [Nealson & Ford, 1980], produces spores that bind and oxidize Mn(II) [Nealson & Tebo, 1980; Rosson & Nealson, 1982]. Mn(II) oxidation by SG-1 spores does not involve metabolic energy: spores rendered nonviable still oxidize Mn [Rosson & Nealson, 1982]. Oxidation occurs over a wide range of concentrations from below nanomolar to millimolar levels, and are active over a wide range of temperature ($<3°C$ to $>70°C$), pH (>6.5) and osmotic

New Developments in Marine Biotechnology, edited by Le Gal and Halvorson.
Plenum Press, New York, 1998.

strength (from distilled water to seawater). Mn oxides can accumulate on the surface of the spores up to approximately 6 times their own weight under ideal conditions. Although Mn oxidation is a thermodynamically favorable process, the rates of Mn oxidation by SG-1 spores at neutral pH are over 4–5 orders of magnitude faster than the abiotic Mn oxidation rates [Hastings & Emerson, 1986]. Using SG-1 spores as a catalyst for Mn mineral formation, and by varying the temperature and Mn(II) concentration at pH 7.4–8.0, a variety of Mn oxides can be formed under environmentally relevant conditions [Mandernack et al., 1995]. Mn oxidation activity has been shown to reside in the outer spore layers [Van Waasbergen et al., 1996; De Vrind et al., 1986].

Radiotracer studies with metals other than Mn with SG-1 spores have demonstrated that the spores will bind appreciable amounts of cadmium, cobalt and zinc, even in the absence of Mn (and therefore Mn oxides). SG-1 has also been shown to oxidize cobalt(II) [Tebo & Lee, 1993; Lee and Tebo, 1994]. Kinetic studies of cobalt(II) oxidation with SG-1 suggests that SG–1 spores have two oxidation systems, a high-affinity–low-rate system ($K_M = 3.3\times10^{-8}$ M; $V_{max} = 1.7 \times 10^{-15}$ $M^{-1}\cdot h^{-1}$) and a low-affinity–high-rate system ($K_M = 5.2 \times 10^{-6}$; $V_{max} = 8.9 \times 10^{-15}$ $M^{-1}\cdot h^{-1}$) [Lee & Tebo, 1994].

Recently we have characterized the surface chemistry and Cu(II) adsorption properties of SG–1 spores [He & Tebo, submitted]. The specific surface area of the spores is fairly high, about 75 m^2 g^{-1}, in a range similar to metal (hydr)oxides and other clay minerals. The spores have a net negative charge on their surface at pH ≥ 4.5, a high density of charged sites, and an excellent affinity and capacity for Cu(II) adsorption. The high surface area and surface site density, comparable to Fe and Al mineral colloids, give SG-1 spores a high capacity for binding metals on their surface. These properties make SG-1 spores well-suited for the removal of metals and radionuclides from contaminated waters.

We have employed molecular approaches to study the biochemical mechanism of Mn(II) oxidation. Using transposon mutagenesis we isolated 27 independent non–oxidizing, but still sporulating, mutants of SG-1 and cloned and sequenced seven genes (*mnxA–G*) that make up a gene cluster (Mnx operon, Fig. 1) that are involved in metal oxidation [Van Waasbergen et al., 1996; Van Waasbergen et al., 1993]. Upstream of the translational start site of the first gene (*mnxA*) are putative –10 and –35 sequences that closely correspond to the mother cell specific σ^K consensus promoter region sequences from the closely related species, *B. subtilis*. σ^K controls genes involved in spore production and this result supports our conclusion that *mnxA-G* are spore proteins involved in Mn(II) oxidation.

The proteins from three of the genes (MnxA, MnxB, and MnxE) showed areas of very hydrophobic nature as expected from spore coat proteins. MnxC is hydrophilic except for a predominantly hydrophobic stretch of approximately 30 amino acids near the start that has characteristics of a bacterial signal sequence, suggesting that *mnxC* may encode a protein that is targeted to a membrane. One of the regions of similarity to other proteins includes an amino acid motif characteristic of thioredoxin, suggesting MnxC may have some redox activity.

mnxG, the last gene in the cluster, codes for a protein that we believe is the Mn oxidase. The predicted MnxG protein is very large (138 kDa), and has amino acid sequence

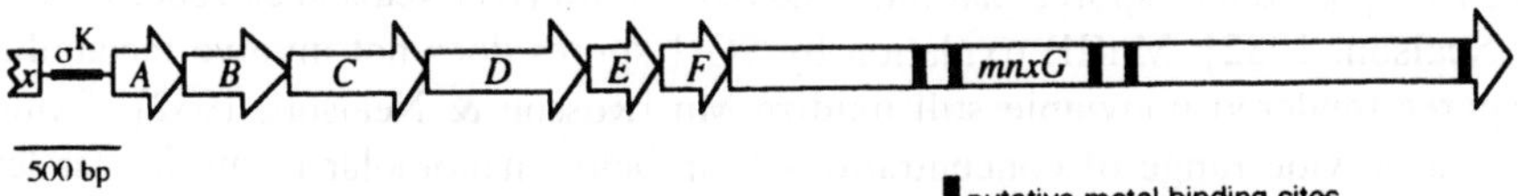

Figure 1. The gene cluster involved in Mn(II) oxidation in SG-1.

similarity to the family of multi-copper oxidases, a fairly divergent group of proteins that use multiple copper ions to oxidize a variety of substrates. The signature regions for copper binding found in the multicopper oxidases have been identified in MnxG. Multicopper oxidases bind copper ions of three spectroscopically distinct types termed +1, +2, and +3 centers. The copper +1 center accepts the electron from the substrate and shuttles it to the +2 and +3 centers, which are believed to bind and reduce molecular oxygen. Members of the multicopper oxidase family include laccase, ascorbate oxidase (cucumber & squash), ceruloplasmin (ferroxidase from plasma), and CopA (a copper resistance protein from *Pseudomonas syringae*). Additions of small amounts of copper(II) enhance manganese oxidation rates further supporting the idea that *mnxG* codes for the Mn oxidizing protein. Interestingly, the Mn(II)-oxidizing protein of another Mn(II)-oxidizing bacterium, *Leptothrix discophora*, has also recently been shown to be related to the multicopper oxidase family of proteins [Corstjens et al., 1997]. Although few Mn(II)-oxidizing bacteria have been examined, perhaps copper plays a more universal role in microbial Mn(II) oxidation.

We have recently cloned and expressed *mnxG* in *E. coli* and have generated MnxG-specific polyclonal antibodies. Utilizing the antibodies in Western blots of French Press–fractionated spores we have localized MnxG to the outermost spore layer, which we have characterized to be the exosporium. The isolated outer layer retains full Mn(II)-

Table 1. SG-1 properties useful for metal removal applications

Property	Consequence
• Spores, not vegetative cells, oxidize and precipitate metals	• Growth does not have to be sustained. Spores are naturally resistant to physical and chemical stresses. Spores may be good delivery vehicles for other proteins/activities.
• The spores directly oxidize several metals (Mn(II), Fe(II), Co(II))	• Several metals can be removed and oxidized simultaneously.
• Nonviable spores still oxidize and precipitate metals	• Can be rendered biologically inert. Live organisms don't have to be released to the environment.
• Metal oxidation and precipitation occurs over a wide range of conditions: metal concentration, temperature, pH and osmotic strength	• SG-1 is versatile and its activities will not be easily interfered with.
• SG-1 spores have a high affinity for metal	• Metal removal will be relatively selective and the residual metal left in solution will be lower than achieved by standard procedures.
• SG-1 spores have a high metal loading capacity	• Large quantities of metals will be removed per unit biomass.
• Metal loading capacity is relatively insensitive to metal concentration	• Applications for metal removal at both high and low metal concentrations
• The metal oxides produced by SG-1 are strong oxidants and good adsorbents	• A variety of other metals will be adsorbed or co-precipitated with the metal oxides and the metal oxides will function at catalytic adsorbent for the oxidation of other compounds.
• SG-1 spores maintain their activity after removal of the metal oxides from the spore surfaces	• The spores can be re-used (recycled).
• Metal oxidation is via an exosporium protein (enzyme) probably coded for by a single gene	• Genetic engineering will allow strain improvement such as overexpression of the protein or modification of the metal specificity or affinity.
• Metal oxidation occurs on an outer layer of the spores, the exosporium, that can be readily removed, has a high surface area, and is resistant to degradation.	• Cell-free systems can be used and will persist under environmental conditions.

oxidizing activity. These results suggest that the high surface area and metal binding and oxidizing capacity of SG-1 is due to the exosporium.

To summarize, the properties of SG-1 that make it an excellent candidate for bioremediation are listed in Table 1. Development of molecular tools in SG-1 now allows the protein-mediated oxidation system to be genetically manipulated to further enhance the utility of the spores for metal removal and recovery applications.

ACKNOWLEDGMENTS

Funding for this research has been provided by the National Science Foundation (OCE94-168944 and MCB94-07776), the University of California Toxic Substances Research and Teaching Program, the National Sea Grant College Program/National Oceanic and Atmospheric Administration/U.S. Department of Commerce (#NA66RG0477 Project R/CZ 123) through the California Sea Grant College System, and the California State Resources Agency.

REFERENCES

Corstjens, P.L.A.M., et al., 1997, *Geomicrobiology Journal*, **14**: 91–108

Hastings, D. and S. Emerson, 1986, *Geochimica et Cosmochimica Acta*, **50**:1819–1824.

He, L.M. and B.M. Tebo, Surface characterization of and Cu(II) adsorption by spores of a marine Bacillus. Applied and Environmental Microbiology, submitted.

Lee, Y. and B.M. Tebo,1994, *Applied and Environmental Microbiology*, **60**:2949–2957

Mandernack, K.W., J. Post, and B.M. Tebo, 1995, *Geochimica et Cosmochimica Acta*, **59**(21): 4393–4408.

Nealson, K.H. and J. Ford, 1980, *Geomicrobiology Journal*,. **2**: 21–37.

Nealson, K.H. and B. Tebo, 1980, *Origins of Life*, **10**: 117–126.

Rosson, R.A. and K.H. Nealson, 1982, *Journal of Bacteriology*, **151**: 1027–1034.

Tebo, B.M., 1983, The ecology and ultrastructure of marine manganese oxidizing bacteria, . University of California, San Diego, p. 220.

Tebo, B.M. and Y. Lee, 1993, Microbial oxidation of cobalt, in Biohydrometallurgical Technologies, A.E. Torma, J.E. Wey, and V.L. Lakshmanan, Editors. The Minerals, Metals, & Materials Society: Warrendale, PA. p. 695–704.

Tebo, B.M., et al., 1997, *Reviews in Mineralogy*, **35**: 225–266.

de Vrind, J.P.M., et al.,1986. *Applied and Environmental Microbiology*, **52**: 1096–1100.

van Waasbergen, L.G., J.A. Hoch, and B.M. Tebo, 1993, *Journal of Bacteriology*, **175**(23): 7594–7603.

van Waasbergen, L.G., M. Hildebrand, and B.M. Tebo, 1996, *Journal of Bacteriology*, **178**(12):3517–3530.

38

THE EFFECTS OF BIOREMEDIATION ON THE OIL DEGRADATION IN OIL POLLUTED ENVIRONMENTS

Kim Sang-Jin,* Jae Hak Sohn, Doo Suep Sim, Kae Kyoung Kwon, and Tae Hyun Kim

Korea Ocean Research and Development Institute
Ansan P.O. Box. 29, 425-600, Korea

1. INTRODUCTION

Oil spill accidents in marine environments occurred increasingly to over 300 times per year in Korea (Kim, 1995). Oil spills cause severe damage to the marine ecosystem and oil residues persist for a long time. The fishery industry and public health are often threatened by oil contamination in the marine environment. Hence it is necessary to promote a study on countermeasures to oil spills including restoration technology of oil polluted environments. Interest in bioremediation has been evoked since the Exxon Valdez accident. There are many study sites for the bioremediation of oil-polluted environments such as open-water or shoreline environments. In comparison with open-water experiments, much more bioremediation field studies have been carried out on shorelines (Swannell et al., 1996). These experiments varied considerably with respect to geographic area, oil type and concentration, study site dimensions, position of oil on the shore, sediment composition, and the analytical methods used to estimate biodegradation. These trials can be classified into two groups; application of fertilizers, and application of specific microbes (seeding or bioaugmentation). However, there have been few evaluations of the effects of seeding on contaminated shorelines or in soil. To evaluate the effect of fertilization and inoculation of oil degrading microorganisms on oil biodegradation, our experiment was performed on the sea shoreline artificially contaminated with crude oil.

* Corresponding author: Marine Microbiology Lab., Korea Ocean Research & Development Institute, Ansan P.O. Box. 29, 425-600, Korea. Phone: 82-345-400-6240; Fax: 82-345-408-5934; E-mail: s-jkim@sari.kordi.re.kr

New Developments in Marine Biotechnology, edited by Le Gal and Halvorson.
Plenum Press, New York, 1998.

2. MATERIALS AND METHODS

2.1. Experimental Test Area

Our bioremediation experiment was carried out on a sandy shoreline of a land reclamation area which located at Siwha, Kyung-Gi-Do, west coast of Korea. Most portion of the sand was consisted of shell particles and hence was relatively coarse.

2.2. Preparation of Test Area and Sampling

Four PVC pipes with 360 mm diameter and 250 mm height, were buried on the shoreline. 150 ml of Arabian light crude oil (final concentration; 3% (v/v)) was mixed with 5L of sand from each mesocosm. Only nutrients or nutrients with oil degrading microorganisms were amended in the mesocosm II, III and IV, respectively. Mesocosm I was not amended as a control site. The treatment conditions of each mesocosm are summarized in Table 1.

A mixture of strains, *Acinetobacter* sp. 2-2, *Moraxella* sp. 12-7 and *Candida lipolytica* 180, which have been selected as oil degrading bacteria with high activity from our previous study, were used as the inoculum (Kim et al., 1994). Each strain was grown in Zobell 2216E liquid medium and equal portions of the three strains were mixed with diatomaceous earth at the concentration of 10^9 cells/g-diatomaceous earth. The diatomaceous earth was added into mesocosm III and IV to make final concentrations of 3×10^6 cells/g-sand and 3×10^8 cells/g-sand, respectively.

Five sub-samples were collected from each mesocosm and were throughly mixed in 100 ml presterilized plastic cups with a stainless steel spoon on 0, 1, 2, 3, 5, 8, 12 and 21 days after treatment. Samples were immediately carried to the laboratory under refrigeration and processed within 2 hours. For maintaining the optimal water content between 40% and 75% seasand of water holding capacity, (Atlas, 1984; King et al., 1992), 1.5–2 L of water (Aged sea water:Deionized water = 1:3) was supplied daily on the surface of each mesocosm and the surface sand was tilled every day for about one min for oxygen supplement.

2.3. Temperature and Water Content

Cumulative average of daily ambient temperature was calculated from the data of daily temperatures, which were measured at the nearest area from the test area by Korea

Table 1. Treatment conditions of mesocosms

Mesocosm	Crude oil[1]	Slow release fertilizer[2]	Seeding with mixed microorganisms[3]
I	+	–	–
II	+	+	–
III	+	+	+4
IV	+	+	+4

[1]Arabian light crude oil was added by 3% (v/v); 150 ml crude oil/5000 cm^3-sand

[2]69 g of mixed form and 17 g of urea form of SRF were added into a mesocosm. Mixed form was consisted of 10% nitrogen and 11% soluble phosphate, and urea form consisted of 46% urea (w/w). These two types of SRF were coated with latex polymer.

[3]*Acinetobacter* sp. 2-2, *Moraxella* sp. 12-7 and *Candida lipolytica* 180 which were previously studied (Kim et al., 1994).

[4]An equal number of each strain was mixed with diatomaceous earth to a final concentration of 10^9 cells/g-diatomaceous earth, and then the diatomaceous earth was added into mesocosms III and IV to make 3×10^6 cells/g-sand and 3×10^8 cells/g-sand, respectively.

Meteorological Station (1994). Water contents were measured using the protocol described in Standard Methods (APHA, AWWA and WPCF, 1985).

2.4. Number of Microorganism

One cm^3 of sample was added to 100 ml of tertiary distilled water and blended with a blender mixing (Waring Co.) at full speed for 1 min to prepare a sand slurry. One ml of the slurry was serially diluted and inoculated onto ZoBell 2216E agar medium for the enumeration of heterotrophic microorganisms, and oil agar medium (Mineral salts agar supplemented with 0.2% crude oil) for petroleum-degrading microorganisms. The number of heterotrophic microorganism was counted after 1 week and PDB were counted after 2 weeks incubation at 25°C.

Acinetobacter sp. 2-2 and *Candida lipolytica* 180 strains in mesocosms III and IV were simultaneously counted from the same agar plates for heterotrophic microorganism by their unique colony color and morphology.

2.5. Extraction and Fractionation of Residual Oil Component

Extraction procedure was followed according to the method of Krahn et al. (1991) with some modification. Briefly, 1 g of anhydrous sodium thiosulfate and 5 mg copper were added to 1 g of sample in 30 ml glass centrifuge tubes. After addition of 250 g of squalane ($C30H32$, Sigma) as an internal standard, 20 ml of chloroform was added and sonicated for 10 min in a sonic bath (Branson 4810). Organic solvent was collected in a 100 ml evaporation flask after centrifugation at 3,000 rpm for 1 min. This procedure was repeated twice with 10 ml and 5 ml of chloroform.

Collected chloroform was evaporated and the oil was dissolved in 1 ml hexane. For fractionation of oil, a silica-alumina column (Wade et al., 1988) was prepared. Silica (70–230 mesh size, Merck) was packed at the bottom of a glass column (diameter, 12 mm) to a height of 3.5 cm and alumina (70–230 mesh size, WB type II, Merck) with 1.5 cm on top of the silica. Concentrated sample was loaded and an aliphatic fraction of oil was eluted with 20 ml hexane.

For quantitative analysis of the aliphatic fraction, gas chromatography (HP 5890) with FID detection was used. Total peak area was converted to the amount of residual aliphatic hydrocarbon by comparing with the peak area of the internal standard, squalane. C17/pristane and C18/phytane ratios were measured as an index of biological degradation (Blumer et al., 1974). Operation conditions of G.C. are described in Table 2.

3. RESULTS AND DISCUSSIONS

Fig. 1 shows the cumulative average of daily ambient temperature and water contents in each mesocosm. Air temperature of the test area fluctuated between 11.8°C and 27.7°C. The highest temperature was observed at the first and second day during the treatment period and after then the temperatures were recorded constantly less than 18°C. The average temperature during the experiment period was 17.9°C . Although a mixture of aged seawater and deionized water (1:3, v/v) was provided after the sampling to keep the optimal water content, water content in each mesocosm was relatively low. Average water content was 3.6% in the control mesocosm and in the range of 5.4 to 5.8% in the mesocosm II, III or IV. The lowest water content in the control mesocosm could be caused by

Table 2. Analysis condition of gas chromatography for the measurement of residual aliphatic hydrocarbon

Instrument	HP 5890 series II (Hewlett Packard)
Integrator	HP 3396 series II (Hewlett Packard)
Column	30 m × 0.32 mm i.d., packed with fused methyl-silicon (Ultra-1, HP)
Detector	FID (Flame Ionization Detector)
Injection volume	1 µl
Temperature	
Injector	260°C
Detector	280°C
Temperature program	
Initial	100°C
Final	280°C
Rate	8°C/min.
Final time	7.5 min.
Carrier gas	High purity nitrogen
Flow rate	42 cm^3/min.
Column head pressure	4 kgf/cm

the higher concentration of residual oil compared to the oil concentration in the other mesocosms. The water contents in the mesocosms were low because the samples were collected just before irrigation except the first day of the treatment. On the first day, water content was highest, because the samples were collected after treatments such as oiling, addition of nutrients and/or oil degraders, tilling and irrigation. The high value of the water content on the third day was due to a heavy rainshower.

The residual concentrations of aliphatic hydrocarbon in each mesocosm are presented in Fig. 2-A. The residual concentrations of aliphatic hydrocarbon in mesocosms II, III and IV decreased drastically from 2.3 to less than 0.3 mg/g dry weight of sand after 3 weeks when compared with those in the control mesocosm. These results showed that the degradation rates of aliphatic hydrocarbon were stimulated by fertilization and/or addition of oil-degrading microorganisms. More than 60% and about 40% reductions in residual oil concentration were observed in mesocosms IV and III after 5 days respectively, and this result indicated that oil degradation can be stimulated by seeding of oil degrader especially in the early phase of bioremediation,however oil degradation is dependant on the concentration of oil degrading inoculum. Therefore oil degrader inoculant can be applied

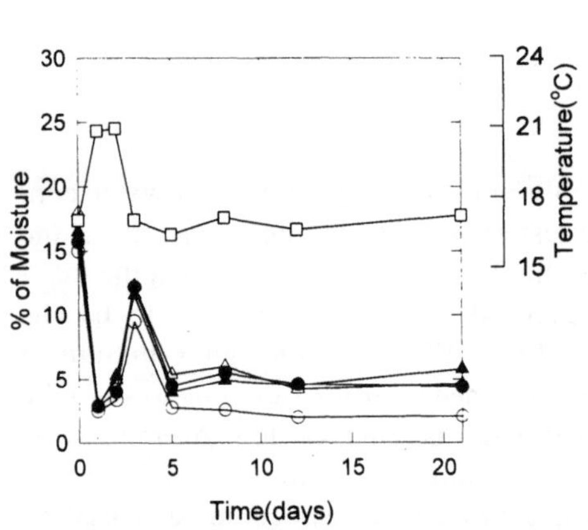

Figure 1. Water content in each mesocosm and cumulative average of daily ambient temperature of test area.

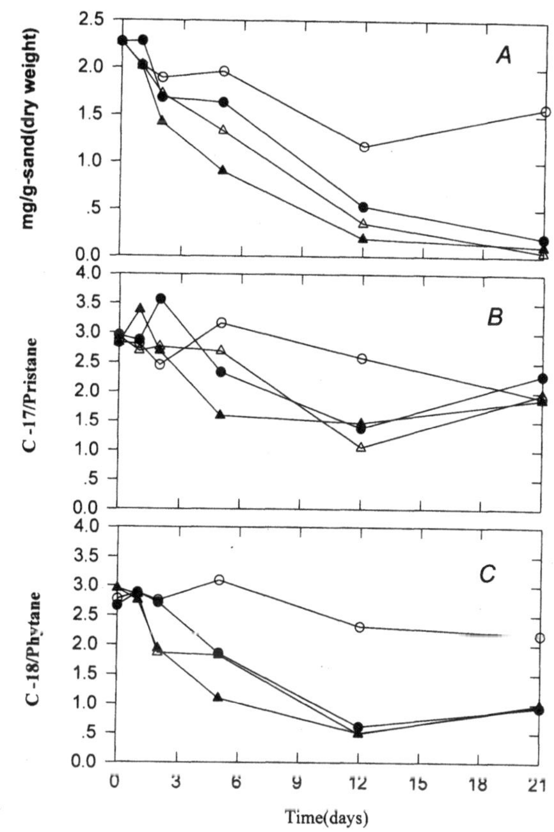

Figure 2. Concentration of residual aliphatic hydrocarbon (A), C-17/Pristane (B), and C-18/Phytane (C) ratios in mesocosms.

for a rapid recovery of oil-polluted environments, especially when they are sensitive environments (e.g. productive environment or ecosystem with abundant biota). According to the study of Jobson et al. (1974), nutrient addition alone had a greater effect on oil degradation than the seeding of oil degraders in soil. Lee and Levy (1987) observed only a limited increase in the biodegradation rates of Scotian Shelf Condensate in sandy beach sediments when seeded with a mixed culture of marine oil-degrading bacteria. They concluded that oil biodegradation was not limited by the need for a microbial inoculum. However, Vecchioli et al. (1990) found that soil amended with hydrocarbon degraders was decontaminated more rapidly than fertilized controls. A similar result was reported by Rosenberg et al. (1992).

The ratios of C17 to pristane (Fig. 2-B) and C18 to phytane (Fig. 2-C) decreased much more drastically in mesocosms II, III and IV compared with those of the control mesocosm I. These results indicated that the aliphatic hydrocarbon compounds in mesocosms II, III and IV were most degraded by the microorganisms. However, both ratios were increased between 12 and 21 days after the treatment indicating that branched alkanes could be preferentially degraded in some cases. This result implied that the C17/pristane and C18/phytane ratios should be interpreted with caution.

The distribution of total heterotrophic microorganism and petroleum degrading microorganisms is shown in Fig. 3-A and 3-B, respectively. The numbers of total heterotrophic microorganisms in the treated mesocosms were much higher than those in the control mesocosm. This result indicated that fertilization is very effective for the optimal

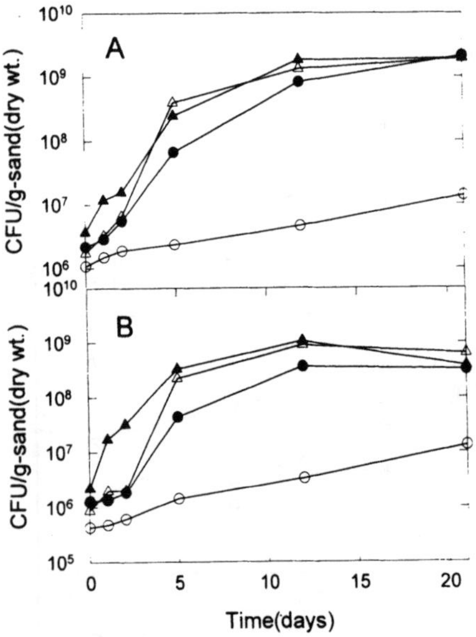

Figure 3. Number of total heterotrophic microorganisms (A) and petroleum degrading microorganisms (B) in mesocosms.

growth of microorganisms in environments contaminated with oil. These findings could also be confirmed in the case of petroleum degrading microorganism. Both numbers of total heterotrophic and petroleum degrading microorganisms in the mesocosms treated with fertilizer and oil-degrading inoculants were higher compared with those in the mesocosm treated only with fertilizer (mesocosm II) during the entire experimental period. The rapid growth rate of petroleum degrading microorganisms can support the fast disappearance of residual aliphatic compound in the mesocosms III and IV. The number of inoculated PDB in mesocosm IV was 100 times higher than that in mesocosm III at the beginning of the experiment. However, the difference in numbers of petroleum degrading microorganism

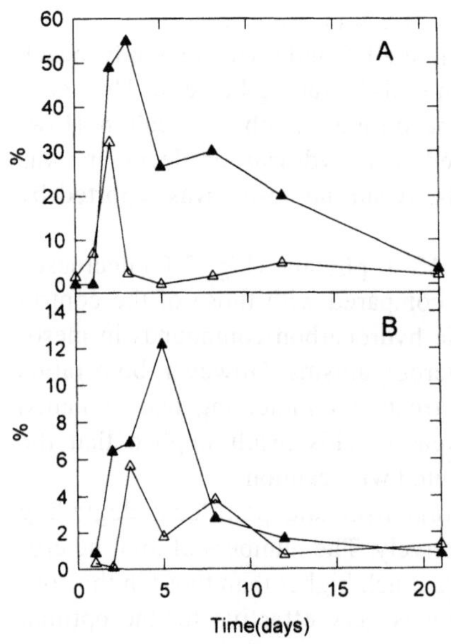

Figure 4. Proportion of *Acinetobacter* sp. 2-2 strain (A) and *Candida lipolytica* 180 strain (B) to total heterotrophic microorganisms in the mesocosm experiment.

between mesocosm III and IV became smaller after 5 days of treatment. The higher inoculum may be much more effective for oil biodegradation in the early phase.

To evaluate the survival of artificially seeded oil degraders, proportions of *Acinetobacter* sp. and *Candida lipolytica* strains to total heterotrophic microorganisms were monitored (Fig. 4). Although *Acinetobacter* sp. was not added in the mesocosm I and mesocosm II, negligible percentage was found (max. <5%). This implys that the result of *Acinetobacter* sp. proportion can be used for monitoring the fate of artificial inoculants. *Acinetobacter* sp. survived more than 20% during the experiment period except for the first and last day of the experiment (Fig. 4-A) in mesocosm IV. It indicated that this strain played an important role for oil biodegradation. A similar result was observed in the case *of Candida lipolytica* strain (Fig. 4-B). However, the percentage of *Candida lipolytica* showed a lower value than that of *Acinetobacter* sp. even though *C. lipolytica* strain was added at the same concentration as *Acinetobacter* sp. strain at the beginning of the experiment.

4. ABSTRACT

To evaluate the effects of bioremediation on oil biodegradation in the oil polluted environment, a study was performed on a sandy seashore. The study area was artificially contaminated with Arabian light crude oil (3%, v/v) and was divided into four parts. The first part (mesocosm I) served as a control without any amendment, the second one (mesocosm II) was amended only with SRF(slow releasing fertilizer) and the third (mesocosm III) and fourth parts (mesocosm IV) were amended with SRF and the precultured oil degrading microorganisms at the concentration of 3×10^6 cells/g-sand and 3×10^8 cells/g-sand, respectively. As oil degrading microorganisms, *Acinetobacter* sp., *Moraxella* sp. and *Candida lipolytica* species were mixed in the same proportion before addition. All study areas were irrigated and tilled in the same way every day. The concentrations of residual aliphatic hydrocarbon in each mesocosm were drastically decreased in the order of the fourth, third, second and first part within 12 days. The numbers of petroleum degrading microorganisms in the SRF amended parts (mesocosm II, III and IV) were much higher compared to those in the control mesocosm. Among the amended parts, the fourth mesocosm showed the highest number of oil degrading microorganism during the entire experimental period. The survival rates of two microbial inoculant strains were monitored during the experimental period. In the third part, 30% of the total heterotrophs number was composed of *Acinetobacter* sp. after 3 days of treatment. In the fourth part, the ratio of *Acinetobacter* sp. to the total number of heterotrophic microorganism sharply increased upto 0.5 until 4 days, and decreased slowly until the final day of the experiment. The ratio of *Candida lipolytica* to the total heterotrophs number showed a similar pattern to *Acinetobacter* sp. however, the proportion *of Candida lipolytica* was much lower than that of *Acinetobacter* sp. These results demonstrate that the addition of oil degrading microorganisms played a major role for the degradation of oil in artificially polluted environment, and suggest a high potential for an effective treatment of oil polluted environments.

ACKNOWLEDGMENTS

This research was granted as one of the G-7 research program under the Ministry of Environment, and Science and Technology, Korea. We thank Dr. Y. S. Oh for manuscript correction and also Hanhwa Group for providing the field test area.

REFERENCES

APHA, AWWA and WPCF., 1985, Standard Methods for the examination of water and wastewater. 16th edition. APHA, Washington, DC.

Blumer, M., Ehrhardt, M. and Jones, J. H., 1974, *Deep-Sea Res.*, **20**:239–259.

Jobson, A. M., Cook, F. D. and Westlake, D. W. S., 1974, *Appl.Microb.*, **27**:166–171.

Kim, S. J. et al., 1994, Korea Ministry of Environment and of Science and Technology, Report No. 9-4-2, Development of marine pollution technology. p. 214 (in Korea).

Kim, S. J., 1995,. Ocean Pollution Technology Seminar, KIMM, KNMP, Inchon, Korea, Nov. 16, 1995, p. 39–53 (in Korea).

King, R. B., Long, G. M. and Sheldon, J. K., (ed.) 1992, Chapter 3;Microbial nutrition and environmental requirements in practical environmental bioremediation. Lewis Publishers, Boca Raton, Florida, p.30–32.

Korea Meteological Station, 1994, Meteological monthly bulletin (In Korea).

Krahn, M. M., Ylitalo, G. M. Joss, J. and Chan, S. L., 1991, *Marine Environmental Research*, **31**:175–196.

Lee, K., and Levy, E. M., 1987, In: Proceedings of the 1987 Oil Spill Conference(Prevention, Behaviour, Control, Cleanup). American Petroleum Institute,Washington, D.C., p. 411–416.

Rosenberg, E., Legmann, R., Kushmaro, A., Taube, R., Adler, E. and Ron, E. Z., 1992, *Biodegradation*. **3**:337–350.

Swannell, R. P. J., Lee, K. and Mcdonagh, M., 1996, *Microbiological Reviews*, **60**:342–365.

Vecchioli, G. I., Del Panno, M. T., and Painceira, M. T., 1990, *Environ. Pollut.*, **67**:249–258.

Wade, T. L., Atlas, E. L., Brooks, J. M., Kennicutt II, M. C., Fox, R.G., Sericano, J., Garcia-Romero, B. and De-Freitas, D. 1988, *Estuaries*, **11**:171–179.

39

HEAVY METAL BINDING PROPERTIES OF WILD TYPE AND TRANSGENIC ALGAE (*CHLAMYDOMONAS* sp.)

Xiao-Hua Cai,[1] Jagat Adhiya,[2] Samuel Traina,[2] and Richard Sayre[1]

[1]Departments of Biochemistry and Plant Biology and Graduate Program in Biophysics
[2]School of Natural Resources
Ohio State University
Columbus, Ohio 43210

Trace metal pollution is among the most pervasive and serious environmental problems facing the biosphere. In 1988, global releases of lead, mercury, cadmium, copper, and zinc were estimated to be: 1,160; 11; 30; 2,150; and 2,340 thousand metric tons/year, respectively (Nriagu & Pacyna, 1988). The largest sources of trace metal pollution were mining and the smelting of non-ferrous metals (Nriagu & Pacyna, 1988). Substantial quantities of trace metals are also released as the result of; coal combustion, refuse and sewage disposal, and manufacturing (steel production and plating, battery manufacturing, and electroplating).

Many trace elements (e.g., copper, cobalt, nickel and zinc) are essential for living organisms and function as enzyme cofactors or in redox-active enzymes (Logan & Traina, 1993). At high concentrations, however, even essential trace elements can be toxic (e.g. copper and Wilson's disease). The heavy metals (e.g. cadmium, lead, and mercury; density > 5 gm/cm^3) and the radionuclides (e.g. uranium) are the most toxic trace metals. Chronic exposure to low levels (1.0 ppm) of heavy metals can result in kidney, bone, nervous and immune system dysfunction, and/or result in cancer (Logan & Traina, 1993).

The recovery of heavy metals from polluted environments poses particular problems. Since toxic trace metals are effective in their elemental form they can not be eliminated from the environment by molecular decomposition (as is the case for many organic pollutants) but must be sequestered from the environment (Logan & Traina, 1993). Traditional methods of sequestering trace elements have involved chemical engineering approaches in which the elements are precipitated, reduced and/or sorbed from contaminated media (Logan & Traina, 1993). These methods are generally non-selective and result in large volumes of waste which must be disposed at some other site. More selective approaches such as chelating resins,

New Developments in Marine Biotechnology, edited by Le Gal and Halvorson.
Plenum Press, New York, 1998.

which concentrate trace elements, have been considered, but large volume use of ion exchange and/or ion selective resins is expensive (Logan & Traina, 1993). Therefore, there is a need for safe, effective, and renewable systems to selectively (without interference from other metals) sequester and concentrate toxic trace metals. Biological trace metal recovery systems meet many of the objectives for safe, effective, and renewable recovery of trace metals from contaminated sites and wastes (Aksu & Kutsal, 1991; Butt & Ecker, 1987; Grill et al., 1985; Harris & Ramelow, 1990; Lindow et al., 1989; Romeyer et al., 1988; Tebo, 1995; Xue et al., 1988). Since bioremediation technologies have only recently been developed, their full potential remains to be determined. Two of the more promising bioremediation strategies which are being developed for trace metal recovery are: 1) the identification of organisms which have enhanced capabilities to sequester trace metals (hyperaccumulators), and 2) the expression of trace metal binding factors in transgenic organisms (Cai et al., 1995; Cai et al., 1997; Minz et al., 1996; Nucifora et al., 1989; Raskin et al., 1994; Romeyer et al., 1988; Romeyer et al., 1990; Silver et al., 1989; Tommey et al., 1991). We report on the trace metal binding properties of the single-celled alga, C*hlamydomonas* sp. Which shows promise for the selective and quantitative recovery of toxic trace metals.

We have characterized the trace metal binding properties of; live C*hlamydomonas* cells, dried algal powders, and transgenic algae expressing foreign trace metal binding proteins. Trace metal binding sites of dried *C. reinhardtii* cells were characterized by potentiometric titration. Three major heavy metal binding groups with pK_as of ~3.5, ~4.5 and ~9.5 were identified with functional group acidities of ~3.5 mol kg^{-1} (pH 3.5 and 4.5), and ~2 mol kg^{-1} (pH 9.5). Under optimal binding conditions (low ionic strength and pHs > 5) dried cells absorbed > 90% of the cadmium in solutions containing 40 mg Cd/L. Significantly, cadmium binding was observed to be reversible as a function of pH. Exposure of cells to low pHs (2.0), causes pre-absorbed cadmium to be quantitatively released from the cells. Upon elevation of the pH to 7.0, however, the cadmium was rebound. In contrast, the binding of gold and uranium (> 90% at 400 ppm) to dried cells was pH insensitive. These results indicate that the chemistry of the cadmium and gold/uranium binding sites is different.

Live *Chlamydomonas* cells may reduce free intra-cellular trace metal levels by a variety of mechanisms. Cells may exclude, sequester, or remove trace metals from the cytoplasm. In the presence of low cadmium concentrations (< 25 μM) *Chlamydomonas* apparently sequesters or excludes cadmium (Howe & Merchant, 1992). In the presence of 10^{-5} M cadmium, however, *C. reinhardtii* cells synthesize phytochelatins or glutathione polymers, which selectively bind heavy metals (Cai et al., 1995; Howe & Merchant, 1992). To determine the contribution of phytochelatins to the total cadmium binding capacity, *C. reinhardtii* cells were grown in 40 μM cadmium plus and minus 200 μM buthionine sulfoximine (BSO), a phytochelatin synthesis inhibitor. Surprisingly, two-fold more cadmium was bound to cells grown with BSO than without BSO. The additional bound cadmium (+BSO) was associated with a membrane/cell wall fraction ($10^6 \times$ G pellet) obtained from ruptured cells. The identity of this fraction remains to be determined.

We have also compared the cadmium binding capacity of *Chlamydomonas* species which are more resistant to trace metals than *C. reinhardtii*. One such species, isolated from trace metal contaminated acid soils near Sudbury, Ontario, is *C. acidophila*. *C. acidophila* has been shown to grow in 100-fold higher copper concentrations than *C. reinhardtii* (Twiss, 1990). When both species were grown in the presence of 25 uM cadmium, however, *C. reinhardtii* bound four-fold (2.5% of total cadmium in medium) more cadmium than *C. acidophila* (0.56% of total cadmium in medium). These results suggest that *C. acidophila* more effectively excludes or removes cadmium from the cell than *C. reinhardtii*.

Table 1. Cadmium binding capacity of wild type (CC-425) and transformed Chlamydomonas (pMTCC-8) expressing a foreign metallothionein gene grown at various cadmium concentrations

	Cadmium concentration in growth medium (μM)		
Strain	5	10	40
Cadmium bound by cells (μg Cd/mg Chl)			
CC-425	0.24 ± 0.1	0.52 ± 0.3	79.6 ± 27.8
pMTCC-8	0.43 ± 0.1	0.68 ± 0.1	86.4 ± 33.1
Cadmium removed from medium (% total Cd)			
CC-425	0.2 ± 0.1	0.2 ± 0.1	5.5 ± 3.5
pMTCC-8	0.4 ± 0.1	0.5 ± 0.1	9.0 ± 5.6

The results are the average of four determinations. Values are ± standard deviations.

To determine whether we could genetically alter the cadmium binding capacity of *C. reinhardtii*, we constitutively expressed a foreign class II metallothionein (MT-II) gene from chicken in transgenic cells (Cai et al., 1997). Cells (pMTCC-8) expressing MT-II grew at significantly higher densities (2X) in the presence of high cadmium (40 μM) concentrations than wild type (strain CC-425). Cultures expressing MT-II also absorbed 2-fold more cadmium than wild type. As shown in Table 1, the cadmium binding capacity of the transformant (pMTCC-8) was two-fold greater than that of the non-transformant (CC-425) when grown at 5 uM cadmium. As the cadmium concentration of the growth medium increased, however, there was little difference between the cadmium binding capacity of the transformant and the wild type, suggesting that cadmium binding by the MT-II was saturated. Overall, these results indicate that the trace metal binding properties of algal cells may be improved through expression of introduced genes encoding trace metal binding proteins.

REFERENCES

Aksu, Z., and Kutsal, T., 1991, A bioseparation process for removing lead (II) ions from waste water by using *Chlorella vulgaris*. *J. Chem. Tech. Biotech.* **52**:109–118.

Butt, T., and Ecker, D., 1987, Yeast metallothionein and applications in biotechnology. *Microbiol. Rev.* **51**:351–364.

Cai, X-H., Logan, T., Gustafson, T., Traina, S. and Sayre, R.T., 1995, Applications of eukaryotic algae for the removal of heavy metals from water. *Molecular Marine Biology and Biotechnology* **4**:338–344.

Cai, X-H., Adihya, J., Traina, S. and Sayre, R.T., 1997, Enhanced heavy metal tolerance in transgenic algae expressing a foreign metallothionein gene. (submitted to Plant Physiology).

Davies, J., Weeks, D. and Grossman, A., 1992, Expression of the arylsufatase gene from the β_2-tubulin promoter in *Chlamydomonas reinhardtii*. Nuc. Acids. Res. **20**:2959–2965.

Grill, E., Winnacker, E-L., and Zenck, M., 1985, Phytochelatins: The principle heavy-metal complexing peptides of higher plants. *Science* **230**:674–676.

Harris, P.O., and Ramelow, G.J., 1990, Binding of metal ions by particulate biomass derived from *Chlorella vulgaris* and *Scenedesmus quadracauda*. *Environ. Sci. Technol.* **24**: 220–228.

Howe, G., and Merchant, S., 1992, Heavy metal-activated synthesis of peptides in *Chlamydomonas reinhardtii*. Plant Physiol. **98**:127–136.

Lindow, S., Panopopulos, N., and McFarland, B., 1989, Genetic engineering of bacteria from managed and natural habitats. *Science* **244**:1300–1307.

Logan, T.J., and Traina, S.J., 1993, Trace metals in agriculture. In: Metals in Groundwater. pp 309–349. Ed. by Allen HE, Perdue EM and Brown DS. Lewis Pub. Chelsea, MI.

Minz, D., Rosenberg, E., and Ron, E., 1996, Cadmium binding by bacteria: screening and characterization of new isolates and mutants. *FEMS Microbiol. Lett.* **135**:191–194.

Nriagu, J.O., and Pacyna, J.M., 1988, Quantitative assessment of worldwide contamination of the air, water and soils by trace metals. *Nature* **333**:134–139.

Nucifora, G., Misra, T., Chu, L., and Silver, S., 1989, Cadmium resistance from *Staphylococcus aureus* plasmid pI258 cadA gene results from a cadmium-efflux ATPase. *Proc. Nat. Acad. Sci.* **86**:3544–3548.

Raskin, I., Kumar, P.N., Dushenkov, S., and Salt, D.E., 1994, Bioconcentration of heavy metals by plants. *Curr. Opin. Biotechnol.* **5**:285–290.

Romeyer, F., Jacobs, F., Masson, J. and Brousseau, R., 1988, Bioaccumulation of heavy metals in *E. coli* expressing an inducible synthetic human metallothionein gene. *J. Biotech.* **8**:207–220.

Romeyer, F., Jacobs, F. and Brousseau, R., 1990, Expression of a *Neurospora crassa* metallothionein and its variants in *E. coli. App. Eviron. Microbiol.* **56**:2748–2754.

Silver, S., Nucifora, G., Chu, L. and Misra, T., 1989, Bacterial resistance ATPases: primary pumps for exporting toxic cations and anions. *Trends. Biol. Sci.* **14**:76–80.

Tebo, B.,1995, Metal precipitation by marine bacteria: potential for biotechnological applications. In: Genetic Engineering and Principles. pp. 231–261, Ed. by Setlow JK. Plenum Press, NY, NY.

Tommey, A., Shi, J., Lindsay, W., Urwin, P. and Robinson, N.,1991, Expression of the pea gene PsMTa in *E. coli. FEBS Lett.* **292**:48–52.

Twiss, M.R., 1990, Copper tolerance of *Chlamydomonas acidophila* isolated from acidic, copper-contaminated soils. *J Phycol.* **26**:655–659.

Xue, H-B., Stumm, W. and Sigg, L., 1988, The binding of heavy metals to algal surfaces. *Wat. Res.* **22**:917–926.

40

DNA REPAIR ENZYMES IN HYPERTHERMOPHILIC ARCHAEA

Jocelyne DiRuggiero and Frank T. Robb

Center of Marine Biotechnology
University of Maryland Biotechnology Institute
701 E. Pratt St.
Baltimore, Maryland

The presence of hyperthermophilic microorganisms in hydrothermal environments at temperatures above 90°C is now well established (Stetter, 1990). In fact, isolates from deep sea marine hydrothermal vents can grow optimally under pressure at temperatures up to 113°C (Blochl et al., 1997). Most of the hyperthermophiles are members of the Domain Archaea as defined by Woese et al. (1990), and their existence raises the question of how metabolic processes are sustained at extremely high temperatures. Not surprisingly, all of the enzymes isolated to date from hyperthermophiles display unusual thermostability (Adams, 1993; DiRuggiero and Robb, 1996), however, temperature in the region of 100°C greatly accelerate the spontaneous chemical degradation of DNA by orders of magnitude (Lindahl, 1993). Despite this, hyperthermophiles such as *Pyrococcus furiosus* are capable of rapid growth at or near to 100°C (Fiala and Stetter, 1986) and apparently maintain a normal level of genetic stability (Brown et al., 1994; Keeling and Doolittle, 1995). We suggest that efficient DNA repair and replication mechanisms are a prerequisite for growth and survival of these microorganisms, yet these processes remain largely unexplored. Further, despite extensive sequencing of Archaeal genomes and the recent publication of the complete genome sequence of the methanogen *Methanococcus jannaschii* (Bult et al., 1996) very little is known of the proteins involved in DNA repair and replication in hyperthermophilic Archaea.

Living cells eliminate DNA lesions by molecular mechanisms such as excision repair and recombination which are ubiquitous and quite similar processes from bacteria to humans. A large body of work is available on DNA repair proteins and mechanisms from both Bacteria and Eukarya (Friedberg et al. 1995), however experimental evidence for repair of damaged DNA in hyperthermophilic and extreme thermophilic Archaea is practically non-existent (DiRuggiero et al., 1997; Grogan, 1997; Jacobs and Grogan, 1997). We recently demonstrated that the hyperthermophile *P. furiosus* is capable of maintaining the integrity of its genome at temperatures which rapidly destroy unprotected DNA

New Developments in Marine Biotechnology, edited by Le Gal and Halvorson.
Plenum Press, New York, 1998.

or DNA in the mesophile, *Escherichia coli* (Peak et al. 1995). We have also discovered exceptional resistance to ionizing radiation in *P. furiosus* cells which showed no loss of viability after irradiation with 2,000 Gy of ^{60}Co, and 75% survival when the cells were exposed to a dose of 2,500 Gy (DiRuggiero et al. 1997). Compared to *E. coli*, *P. furiosus* cells are 7-fold more resistant to ionizing radiation. In addition, using pulsed field electrophoresis, we showed that the *P. furiosus* 2 Mb chromosome was fragmented into pieces from 500 to 30 kb after irradiation at a dose of 2,500 Gy, and was fully reassembled upon incubation at 95°C (DiRuggiero et al., 1997). Recently, Jacobs and Grogan (1997) determined that the rates of forward mutation in the extremely thermophilic archaeon *Sullolobus acidocaldarius* (optimal growth temperature 80°C) were similar to the rates for protein-encoding genes in *E. coli*, suggesting the presence of relatively precise mechanisms for repair of spontaneous damage in *S. acidocaldarius* cells. In addition, Grogan (1997) reported physiological evidence of light-dependent repair (photoreactivation) by *S. acidocaldarius* to reverse lethal effects of UV-irradiation exposure.

At the molecular level, only a few genes with homology to eukaryotic and bacterial repair genes have been described for hyperthermophiles. In studies published to date, recognition of these genes has been deduced from amino acid sequence homology with existing proteins from eukaryotes and bacteria, except in the case of the RecA-like genes, where preliminary functional studies have been done (McCready, 1996; Rashid et al., 1996). Among the bacteria-like repair gene described so far is the putative *dinF*-like SOS repair gene (damage inducible) found in *P. furiosus*, and whose function remains enigmatic even in *E. coli.* (Bouyoub et al., 1996). The homolog of another *E. coli* damage-inducible gene has been found in *S. solfataricus* (Kulaeva et al., 1996). The *Sulfolobus* protein sequence shows high sequence identity to bacterial DinB, UmuC and to the *Saccharomyces cerevisiae* REV1 proteins. All of these proteins are involved in the correction of damage-induced mutagenesis. The existence of the *dinF* and *dinB* genes in hyperthermophiles suggests that adaptive repair systems similar to the bacterial SOS DNA repair systems (Radman 1975) could exist in Archaea, although there is no experimental evidence for this hypothesis.

Homologs of RecA-like proteins (RadA and Pk-REC) have been reported in four species of Archaea, including hyperthermophilic Archaea (Rashid et al., 1996; Sandler et al., 1996). Woods and Dyall-Smith (1997) showed that a deletion mutation in the mesophilic archeaon *Haloferax volcanii radA* gene resulted in an UV-sensitive phenotype, and Rashid et al. (1996) demonstrated the complementation of an *E. coli recA* mutant to UV resistance with a recombinant RecA-like protein (Pk-REC) from the hyperthermophilic archaeon *Pyrococcus* sp. KOD1, establishing the role of archaeal RecA-like proteins in DNA repair. We have cloned and sequenced two RecA-like genes in *P. furiosus,* namely *rad A* and *rad B* (DiRuggiero, unpublished). The deduced amino acid sequence of the *arr* gene is similar in length and composition to the sequences of Pk-REC and Rad51 of *Pyrococcus sp.* KOD1 and *M. jannaschii* respectively, whereas the translated sequence of the *radA* gene is homologous to the archaeal RadA protein sequences (DiRuggiero, unpublished; Sandler et al., 1996). These results clearly show the existence of two RecA-like proteins in *P. furiosus.* Furthermore, the two corresponding proteins also exist in *M. jannashii* but annotated with different names (Bult et al., 1996; Sandler et al., 1996), and our searches of the complete genomes of *Methanobacterium thermoautotrophicum* (optimal growth temperature 65°C) and *Archaeoglobus fulgidus* (optimal growth temperature 80°C) also revealed the presence of two RecA-like proteins similar to those found in *M. jannaschii* and *P. furiosus.* Our phylogenetic analysis of RecA-like proteins shows that both type of archaeal proteins are more similar to the eukaryotic Rad51/DMC1 proteins than to the bacterial RecA proteins (DiRuggiero, unpublished; Sandler et al., 1996).

In addition, we have cloned and sequenced the gene encoding a structure specific endonuclease in *P. furiosus*. The deduced amino acid sequence is highly homologous to the *M. jannaschii* Rad2 (55% identity), the human and murine FEN-1 (43% and 44% identity respectively) and *S. cerevisiae* Rad2 (30% identity). Our searches of the complete genomes of *M. thermoautotrophicum* and *A. fulgidus* also revealed the presence of *fen-1* genes in these organisms. All Rad2 and FEN-1 proteins present the same two regions of high homology: (i) the N region of about 100 amino acids and, (ii) the I region of about 140 amino acids (Figure 1). However, unlike eukaryotic Rad2 and FEN-1 proteins, the archaeal FEN-1 do not have a C region, which is constituted of a highly basic carboxy-terminal tail, and is conserved in structure rather than sequence (Figure 1). The N and I regions are separated by a region of about 15 amino acids in the FEN family and about 600 amino acids in the Rad2 family and it is called the S-region for spacer. Because both FEN-1 and RAD2 have been shown to be nucleases, it is likely that the DNA-binding domain and active site are within one of the regions of homology. From the combined perspectives of gene structure, primary sequence similarity and phylogeny, the archaeal FEN-1 proteins appear to be orthologs of eukaryotic FEN proteins and paralogs of eukaryotic Rad proteins.

The *P. furiosus fen-1* gene was expressed in *E. coli* and purified to homogeneity by heat treatment and ion exchange chromatography (DiRuggiero, Kaiser and Robb, unpublished). We found that the thermostable recombinant protein has the same structure-specific endonuclease activity as the human and murine FEN-1 (Harrington and Lieber 1994). The enzyme cleaves 5′-3′ single-stranded DNA flap structures and possesses a 5′-3′exonuclease activity. In eukaryotes, FEN-1 is essential for the removal of Okazaki fragments during chromosome replication, interacting with the replicative helicase Dna2 and the proliferating cell nuclear antigen (PCNA) (Lieber, 1997). FEN-1 is also involved in DNA repair and recombination. DeMott et al. (1996) and Barnes et al. (1996) showed that

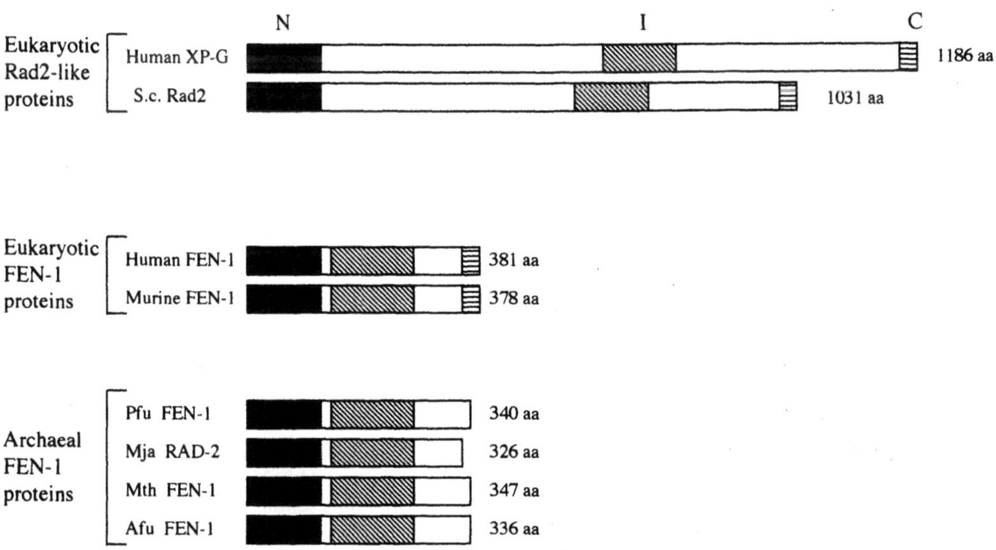

Figure 1. Schematic representations of eukaryotic and archaeal FEN and eukaryotic Rad proteins. N, I, and C represent conserved regions. S.c., *Saccharomyces cerevisiae*; Pfu, *Pyrococcus furiosus* (Genbank accession number AFO13497); Mja, *Methanococcus jannaschii* (Genbank accession number U67586); Mth, *Methanobacterium thermoautotrophicum* (Genome Therapeutics Co. and Ohio State University, 1997); Afu, *Archaeoglobus fulgidus* (The Institute for Genome Research, 1997).

mammalian FEN-1 can cleave 5′ terminal abasic lesion and adduct-damaged DNA efficiently suggesting a role in the base excision repair pathway. FEN-1 may also play a role in DNA end-joining mechanisms in V(D)J recombination as well as double-strand break repair in response to ionizing radiation (Lieber, 1997). In addition, genetic analysis of yeast FEN-1 mutants also implicate this enzyme in DNA repair (Sommers et al. 1995). Consequently, these results strongly suggest a dual role of the *P. furiosus* FEN-1 in replication and in repair of double-stranded breaks and other DNA damages.

ACKNOWLEDGMENTS

This research was supported by grants from the Department of Energy (DE-FG02-92ER20083), the National Science Foundation (BES 9410687), NATO (CRG. 960211) and the U.S. Department of Commerce, Advanced Technology Program. We thank Dr. M. Kaiser for his help with the FEN-1 assays, and Dr. J. Brown for his assistance with the phylogenetic analysis.

REFERENCES

Adams, M.W.W. 1993, *Ann. Rev. Microbiol.* **47**:627–658

Barnes, C.J., Wahl, A.F., Shen, B., Park, M..S.. and Bambara, R.A,. 1996, *J. Biol. Chem.*, **217**:29624–29631

Blochl, E. Rachel, R. Burgraf, S. Hafenbradl, D., Jannasch, H.W. Stetter, K.O., 1997, *Extremophiles* **1**:14–21

Bouyoub, A., Barbier, G., Querellou, J. and Forterre, P. 1996, *Gene*, **167**:147–149

Brown, J.R., Masuchi, Y., Robb, F.T. and Doolittle, W.F. 1994, *J. Mol. Evol.*, **38(6)**:566–576

Bult, C.J., White, O., Olson, G.J., Zhou, L., Fleishmann, R.D., Sutton, G.G., Blake, J.A., FitGerald, L.M., Clayton, R.A., Gocayne, J.D. and et al. 1996, *Science*, **273**:1058–1073

DeMott, M.S., Shen, B., Park, M.S. and Bambara, R.A. 1996, *J. Biol. Chem.*, **271**: 30068–30076

DiRuggiero, J. and F.T. Robb, 1996, In Enzymes and Proteins from Hyperthermophilic Microorganisms. Adams, M.W.W. (ed.). Academic Press.pp 311–339.

DiRuggiero, J., Santangelo, N., Nackerdien, Z., Ravel, J. and Robb, F. T. 1997, *J. Bacteriol.* **179**: 4643–4645

Fiala, G. and Stetter, K.O., 1986, *Arch. Microbiol*, **145**: 56–61

Friedberg, E.C., Walker, G.C. and Siede, W. 1995, DNA Repair and Mutagenesis. Washington, D.C. ASN Press

Grogan, D.W. 1997, Microbiology. **143**:1071–1076

Harrington, J.J. and Lieber, M.R. 1994, *EMBO J.*, **13**:1235–1246

Jacobs, K.L. and Grogan, D.W. 1997, *J. Bacteriol.*, **179**:3298–3303

Keeling, P.J. and Doolittle, W.F. 1995, *Mol. Microbiol.*, **17**:39–400

Kulaeva, O.I., Koonin, E.V., McDonald, J.P., Randall, S.K., Rabinovich, N., Connaughton, J.F., Levine, A.S. and Woodgate, R. 1996, *Mutation Res.*, **357**:245–253

Lieber, M.R. 1997, *BioEssays*, **19**:233–240

Lindahl,T., 1993, *Nature*, **362**:709–715

McCready, S. 1996, *Mutat. Res.* **364**:25–32

Peak, M.J., Robb, F.T. and Peak, J.G. 1995, *J. Bacteriol.*, **177**:6316–6318

Radman, M. 1975, In A. P. Hanawalt and R.B. Setlow (Eds.), Molecular Mechanisms for Repair of DNA, (pp. 355–367). New York: Plenum Publishing Corp.

Rashid, N., Morikawa, M. and Imanaka, T. 1996, *Mol. Gen. Genet.*, **253**:397–400

Sandler, S.J., Satin, L H., Samra, H.S. and Clark, A.J. 1996, *Nucleic Acids Res.*, **24**:2125–2132

Sommers, C.H., Miller, E.J.. Dujon, B., Prakash, S. and Prakash, L 1995, *J. Biol. Chem.*, **270**:4193–4196

Stetter, K.O., Fiala, G., Huber, R. and Segerer, A. 1990, *FEMS Microbiol. Rev.*, **75**:117–124

Woese, C.R., Kandler ,O. and Wheelis, M.L. 1990, *Proc. Nat. Acad. Sci. USA*, **87**:4576–4579

Woods, W.G. and Dyall-Smith, M.L. 1997, *Mol. Microbiol*, **23**:791–797.

41

CHAPERONIN IN A THERMOPHILIC METHANOGEN, *METHANOCOCCUS THERMOLITHOTROPHICUS*

Masahiro Furutani,[1] Toshii Iida,[1] and Shigeyuki Yamano,[2] and Tadashi Maruyama[1]

[1]Marine Biotechnology Institute
Kamaishi, Iwate, Japan
[2]Marine Biotechnology Institute
Shimizu, Shizuoka, Japan

1. INTRODUCTION

Molecular chaperones play an important role in the protein foldings in vivo. Chaperonin is a 60 kDa major member of molecular chaperones and has two types, Group I and Group II. The Group I chaperonin is GroEL-like complex in eubacteria, mitochondria and chloroplasts. Whereas archaeral chaperonin which is colled thermosome and eukaryotic cytosol TCP- I protein belong to Group II (Trent et al., 1991). While a chaperonin was purified from a hyperthermophilic methanogen, *Methanopyrus kandleri* (Andrä et al., 1996), biochemical and functional characters of chaperonins in methanogens have remained to be clarified. We report here characteristics of chaperonin of a thermophilic methanogen, *Methanococcus thermolithotrophicus*.

2. MATERIALS AND METHODS

2.1. Cloning of a Chaperonin Gene from *M. thermolithotrophicus*

Genomic DNA of *Methanococcus thermolithotrophicus* DSM2095 was prepared as described previously (Hoaki et al., 1994). A partial chaperonin gene was amplified by PCR from the genome of *M. thermolithotrophicus* with various primers designed from the consensus amino acid sequences of other archaeral chaperonins. After determination of the partial nucleotide sequence (about 740 bp) of the chaperonin, a probe DNA (730 bp) for the screening of the whole chaperonin gene from the genomic library was prepared. The positive clones were detected using the probe labeled with a DIG DNA labeling and detection

New Developments in Marine Biotechnology, edited by Le Gal and Halvorson.
Plenum Press, New York, 1998.

kit (Boehringer Mannheim Co., Mannheim, Germany). The hybridization and washing were carried out at 65°C in 0.2 × SSC, 0.5% (w/v) SDS. The inserted fragments from positive clones were sequenced using a Dye termination cycle sequencing kit (Perkin Elmer Co.).

2.2. Expression of Chaperonin Gene in *E. coli*

The chaperonin gene of *M. thermolithotrophicus* was expressed in *E. coli* using a pET-11d expression vector (Novagen Co., Madison, WI). The chaperonin gene having *Nco*I and *Bam*HI sites at its N- and C-terminals, respectively, was prepared by PCR using the forward primer, GGCCATGGCAGCTAACCAGCCA and the reverse primer, CCGGATCCTTACATCATTCCGCCCATAC. After amplification, the DNA fragment was recovered and then ligated into *NcoI-Bam*HI sites of pET-11d vector. The resultant methanogen chaperonin expression vector, pETTS was introduced into *E. coli* BL 21 (DE3) (Novagen Co., Madison, WI). *E. coli* BL 21 (DE3) harboring pETTS was grown in 2 X LB (yeast extract 10 g, bactotrypton 20 g, NaCl 10 g/L, pH 7.5) at 35°C. When the OD650 of the culture reached 1.2–1.5, 1 mM isopropy 1-β-D-thiogalactoside was added. The culture was further incubated for additional 4 hours at 35°C.

2.3. Purification of Recombinant Chaperonin from *E. coli*

The recovered *E. coli* cells with pETTS were sonicated in 50 mM K-phosphate buffer (pH 7.0) and was incubated at 65°C for 90 minutes. The supernatant was collected by centrifugation and then $(NH_4)_2SO_4$ was added at the final concentration of 1.3 M. The solution was centrifuged and the supernatant was applied to a TSKgel Ether-5PW column (Tosoh, Tokyo, Japan). The proteins were separated using a linear gradient from 1.5 to 0 M $(NH_4)_2SO_4$ in 25 mM Hepes-KOH buffer (pH 7.0). The chaperonin fractions were collected and dialyzed against 25 mM Hepes-KOH buffer (pH 7.3) overnight at 4°C and then applied to a TSKgel Super Q-5PW column (Tosoh, Tokyo, Japan). The proteins were separated with a linear gradient from 0–0.5 M NaCl in 25 mM Hepes-KOH (pH 7.3). The chaperonin fractions were collected, EDTA was added at 10 mM and incubated for 1 hour at room temperature. The solution was dialyzed against 25 mM Hepes-KOH (pH 6.8) overnight at 4°C. The inner solution was treated with 50 mM $MgCl_2$, 0.25 mM ATP and 0.3 M $(NH_4)_2SO_4$ at 30°C for 6 hours and then was concentrated by ultrafiltration with a Ultrafree PLTK filter (MILLIPORE JAPAN Co., Tokyo). The concentrated solution was applied to a TSKgel G3000SWXL column (Tosoh, Tokyo, Japan) in 25 mM Hepes-KOH (pH 6.8) containing 50 mM $MgCl_2$, 0.2 mM ATP and 0.3 M $(NH_4)_2SO_4$. The purified chaperonin was analyzed by SDS-PAGE.

2.4. Electron Microscopy

An aliquot of chaperonin solution was negatively stained with 2.0% uranyl acetate and observed in a Hitachi H-7000 electron microscope operated at 75 or 100 kv.

2.5. ATPase Assay

The assay mixture (100 μl) of ATPase activity of chaperonin contained 50 mM $MgCl_2$, 2 mM ATP, 300 mM KCl and 0–5 ug of chaperonin in 50 mM Hepes-KOH (pH 6.8). The mixture was incubated at 40–90°C for 30 minutes. The reaction was finished by addition of 10% (w/v) perchloric acid. After centrifugation released Pi in supernatant was measured by malachite green method (Lanzetta et al., 1979) in 96-wells microplate.

3. RESULTS AND DISCUSSION

3.1. Cloning of Chaperonin Gene from *M. thermolithotrophicus*

The chaperonin gene of *M. thermolithotrophicus* encoding 544 amino acids was cloned. We call this chaperonin MTTS (*M.thermolithotrophicus* thermosome). The protein sequence of MTTS shows 78.7, 66.5 and 20.0% identities with those of *Methanococcus jannaschii, Methanopyrus kandleri* and *E. coli*, respectively. A highly conserved ATP-binding site (GDGTTT) was found in MTTS protein sequence. The unique Gly and Met rich sequence was found in its C-terminal region. This sequence is conserved also in Group I chaperonin, GroEL of *E. coli*. While other archaeral chaperonins usually have a α and β subunits, only one chaperonin gene was cloned in our study. Only one chaperonin gene was found in also *M. jannaschii* genome (Bult et al., 1996). A chaperonin which is composed of one kind of subunit was purified from a hyperthermophilic methanogen, *Methanopyrus kandleri* (Andrä et al., 1996). These suggest that chaperonins of methanogen may be composed of only one kind of subunit.

3.2. Purification of Recombinant MTTS from *E. coli*

The MTTS gene was expressed in *E. coli* BL2 1 (DE3). The recombinant MTTS was purified to homogeneity by heat treatment, hydrophobic and anion exchange chromatographies. Almost all of the recombinant chaperonin was eluted in monomeric fraction on gel filtration. The monomeric MTTS was incubated in 50 mM $MgCl_2$, 0.25 mM ATP and 0.3 M $(NH_4)_2SO_4$ at 30°C for 6 hour. The solution was concentrated by ultrafiltration and then applied to a gel filtration column in the presence of 50 mM $MgCl_2$, 0.2 mM ATP and 0.3 M $(NH_4)_2SO_4$. A new peak which corresponded to oligomeric MTTS (MTTS complex) appeared at molecular mass above 669 kDa (Figure 1). The ring structure of MTTS in this fraction was confirmed in electromycroscopy. The ring structures seemed to be composed of eight subunits. The reconstructed MTTS complex was stable for at least one month in the presence of 50 mM $MgCl_2$, 0.2 mM ATP and 0.3 M $(NH_4)_2SO_4$ at 4°C. However, MTTS complex further formed the filaments at higher protein concentration above 2.0 mg/ml during the storage at 4°C. It had recently been reported that the purified chaperonin from *Sulfolobus shibatae* forms filament in the presence of Mg/ATP (Trent et al., 1997). Our study showed that the chaperonin composed of one kind of subunit is able to form the filament in the presence of Mg/ATP. The electromicroscopic study revealed that the chaperonin filament rapidly dissociated to the chaperonin complex by the dilution to the

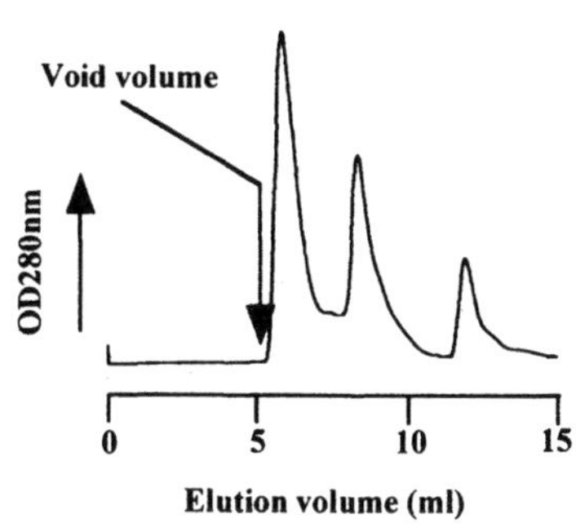

Figure 1. Gel filtration of *M. thermolithotrophicus* chaperonin.

Table 1. ATPase activity of *M. thermolithotrophicus* chaperonin

Chaperonin (μg/assay mixture)	ATP hydrolysis (n mole/assay mixture)
12.5	32.0
25.0	63.9
50.0	99.5

protein concentration of 0.5 mg/ml in the presence of Mg/ATP. The physiological meanings of the chaperonin filament have not been understood yet.

3.3. ATPase Activity of MTTS

The ATPase activity of the reconstructed MTTS complex was assayed (Table 1). MTTS exhibited ATPase activity of 62.4 mole ATP hydrolyzed/mole complex/min at 65°C. However, monomeric MTTS showed undetectable ATPase activity. This is the first time that ATPase activity of chaperonin from methanogen was confirmed. The optimum temperature of the ATPase activity was 60–70°C. Significant inactivation was observed at 90°C. Further biochemical characterization of MTTS is being under investigation.

4. SUMMARY

A gene encoding 544 amino acids of a subunit of chaperonin was cloned from a thermophilic methanogen, *Methanococcus thermolithotrophicus*. The deduced amino acid sequence showed 66.5, 56.1 and 20.1% similarity to those of *Methanopyrus kandleri*, *Thermoplasma acidophilum* and *E. coli*, respectively. The chaperonin gene was expressed in *E. coli* by using a pET expression system. The expressed recombinant chaperonin purified from *E. coli* appeared to be monomeric on gel filtration. The monomer assembled to an oligomer with apparent mass above 669 kDa in the presence of 50 mM $MgCl_2$, 0.25 mM ATP and 0.3 M $(NH_4)_2SO_4$. The oligomeric chaperonin exhibited ATPase activity at 40–90°C with the optimum temperature of 60–70°C. This is the first report on ATPase activity of the chaperonin in methanogens.

REFERENCES

Andrä, S., Frey, G., Nitsch, M., Baumeister, W., and Stetter, K.O., 1996, *FEBS lett.* **379**:127–131
Bult, C.J., et al., 1996, *Science*, **273**:1058–1073
Hoaki, T., Nishijima, M., Kato, M., Adachi, K., Mizobuchi, S., Hanzawa, N., and Maruyama, T., 1994, *Appl. Environ. Microbiol.* **60**:2898–2904
Lanzetta, P.A., Alvarez, L.J., Reinach, P.S., and Candia, O.A., 1979, *Anal. Biochem.* **100**:95–97
Trent, J.D., Kagawa, H., Yaoi, T., Olle, E., and Zaluzec, N.J., 1997, *Proc. Natl. Acad Sci. USA* **94**: 5383–5388
Trent, J.D., Nimmesgerm, E., Wall, I.S., Hartl, F.U., and Horwich, A.L., 1991, *Nature* **354**:490–493

42

PRODUCTION AND APPLICATION OF NATURAL STABILIZING COMPOUNDS FROM HALOTOLERANT BACTERIA

Erwin A. Galinski[1] and Thomas Sauer[2]

[1]Institute for Biochemistry
Westfälische Wilhelms-Universität
Wilhelm-Klemm-Straße 2
48149 Münster, Germany
[2]Chiron Behring GmbH & Co.
Preclinical Development Vaccine
P.O.Box 1630
35006 Marburg, Germany

Marine, halotolerant microorganisms have great potential for the production of organic osmolytes (compatible solutes), which typically attract water and stabilize biological structures in vivo and in vitro [Galinski, 1995]. In this study we describe the exploitation of the marine, extremely halotolerant *Halomonas elongata* for the semi-continuous production of the tetrahydropyrimidines ectoine and S,S-β-hydroxyectoine, both novel compatible solutes with applications as stress protectants for enzymes, DNA and whole cells [Lippert & Galinski, 1992; Lapidot et al., 1993; Louis et al., 1994] (Fig. 1).

The novel biotechnological process named "bacterial milking" is presented in Fig. 2. It is characterized by high cell-density fermentation of the production strain at elevated salinity (15–20% NaCl) followed by a rapid dilution step, which triggers immediate release of compatible solutes into the medium [Galinski et al., 1993; Sauer & Galinski, 1997]. In practice, a complete "bacterial milking" cycle is achieved by separating the cells from the high-salt medium (step 2), dilution of the dense cell suspension with water (step 3), a second separation step to collect the product solution (step 4) and subsequent up-shock treatment with the original high-salt medium to trigger biosynthesis of new solutes.

Surprisingly, the cells were not harmed by this *in vivo* extraction technique and upon transfer to the original high salinity medium rapidly resynthesized 90% of the lost solutes within 6 hours (Fig. 3). By repeatedly performing "bacterial milking" cycles we were able to supply large amounts of ectoines with a productivity of 150 mg per g cell dry weight and "milking" cycle [Sauer & Galinski, 1997]. The described production process may pos-

New Developments in Marine Biotechnology, edited by Le Gal and Halvorson.
Plenum Press, New York, 1998.

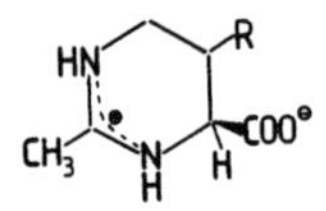

Figure 1. Structure of L-ectoine (R=H) and S,S-β-hydroxyectoine (R=OH).

sibly be applied to natural open-pond systems under non-sterile conditions (e.g. salinas). It represents a novel approach for the exploitation of marine microbial resources by biotechnology industries.

Besides their application as moisturizers in skin-care products, compatible solutes (and in particular ectoines) have a wide range of applications for the stabilization of biomolecules in the dry state and in aqueous solution, namely under adverse conditions such as high temperature, freeze-thaw treatment and in the presence of denaturants such as urea

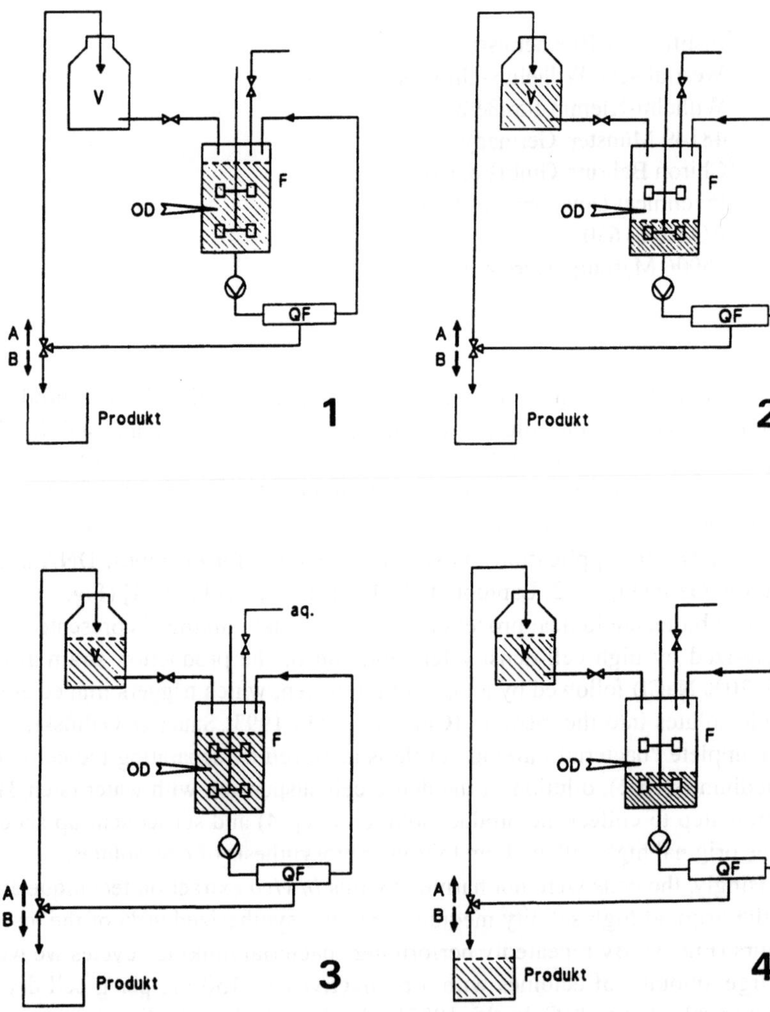

Figure 2. Schematic presentation of the "bacterial milking" process (4 stages). V: reservoir for high-salt medium; F: fermenter; OD: unit for optical density measurements; QF: crossflow filtration; A,B: two-way valve.

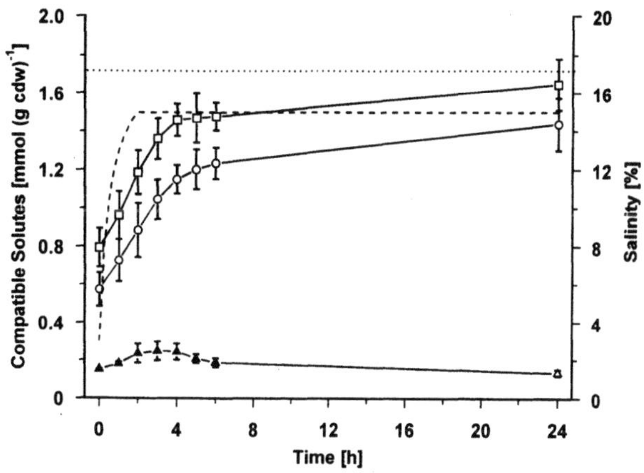

Figure 3. Resynthesis of compatible solutes by *Halomonas elongata* to complete a full "bacterial milking" cycle. Salinities of hyper- and hypoosmotic shocks were 15% (w/v) and 3% (w/v), respectively. s: glutamate, O: ectoine, □: total solutes; dotted line: total solute content before osmotic downshock; broken line: medium salinity profile of hyperosmotic shock (from Sauer & Galinski, Biotechnol. Bioeng. (in press)).

and guanidinium chloride. It is now understood that ectoines like other compatible solutes influence extrinsic stability factors such as water structure, water activity and solution properties of the molecular environment. The presence of natural stabilizing compounds has a marked effect on all thermodynamic parameters (ΔH, ΔS and ΔCp) governing the stability curve of biomolecules in solution. Compatible solutes are, therefore, nature's answer to stability problems during stress adaptation in an ever changing environment.

REFERENCES

Galinski, E.A., 1995, Osmoadaptation in bacteria. *Adv Microb Physiol* **37**: 273–328

Galinski, E.A., Trüper, H.G., Sauer, T., 1993, Eur Pat Appl EP93/03687 (Cl. C12P1/00)

Lapidot, A., Inbar, L., Aloni, Y., Ben-Asher, E. 1993,. Eur Pat Appl. EP 553,884 (Cl. C12P17/12)

Lippert, K., Galinski, E.A., 1992, Enzyme stabilization by ectoine-type compatible solutes: protection agains heating, freezing and drying. *Appl Microbiol Biotechnol* **37**: 61–65

Louis, P., Trüper, H.G., Galinski, E.A., 1994 Survival of *Escherichia coli* during drying and storage in the presence of compatible solutes. *Appl Microbiol Biotechnol* **41**:684–688

Sauer, T., Galinski, E.A., 1997, Bacterial milking: a novel bioprocess for the production of compatible solutes. *Biotechnol Bioeng.* (in press)

43

MOLECULAR DETECTION OF MAGNETIC BACTERIA IN DEEP-SEA SEDIMENTS

Kaori Inoue[1] and Harald Petermann[2]

[1]Department of Biotechnology
Tokyo University of Agriculture and Technology
Koganei, Tokyo 184, Japan
[2]Fachbereich Geowissenschaften
University Bremen
Postfach 330440
2800 Bremen 33, Germany

Bacterial production of magnetite (Fe_3O_4) (Frankel et al., 1979) contributes significantly to the natural remanent magnetism of deep-sea and other sediments (Kirschvink et al., 1984; Stolz et al., 1986). Accurate magnetostratigraphy, reconstruction of past magnetic field behavior and extraction of environmental information from the sedimentary record depend on an understanding of the biology of deep-sea magnetic bacteria. Although magnetofossils (Kirschvink et al., 1984) have been found in deep-sea sediments, there have been no reports of the recovery of magnetic bacteria from the deep ocean.

Detection and diversity of bacteria from the environment are routinely studied by using 16S rDNA methods which are particularly suitable for uncultured bacteria (Pace et al., 1986; Vandamme et al., 1996). It is now generally accepted that rDNA is one of the most useful targets for studying phylogenetic relationships because it presents in all bacteria, is functionally constant, and is composed of highly conserved as well as more variable domains (Woese, 1987). However for the detection of specific bacteria the use of probes based on structural genes is also very specific. Here, we have used the structural gene *magA* as a molecular marker to detect magnetic bacteria in deep-sea sediments.

A gene required for magnetite biomineralization, *magA*, which codes for an iron transport protein, was cloned from the magnetic bacterium *Magnetospirillum* sp. AMB-I (Fig. 1) (Nakamura et al., 1995; Matsunaga et al., 1992). Its sequence has been determined. The *magA* specific PCR primers can amplify an approximately 1.3 kb segment, which represent the whole *magA* gene. The forward primer is from nucleotides at position 879–899 with a *Sca*I linker (5′-TAAAGTACTGTCCATGGAACTGCATCATCC-3′), the reverse primer is from nucleotides at position 2164–2184 also with a *Sca*I linker (5′-CTAAGTACTAATTCCAGAGCCAGGTCCGGC-3′).

New Developments in Marine Biotechnology, edited by Le Gal and Halvorson.
Plenum Press, New York, 1998.

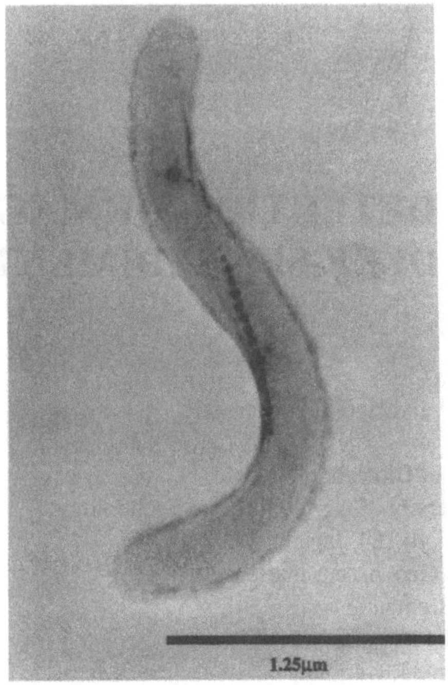

Figure 1. Transmission electron micrograph of *Magnetospirillum* sp. AMB-1.

Highly homologous genes with *mag*A has been amplified and sequenced in *Magnetospirillum magnetotacticum* MS-I and *Magnetospirillum* sp. MCT-I. There are no genes amplified by PCR with *mag*A primers in *Magnetospirillum gryphiswaldense* MSR-1 (Sakaguchi et al., 1993) and sulphate-reducing bacterium RS-1 (Kawaguchi et al., 1995). MS-1 and MGT-1 are more closely related to AMB-1 than to MSR-l by phylogenetic tree constructed using 16S rDNA (Burgess et al., 1993). Although many different size DNA fragments were amplified from *E. coli* and other kinds of bacterial genomic DNA using *mag*A primers, they did not show high homology with *magA* sequence. These results show that *magA* gene is specific to magnetic bacteria and detection of AMB-type magnetic bacteria using *magA* as a probe is possible.

Deep-sea samples were collected from twelve sites. The water depth of all sites was over 1000 m. Extraction and purification of DNA was carried out by previously described methods (Thornhill et al., 1995; Herrick et al., 1993). The *magA* genes were amplified using these primers, cloned and sequenced by describing above. Homology searches were also carried out against the DDBJ with BLASTN program.

Forty clones were obtained after digestion of PCR products into the vector pCR2.1 (Invitrogen Co., San Diego, USA). Sequence analysis showed that four out of forty cloned inserts were highly homologous with the *magA* gene Two of them were from samples collected at Namibia. The Namibia sample GeoB1719 contained living magnetic bacteria as observed using a light microscope (Petermann et al., 1993). Another two samples were collected from the Atlantic Ocean near Belem on the Northern Coast of Brazil. These sequenced genes of the Atlantic Ocean samples (GeoB1508 and GeoB1501) were partial region of the *magA* gene. No living magnetic bacteria were found in the GeoB1508 and GeoB1501. However the sample GeoB 1508 contained magnetosome.

Table 1. Depth distribution of the number of magnetic bacteria, nitrate depth profiles, and oxygen depth profiles

Sediment depth (cm)	PCR (fluorescence intensity[a])	Number of magnetic bacteria	NO_3 (mmol/l)	O_2 (mmol/l)
0–1	+	0	83.1	160
1–2	+	50	88.1	40
2–3	+	100	40.6	0
3–4	+	1000	4.0	0
4–5	+	500	4.0	0
5–6	+++	400	0.6	0
6–7	+	50	0.2	0
7–8	+	0	0.3	0
8–9	+	0	0	0
9–10	+	0	0	0

[a]The relative fluorescence intensity: 70% +, 70–95% ++, 95% +++.

Furthermore, PCR using *magA* primers could amplify the *magA* gene at the depth containing no magnetotactic bacteria. There were no oxygen less than 2 cm depth sediments. There were a very small amount of nitrate concentration less than 3 cm and nitrate were rarely detected in less than 6 cm depth sediments. Magnetic bacteria are sometimes non-magnetic, non-motile and weakly magnetotactic bacteria as in natural environment or no good growth conditions (Frankel et al., 1979; Blakemore et al., 1979; Matsunaga et al., 1991; Matsunaga et al., 1992; Schleifer et al., 1991). It is impossible to detect magnetic bacteria which can not synthesize magnetosome or have no motile using a light microscope at magnetic field. Using structural *magA* gene, although no live magnetic bacteria, could be also detected in samples GeoB 1508 and GeoB 1501 magnetic bacteria.

The *magA* genes of deep-sea magnetic bacteria were cloned and sequenced, confirming the presence of magnetic bacteria in sediments as deep as 4300 m. These results are surprising since they suggest that, magnetic bacteria commonly found in freshwater ponds, are much more widely distributed than previously thought. These results contribute to the understanding of the ecology and distribution of magnetic bacteria as well as the processes by which magnetite is formed in the deep-sea.

REFERENCES

Blakemore, R.P., Maratea, D. & Wolfe, R.S., 1979, *J. Bacteriol.*, **140**:720–729.

Burgess, J.G., Kawaguchi, R., Sakaguchi, T., Thornhill, R.H. & Matsunaga, T., 1993, *J. Bacteriol.*, **175**:6689–6694.

Frankel, R.B., Blakemore, R.P. & Wolfe, R.S., 1979, *Science*, **203**:1355–1356.

Herrick, J.B., Madsen, E.L., Batt, C.A. & Ghiorse, W.C., 1993, *Appl. Environ. Microbiol.*, **59**:687–694.

Kawaguchi, R., Burgess, J.G., Sakaguvhi, T., Takeyama, H., Thornhill, R.H., & Matsunaga, T.,1995, *FEMS Microbiol. Lett.*, **126**:277–282.

Kirschvink, J.L. & Chang, S.R., 1984, *Geology*, **12**:559–562.

Matsunaga, T. & Sakaguchi, T., 1992, Proc. the sixth Interntl. Conf. Ferrites.

Matsunaga, T., Nakamura, C., Burgess, J.G. & Sode, K., 1992, *J. Bacteriol.*, **174**:2748–2753.

Matsunaga, T., Sakaguchi, T. & Tadokoro, F., 1991, *Appl. Microbiol. Biotechnol.*, **35**:651–655.

Nakamura, C., Burgess, J.G., Sode, K. & Matsunaga, T.J., 1995, *J. Biol. Chem.*, **270**:28392–28396.

Pace, N.R., Stahl, D.A., Lane, D.J. & Olsen, G., 1986, *J. Ach. Microbiol. Ecol.*, **9**:1–55.

Petermann, H. & Bleil, U., 1993, *Earth Planet. Sci. Lett.*, **117**:223–228.

Petersen, N. & Dobeneck, T., 1986, *Nature*, **320**:611–615

Sakaguchi, T., Burgess, J.G. & Matsunaga, T., 1993, *Nature*, **365**:47–49.
Schleifer, K.H., Schuler, D., Sping, S., Weizenegger, M., Amann, R., Ludwig, W. & Kohler, M., 1991, *System. Appl. Microbiol.*, **14**:379–385.
Stolz, J.F., Chang, S.R. & Kirschvink, J.L., 1986, *Nature*, **321**:849–851.
Thornhill, R.H., Burgess, J.G. & Matsunaga, T., 1995, *Appl. Environ. Microbiol.*, **61**:495–500.
Vandamme, P., Pot, B., Gillis, M., Vos, P.D., Kersters, K. & Swings,J., 1996, *Microbiol. Rev*, **60**:407–438.
Woese, C.R., 1987, *Microbiol. Rev.*, **51**:221–271.

44

STRUCTURE AND REACTION MECHANISM OF THE β-GLYCOSIDASE FROM THE ARCHAEON *SULFOLOBUS SOLFATARICUS*

Marco Moracci,[1] Maria Ciaramella,[1] Laurence H. Pearl,[2] and Mosé Rossi[1,3]

[1]Institute of Protein Biochemistry and Enzymology
Naples, Italy
[2]Structural Biochemistry Section
Department of Biochemistry and Molecular Biology
University College London, United Kingdom
[3]Dipartimento di Chimica Organica e Biologica
University of Naples
Naples, Italy

Hyperthermophilic Bacteria and Archaea represent the organisms at the upper-temperature border of life. They have been almost exclusively isolated from environments with temperatures between 80 and 115°C, such as hot springs, solfatara fields, submarine hydrothermal areas or abyssal hot-vent systems (Brock, 1986). In particular, hyperthermophilic microorganisms living in marine abyssal biotopes grow optimally in presence of combined stress factors, for istance high pressure and temperatures above 100°C. Most of the microorganisms growing around 100°C belong to Archaea (formerly the archaebacteria) one of the three domains of the living world. On the basis of the recent hypothesis that life appeared at high temperatures, it has been proposed that archaeal hyperthermophiles could be the most 'primitive' living organisms and the closest to the universal ancestor.

The enzymes from hyperthermophiles are of large interest to understand the molecular origins of enzymes thermostability for both fundamental and applied research. The β-glycosidase from the hyperthermophilic Archaeon *Sulfolobus solfataricus* (Ssβ-gly) is a glycohydrolase with broad substrate specificity, true exo-glucosidase activity and able to promote transglycosylation (Trincone *et al.*, 1991; Nucci *et al.*, 1993). This enzyme shows optimal activity at temperatures increasing 85°C and no denaturation in the temperature range 30–100°C (Moracci *et al.*, 1995). For these reasons, we choosed Ssβ-gly as a model system to study catalysis at high temperatures and the molecular origins of thermostability.

On the basis of the amino acid sequence Ssβ-gly enzyme was placed in family 1 of the glycosyl hydrolases along with archaeal, bacterial and eukaryal enzymes. Ssβ-Gly has

New Developments in Marine Biotechnology, edited by Le Gal and Halvorson.
Plenum Press, New York, 1998.

been recently crystallized and the 3-D structure has been resolved at 2.6 Å (Pearl *et al.*, 1993; Aguilar *et al.*, 1997). Ssβ-Gly crystallyses as a tetramer with a dimer in the asymmetric unit, it has the classic (βa)8 barrel (TIM barrel) that has been also observed in the structure of two others family-1 glycosyl hydrolases.

Ssβ-Gly hydrolyses β-glycosides maintaining the anomeric configuration of the substrate, and for this reason was classified as a 'retaining' glycosyl hydrolase. Such enzymes operate by a two-step reaction involving a glycosyl-enzyme intermediate, supported by two acidic residues in the active site: one acting as general acid/base catalyst and the other as nucleophile. The inspection of the structure and the sequence alignments provide the molecular basis of the exo-glucosidase activity of Ssβ-gly and suggested that the residues Glu206 and Glu387 are good candidates for being the catalytic diad. An extended substrate binding tunnel (30 Å in length) run on the surface of the tetramer and can accomodate up to five β-1,4 linked sugar units. Furthermore, the glucosidic scissile bond is in close proximity to residues Glu206 and Glu387 which are 4.5 Å apart, a distance compatible with the retaining mechanism proposed for Ssβ-gly. In order to analyse the role played by these two residues in the reaction mechanism, we have mutated Glu206 and Glu387 to glutamine by site-directed mutagenesis (Moracci *et al.* 1996). Table 1 compares the specific activity of the wild type and the two mutants on several substrates at 65°C. The Glu387Gln mutant is totally inactive against all the substrates tested, as expected for the residue acting as the nucleophile of the reaction. The Glu206Gln mutant showed 10-fold and 60-fold reduced activity on aryl-galacto and glucosides, respectively, if compared to the wild type. Moreover, the mutant enzyme is completely inactive on disaccharides and lacks any transglycosylating activity (not shown).

Kinetic constants were determined for the Glu206Gln mutant at 65°C and compared with the values obtained for the wild-type (Table 2). The k_{cat}/Km ratio decreased to different extents using different substrates: whereas aryl-gluco and fucosides were little affected (1.6- and 6-fold reduction), a 14- and 33-fold reduction occurred for aryl-galactosides. Moreover, a significant Km decrease with 2/4-nitrophenyl-β-D-glucosides was observed. These results suggest that 2/4-nitrophenyl-β-D-gluco and galactoside substrates make different interactions with the Glu206 residue in the Ssβ-gly active site. This might reflect the different enantioselectivity toward secondary hydroxyl groups of 1,2-diols shown by Ssβ-gly in the transglycosylation reaction using phenyl-β-D-gluco and galactosides as glycoside donors, respectively. The residual activity of the Glu206Gln mutant on 4-nitrophenyl-β-D-galactoside substrate loosed the typical pH dependence shown by the wild type. These data sug-

Table 1. Specific activity at 65°C of wild-type and mutant Ssβ-gly

	Wild-type	Glu206 Gln	Glu387 Gln
4-NpGal	55.4	3.8	NA[a]
2-NpGal	60.0	6.8	NA
4-NpGlu	54.9	0.95	NA
2-NpGlu	48.1	0.93	NA
4-NpFuc	58.1	8.6	NA
4-Np-tio-Gal	NA	NA	NA
4-Np-tio-Glu	3.0	0.5	NA
Cellobiose	27.0	NA	NA
Laminaribiose	37.8	NA	NA
Gentiobiose	36.4	NA	NA

[a]NA (no measurable activity) means the rates of changes in absorbance did not vary by varying the substrate concentrations, and where approximately the same as in the control without substrate (typically <0.001 DA_{405}/min).

Table 2. Kinetic constants of hydrolysis at 65°C by wild-type and mutant Glu206 Gln

	Km (mM)		kcat (s^{-1})		kcat/Km (s^{-1} mM^{-1})	
Substrate	Wild type	Mutant	Wild type	Mutant	Wild type	Mutant
2-NpGal	0.94 ± 0.08	1.53 ± 0.2	296 ± 6	34 ± 0.9	309	21
4-NpGal	1.17 ± 0.06	4.80 ± 0.17	275 ± 5	31 ± 0.3	233	7
2-NpGlu	1.01 ± 0.23	0.03 ± 0.007	253 ± 12	4.46 ± 0.14	252	161
4-NpGlu	0.31 ± 0.04	0.03 ± 0.004	240 ± 6	4.40 ± 0.07	829	133
4-NpFuc	0.44 ± 0.11	0.09 ± 0.006	264 ± 12	35 ± 0.5	585	381

gest that Glu206 acts as the general acid/base catalyst in the hydrolysis reaction. However, the fact that the enzyme activity, although significantly reduced, was not completely abolished by the mutation Glu206Gln suggested that additional residue(s) could cooperate with Glu206 in supplying the acid/base catalist function. Alternatively, the residual activity could be related to the nature of the mutation: Gln residue could orient a water molecule such as it could promote the deglycosilation step of the reaction. Studies are in progress to clarify the role of Glu206 residue in the catalytic mechanism.

The resolution of the 3D structure of Ssβ-gly allowed the analysis of the possible mechanisms of stabilization. Among all the structural features which might contribute to thermostability only two stand out in Ssβ-gly to be probably significant: the content of ion-pair networks and of buried solvent molecules is higher than is usually observed in mesophilic enzymes. The active tetramer of Ssβ-gly contains 524 charged groups; tipically in mesophilic proteins, around one third of these groups are involved in ion-pair interactions. The cyanogenic β-glycosidase from clover, that is homologous to Ssβ-gly, has around 41% of its charged residues involved in ion-pairs, over 65% of which are isolated pairs. In contrast, in Ssβ-gly, a significant higher proportion (58%) of the charged groups are involved in ion-pair interactions and nearly 60% of these occurs as part of multiple ion-pair networks involving 3 or more charged centres. Ionic networks as stabilisation elements have been found so far only in thermophilic proteins, suggesting that it could represent a general strategy for the stabilisation of enzymes from hyperthermophiles (for a review see Goldman, 1995). In Ssβ-gly, the largest network is located at the tetrameric interface between the C-termini and involves 16 individual ion-pair interactions among the residues 464–489 of the enzyme.

Within the tetramer, more than 172 of the identified solvent molecules occupy internal cavities which are inaccessible to bulk solvent (one buried solvent molecule : 11.4 amino acid residues). This degree of solvent burial is significantly higher than observed in protein in general (1:27). Around 40% of the solvent molecules buried in the core of Ssβ-gly occur in clusters of two more. This finding has only been observed, among hyperthermophilic enzymes, in Ssβ-gly.

On the basis of the inspection of the Ssβ-gly 3D-structure we suggest that among the forces that stabilize proteins, ion-pair and buried solvent molecules could contribute to the thermal stability of this enzymes. However, the interpretation of these observations is difficult without the support of site directed mutagenesis and protein denaturation experiments. It is tempting to speculate that, during heat denaturation, polar interactions, hydrogen bonds and especially ion-pairs could maintain residual attraction between interacting structures even after relatively large displacements. In contrast, in case of the van der Waals interactions transient displacements breaks the interactions almost completely. Hence, in enzymes from hyperthermophiles that promote catalysis at high temperatures (>75°C), ion-pairs networks and buried solvent could be a general strategy of stabilization.

ACKNOWLEDGMENTS

This work was supported by CNR Target Project Biotechnology and Bioinstrumentation, by CNR Progetto Finalizzato Ingegneria Genetica, by EC project: "Biotechnology of extremophiles" contract n. BIO2-CT93-0274 and by EC FAIR project: "Enzymatic Lactose Valorization" contract n. 1048/96

REFERENCES

Aguilar, C., Sanderson, I., Moracci, M., Ciaramella, M., Nucci, R., Rossi, M. and Pearl, L.H., 1997,. *Journal of Molecular Biology,* **271**:789–802

Brock, T.D.,1986, *Thermophiles: General, Molecular and Applied Microbiology.* New York: John Wiley &Sons.

Goldman, A.,1995, *Structure,* **3**:1277–1279

Moracci, M., Nucci, R., Febbraio F., Vaccaro C., Vespa N., La Cara F. and Rossi M.,1995, *Enzyme and Microbial Technology,* **17**:992–997

Moracci, M., Capalbo, L., Ciaramella, M., and Rossi, 1996, *Protein Eng.* 9, 1191–1195

Nucci, R., Moracci, M., Vaccaro, C., Vespa, N. and Rossi, M. 1993, *Biotechnol. Appl. Biochem.* **17**:239–250

Pearl, L. H., Hemmings, A. M., Nucci R. and Rossi, M.1993, *J. Mol. Biol.* **229**:558–560

Trincone, A., Nicolaus, B., Lama, L., and Gambacorta, A., 1991, *J. Chem. Soc. Perkin Trans.*, **1**:2841–2844

45

IMMUNOLOGICAL INVESTIGATIONS ON ANTARCTIC FISH PARASITISM BY NEMATODES

Maria Rosaria Coscia and Umberto Oreste

Institute of Protein Biochemistry and Enzymology
CNR, Naples, Italy

The parasitism of fish by nematodes has been described in both temperate and polar seas (Orecchia & Paggi, 1978). The nematode species most frequently found infesting antarctic fish, are *Anisakis simplex* (Rudolphi, 1809; Nascetti et al., 1986), *Contracaecum osculatum* (Rudolphi, 1802; Koie & Fagerholm, 1995) and *Pseudoterranova decipiens* (Krabbe, 1878; Palm et al., 1994), all belonging to Anisakidae family. The nematode proteins, mostly including the components of the cuticle identified as Collagens, Cuticlins and Surface-associated proteins (Politz & Philipp, 1992), represent intense and persistent natural stimuli eliciting an immune response in fish.

A detailed examination of the nature of the antibody response and the properties of the immunoglobulins of fish leads to a number of generalisations, though several controversial or missing informations: upon immunisation, fish produce antibodies of moderate affinity and limited heterogeneity; there is a lack of the typical secondary antibody response seen in mammals; antibodies can be found not only in the blood but also in secretions, such as bile and skin mucus (Wilson & Warr, 1992).

The predominant immunoglobulin (Ig) of fish is considered to be an IgM-like molecule arranged in a basic unit consisting of ~70 kDa heavy (H) chains and ~22–26 kDa light (L) chains in equimolar amounts. Fish Ig are present in polymeric forms, (tetramer $(H_2L_2)_4$ in teleosts and pentamer $(H_2L_2)_5$ in cartilaginous fish), or in monomeric forms.

Fish immune response against parasite has been poorly investigated (Buchmann et al., 1992; Buchmann, 1993). We have sought to address this issue by investigating the molecular bases of the parasitism of antarctic fish by nematodes, as this is a valuable model for studying the humoral response at very low temperature (−1.9°C). Fish are ectothermic vertebrates whose physiological functions, including the immunological processes, are dependent on the ambient temperature. It is known that cold temperatures depress life functions, e.g. the proliferative response of lymphocytes of *Tinca tinca* is lower at 12°C than that at 22°C (Collazos et al., 1995).

We isolated and characterised Ig from plasma and bile of *Trematomus bernacchii* and *Nothotenia coriiceps*, and from cutaneous mucus of *Nothotenia gibberifrons* (Coscia

New Developments in Marine Biotechnology, edited by Le Gal and Halvorson.
Plenum Press, New York, 1998.

& Oreste, 1997a,b). Thiophilic adsorption chromatography on T-gel and size-exclusion FPLC on Superose 6 column were used in the purification of Ig.

To investigate the antibody specificity of the purified Ig, nematodes were collected from the liver of specimens of different teleostean species: those recovered from *N. coriiceps*, *C. aceratus*, *C. rastrospinosus*, *P. georgianus* and *C. antarcticus* were identified as third stage larvae of *Pseudoterranova decipiens*, whereas those from *C. hamatus* as third stage larvae of *Contracaecum osculatum*. Both the nematode species belong to the superfamily of Ascaridoidea, family Anisakidae, subfamily Anisakinae.

The ability of antarctic fish plasma Ig to recognise the nematode proteins was tested. The binding of *T. bernacchii* Ig to *P. decipiens* surface associated proteins (*Pd*AP) was found, by ELISA, higher than that to *P. decipiens* somatic proteins (*Pd*SP). Consistent results were obtained by affinity chromatography when *Pd*AP and *Pd*SP were eluted through a *N. coriiceps* Ig coupled column: the fraction of *Pd*AP retained by fish antibodies was more abundant than that of *Pd*SP, due to the presence of antigens at higher concentration in *Pd*AP or antibodies with higher affinity to these antigens (Coscia & Oreste, 1997a).

To compare the *Pd*SP binding activity of antibodies from plasma and secretions, the same amount of *T. bernacchii* plasma and bile Ig, and of *N. gibberifrons* cutaneous mucus Ig was tested by dot-blot immunoassay: in autoradiography, the spot incubated with bile Ig gave a signal clearly higher than that with plasma Ig; in addition, the skin Ig gave a detectable signal (Coscia & Oreste, 1997b). These results are compatible with the conclusion that a mucosal immune response against parasites exists in antarctic fish.

To identify the antigenic components, nematode proteins were separated by SDS-PAGE, blotted onto nitrocellulose sheet and then incubated with *T. bernacchii* plasma Ig; bound antibodies were revealed by ^{125}I-rabbit IgG (anti-*T. bernacchii* Ig H chain). Five out

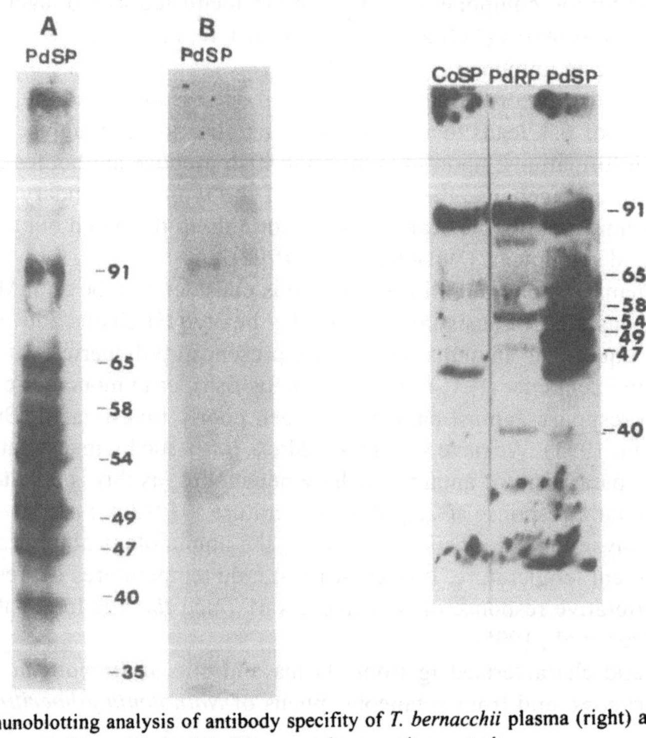

Figure 1. Immunoblotting analysis of antibody specifity of *T. bernacchii* plasma (right) and bile (left, A) Ig for nematode antigens. *D. labrax* bile Ig (left, B) was used as negative control.

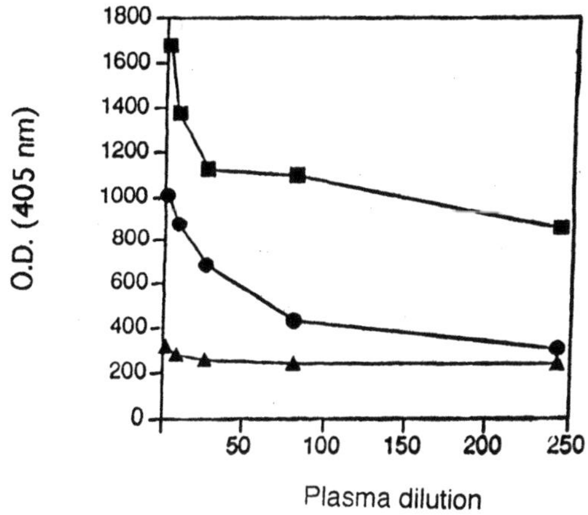

Figure 2. ELISA analysis of the temperature dependence of the interaction between *Pd*SP antigens and *T. bernacchii* antibodies at +20°C (■-■), +4°C (O-O) and −2°C (Δ-Δ).

of 6 *P. decipiens* proteins secreted/excreted by live nematodes (*Pd*RP) and only 7 out of 26 *Pd*SP components were identified as antigens by autoradiography, indicating that *Pd*RP is more antigenic; at least 3 antigens of *C. osculatum* somatic proteins, two of them in common with *Pd*SP, were recognized by plasma Ig. A higher number (thirteen) of *Pd*SP bands was found positive when *T. bernacchii* bile Ig was used as first antibody, whereas no bands were detected when *Dicentrarchus labrax* bile Ig was used as negative control (Fig. 1). These findings confirm the presence of a mucosal immune response and suggest differences in specificity between plasma and mucosal Ig.

Moreover, *Pd*SP specific antibodies were purified from *N. coriiceps* plasma by affinity chromatography with *Pd*SP coupled to CNBr-activated Sepharose 4B; the retained fraction, run in SDS-PAGE under reducing conditions, gave the typical Ig pattern. The isoelectrofocusing pattern of the purified specific antibodies consisted of at least 13 focused bands, giving a measure of the antibody diversity.

To investigate the antigen/antibody interaction and its temperature dependence, ELISA was performed at −2°C, +4°C and +20°C (Fig. 2). *D. labrax* plasma was used as negative control. The antibody recognition, though lower, was detected also at −2°C, indicating that *T. bernacchii* Ig can carry out its function at the temperature of the aquatic environment.

Preliminary investigations were done on the prevalence of the immune response to nematode parasites in different species of antarctic fish: plasma of individual fish of four species were tested for the presence of anti-*C. osculatum* antibodies in double immunodiffusion assay. Immunoprecipitates were observed with 2 out of 10 *T. bernacchii* plasma, 2 out of 4 *N. gibberifrons* plasma, 3 out of 3 *N. coriiceps* plasma, 5 out of 6 *C. hamatus* plasma and none of 2 *P. georgianus* plasma. However, the paucity of the data collected so far, does not allow to evaluate the susceptibility to the parasitic infestation among species.

In conclusion, we demonstrated the presence, in plasma and secretions of infested fish, of antibodies specific for nematode antigens and in particular, of bile antibodies with a relatively high affinity and heterogeneity, compared with that of plasma antibodies.

ACKNOWLEDGMENTS

This research was supported by a Grant from the National Antarctic Project. We thank Dr. G. di Prisco, Institute of Protein Biochemistry and Enzymology, CNR, Naples, for providing us with nematodes and fish plasma and secretions; Dr. S. Mattiucci, Institute of Parasitology, University "La Sapienza", Rome, for the identification of nematode larvae.

REFERENCES

Buchmann, K., 1993, *Fish & Shellfish Immunol* **3**: 397–399
Buchmann, K., Østergaard, L., Glamann, J., 1992, *Scand J Immunol* **36**: 89–97
Collazos, M.E., Ortega, E., Barriga, C., 1995, *Comp Immunol, Microbiol and infectious diseases* **18**: 209–214
Coscia, M.R.,, Oreste, U., 1997a, Proceedings of the third meeting on "Antarctic Biology" Camerino University Press, pp 61–67.
Coscia, M.R., Oreste, U., 1997b, Fishes of Antartica. A Biological overview. Springer-Verlag Publishers. In press.
Koie, M., Fagerholm, H.P., 1995, *Parasitol Res* **81**: 481–489
Nascetti, G., Paggi, L., Orecchia, P., Smith, J.W., Mattiucci, Bullini, L., 1986, *Int J Parasitol* **16**: 633–640
Orecchia, P., Paggi, L., 1978, *Parassitologia* **20**:73–89
Palm, H., Andersen, K., Klöser, Plötz, J., 1994, *Polar Biol* **14**: 539–544
Politz, S.M., Philipp, M., 1992, *Parasitology Today* **8**: 6–12
Wilson, M.R., Warr, G.W., 1992, *Annual Rev of Fish Diseases* 201–221

THE IDENTIFICATION AND CHARACTERISATION OF *GRACILARIA GRACILIS* DEFENCE GENES EXPRESSED IN RESPONSE TO A BACTERIAL INFECTION

Ann E. Jaffray and Vernon E. Coyne

Department of Microbiology
University of Cape Town
Private Bag, Rondebosch 7701, South Africa

The red alga *Gracilaria gracilis* is a commercially important source of food-grade agar world-wide (Armisen, 1995). In South Africa, it is harvested on a commercial basis from natural populations occurring in Saldanha Bay on the west coast of South Africa (Anderson *et al.*, 1989), and is also cultivated commercially on rafts in this area.

G. gracilis is colonised by large populations of agarolytic bacterial epiphytes, many of which are pathogenic to the alga (Jaffray and Coyne, 1996). It is thought that bacteria may have played a role in the total collapse of the natural population of *G. gracilis* in 1989, and in subsequent partial die-offs (*e.g.*, 1993) which have occurred predominantly during the summer months. The alga showed symptoms of bleaching and lesions (Anderson, pers. comm.). It is thought that under conditions of stress such as elevated temperatures and nutrient depletion in the surrounding seawater, bacteria exploit the agar comprising the cell wall of the algal thallus as a nutrient source (Jaffray *et al.*, in press).

Terrestrial plants produce one or more proteins as part of the hypersensitive response to infection by pathogenic micro-organisms (van Loon, 1985; Stinzi *et al.*, 1993). These proteins are called pathogenesis-related (PR) proteins and are accompanied by symptom development on the plant. The function of PR proteins is to protect the plant from sustaining extensive damage when infected by microbial pathogens. Many PR proteins are enzymes or compounds involved in antimicrobial activity, such as glucanases, chitinases, phytoalexins, proteinases, and phenolics.

In our study, a previously-isolated agarolytic, bacterial pathogen of *G. gracilis* (designated strain LS2i), was used to investigate the algal response to infection. This strain induces bleaching of the algal thalli and subsequent thallus degradation. Algal thalli were

New Developments in Marine Biotechnology, edited by Le Gal and Halvorson.
Plenum Press, New York, 1998.

either injected with seawater (healthy control), or the agarolytic bacterial strain (infected). Following incubation at 30°C for 5 days, total algal proteins were extracted and analysed by SDS polyacrylamide gel electrophoresis. Comparisons between healthy and unhealthy *G. gracilis* total protein extracts showed the presence of at least 3 differentially synthesised proteins, which ranged in size from 30kD to 60kD. It is possible that these are PR proteins of *G. gracilis*, as they were produced in algal thalli infected with LS2i, but not by healthy algae. Thus, it seems that *G. gracilis* expresses PR proteins in response to infection by a pathogenic bacterium.

A cDNA-AFLP protocol (Vos *et al.*, 1995) was used to isolate and characterise differentially expressed genes in healthy and unhealthy *G. gracilis*. Messenger RNA extracted from healthy and LS2i-infected *G. gracilis* was reverse-transcribed to cDNA. The DNA was digested with a pair of restriction enzymes (a rare and a frequent cutter) and the fragments ligated to adaptors of known sequence. These fragments were then pre-amplified by polymerase chain reaction (PCR) under low stringency conditions using primers complentary to the adaptor sequences. The resulting products were re-amplified under high stringency conditions by PCR using radioactively-labelled primers with a two-base pair degeneracy at their 3′ ends. The PCR products were resolved by polyacrylamide gel electrophoresis. DNA fragments of up to 500 base pairs in size were detected by autoradiography. In addition, the banding pattern of DNA fragments obtained from healthy *G. gracilis* differed to that obtained from infected *G. gracilis*. Additional DNA fragments were apparent in lanes containing the PCR products from infected *G. gracilis*, suggesting that differential gene expression occurs in response to bacterial infection.

DNA fragments representing nine differentially expressed genes were identified which ranged in size from 100 to 450 base pairs. These bands were excised from the polyacrylamide gel and re-amplified by PCR using primers from the pre-amplification reaction. Of the nine original fragments, five were labelled with [a-^{32}P dCTP] and used to probe genomic *G. gracilis* DNA in order to verify that they originated from *G. gracilis*. Three fragments showed homology to total genomic *G. gracilis* DNA. Similarly, mRNA was isolated from both healthy and infected *G. gracilis* and probed with the same labelled DNA fragments in order to confirm that the DNA fragments represented genes which were expressed solely in infected algal thalli. Two of the fragments hybridised with mRNA from the infected alga, indicating that their expression is induced by the presence of the bacterial pathogen.

Five DNA fragments were cloned into the pBluescript SK (Stratagene) cloning vector and their nucleotide sequences determined. Sequence homology searches of the Genbank database using the BlastX protocol, showed that three of the *G. gracilis* cDNA fragments possessed amino acid sequence similarities to particular enzymes isolated from terrestrial plants. For example, cDNA fragment 2 has a 56% amino acid sequence similarity to 3-hydroxy-3-methylglutaryl coenzyme A reductase from *Solanum tuberosum*. This enzyme is important in isoprenoid biosynthesis regulation. Antimicrobial phytoalexins, are produced in response to pathogenic infection, are a product of this isoprenoid biosynthesis pathway (Korth *et al.*, 1997).

In conclusion, it has been shown that *G. gracilis* responds to a bacterial infection by producing PR proteins. Pathogenesis-related protein expression in *G. gracilis* seems to be regulated at the level of transcription as indicated by the presence and absence of mRNA in infected and healthy algae, respectively. From the results of this study, it seems that *G. gracilis* synthesises a variety of enzymes which may serve as a defence mechanism against infection by bacterial pathogens. Future work will seek to characterise this system in this commercially important seaweed.

REFERENCES

Anderson, R. J., Simons, R. H. and Jarman, N. G. 1989,. *S. Afri. J. Mar. Sci.* **8**: 277–289.

Armisen, R. 1995, *J. Appl. Phycol.* **7**: 231–243.

Jaffray, A. E. and Coyne, V. E. 1996, *J. Appl. Phycol.* **8**: 409–414.

Jaffray, A. E., Coyne, V. E. and Anderson, R. J. *Bot. Mar.* (In press).

Korth, K. L., Stermer, B. A., Bhattacharyya, M. K. and Dixon, R. A. 1997*Plant Mol. Biol.* **33**: 545–551.

Stinzi, A., Heitz, T., Prasad, V., Wiedemann-Merdinoglu, S., Kauffman, S., Geoffroy, P., Legrand, M. and Fritig, B. 1993*Biochemie* **75**:687–706.

Van Loon, L. C. 1985, *Plant Mol. Biol.* **4**: 111–116.

Vos, P., Hogers, R., Bleeker, M., Reijans, M., van de Lie, T., Vornes, M., Frijters, A., Pot, J., Peleman, J., Kuiper, M. and Zabeau, M. 1995,. *NAR.* **23**: 4407–4414.

47

IMPROVING ENZYME THERMOSTABILITY

The *Thermococcus litoralis* Glutamate Dehydrogenase Model

Costantino Vetriani,[1] Dennis L. Maeder,[1,2] Nicola J. Tolliday,[1]
Horst H. Klump,[2] Kitty S. P. Yip,[3] David W. Rice,[3] and Frank T. Robb[1]

[1]Center of Marine Biotechnology
Columbus Center
University of Maryland Biotechnology Institute
Baltimore, Maryland
[2]Department of Biochemistry
University of Cape Town
Rondebosch, South Africa 7700
[3]Krebs Institute
University of Sheffield
Sheffield, United Kingdom

1. INTRODUCTION

The identification of the structural determinants that confer thermostability to a protein remains a major unsolved problem in molecular biology. Trends commonly associated with elevated thermostability in proteins include relatively small solvent-exposed surface area (Chan et al., 1995), decreased length of surface loops (Russel et al., 1994), increased packing density that reduces cavities in the hydrophobic core (Russel et al., 1994; Britton et al., 1995), optimization of hydrophobic interactions (Spassov et al., 1995), extended ion-pair networks (Yip et al., 1995; Chan et al., 1995; Korndörfer et al., 1995; Henning et al., 1995) and hydrogen bonds between charged-neutral residues (Tanner et al., 1995). Overall these characteristics lead to decreased flexibility of the polypeptide chain, a required feature that compensates for increased thermal fluctuations at high temperatures (Vihinen, 1987).

In this work we have selected the GluDH from *T. litoralis* as a model experimental system to investigate the role of intersubunit ionic interactions in protein thermostability at extremely high temperatures. We have cloned, sequenced and expressed the extremely thermostable GluDHs from the closely related hyperthermophilic archaea *Pyrococcus furiosus* (T_{opt} = 100°C) (Di Ruggiero & Robb, 1995) and *Thermococcus litoralis* (T_{opt} = 88°C) (Di Ruggiero et al., submitted) and have found that the stability of the recombinant

New Developments in Marine Biotechnology, edited by Le Gal and Halvorson.
Plenum Press, New York, 1998.

GluDHs reflects the optimal growth temperatures of the source organisms. Differential scanning microcalorimetry (DSC) was used to establish the thermodynamic properties of these enzymes which have melting temperatures (Tm) of 114.5°C and 109°C respectively (Di Ruggiero et al., 1997; Klump et al., 1992).

The GluDH crystal structures from *P. furiosus* (Yip et al., 1995) and *T. litoralis* (Sedelinkova et al., 1996) have been recently solved. The monomeric form of the GluDHs is a polypeptide of 420 amino acids. Three monomers self assemble into a trimer with a three-fold rotational axis of symmetry. Two trimers assemble into a hexamer with a dyad axis of symmetry to form the native enzyme. The comparison of the *P. furiosus* three-dimensional structure to that from a mesophilic organism, *Clostridium symbiosum*, has revealed one major difference: the structure of the hyperthermophilic enzyme contains a striking series of ion-pair networks on the surface of the protein subunits and partially buried at both the inter-domain and inter-subunit interfaces (Yip et al., 1995). Further analyses of a range of GluDH structures obtained from species which span a wide range of optimal growth temperatures have revealed that the number of ion-pair networks decreases as we move down the temperature spectrum towards the less thermostable enzymes (K. Yip, unpublished results). This observations have led to the suggestion that ionic interactions may play a critical role in decreasing the flexibility of the polypeptide chain and maintaining enzyme stability at high temperatures.

To date the evidence linking protein thermostability to reduced flexibility is based on comparative sequence and structural data but not on predictive modeling and site-directed mutagenesis. By analyzing the structures of GluDH from *P. furiosus* (Yip et al., 1995) and *T. litoralis* (K. Yip, unpublished results), we have been able to identify structural features that accompany extreme thermostability. We have predicted stabilizing mutations in the recombinant *T. litoralis* GluDH and we have used site-directed mutagenesis to test these predictions.

2. RESULTS

Identification of target residues for site-directed mutagenesis by sequence comparison and predictive modeling. We have undertaken a detailed examination of both the amino acid sequences and the three-dimensional structures of the GluDHs from *P. furiosus* and *T. litoralis* in order to identify the key structural determinants that play a crucial role in the stabilization of these enzymes at high temperatures. The sequences have 87.4% identity and can therefore be unambiguously aligned allowing the structure of *P. furiosus* GluDH to be used as a template to design mutations for the *T. litoralis* structure.

One of the prospective sites revealed by our comparative analysis of the GluDHs from *P. furiosus* and *T. litoralis* is the glutamic acid at position 138 in the GluDH from *P. furiosus*, which is part of a six residues network of ion pairs involved in intersubunit interactions (Yip et al., 1995). The conserved basic string Arg164–Arg165–Lys166 is part of the intersubunit ionic association with Glu138. Arg35 is also part of the network, making an intrasubunit contact with Glu138; Asp132, located on the second trimer, contributes to the ion pair network on assembly of the two trimers into the hexamer. With the exception of Glu138, which is replaced by a threonine in the GluDH from *T. litoralis*, all the residues in this cluster are conserved between the two structures. Our working hypothesis was that the lack of a charged side-chain at position 138 in *T. litoralis* would lead to the loss of the six residues ion-pair network and therefore affect the stability of the enzyme at high temperature. We speculated that the replacement of the threonine at position 138 with a glutamic acid residue would make the cluster identical to the *P. furiosus* enzyme and restore the network of ionic interactions.

The second prospective sites revealed by our comparative analysis of the two GluDHs is Thr167 in the *P. furiosus* GluDH. Thr 167 is located at the contact surface between subunits and in close proximity to the six residues cluster across the two-fold axis of the molecule. Thr167 is substituted by an aspartic acid in *T. litoralis*, thereby maintaining the charge balance across the two-fold axis. It is possible that Asp167 is ionically associated with the adjacent basic cluster (164–166). However the interaction between Asp167 and the basic cluster is intrasubunit rather than intersubunit, and there may be unfavorable ionic interactions across the 2-fold axis.

We constructed mutants bearing the single Thr138–Glu and the double Thr138–Glu/Asp167–Thr substitutions. We overexpressed the mutant *gdhA* genes in *E. coli*, purified the recombinant GluDHs and used DSC and thermoinactivation experiments to assess their behavior at different temperatures.

Characterization of T. litoralis GluDH mutants: thermoinactivation and DSC. The specific activity at 80°C of the two mutant GluDHs (150 U/mg) was equivalent to that of the recombinant wt *T. litoralis* enzyme (Korndörfer et al., 1995). Thermostability at 104°C is consistent with the DSC data: the double mutant is four-fold more stable than the wild type while *P. furiosus* GluDH is sixteen-fold more stable than wild type (Table 1).

Thermal denaturation of the T138E mutant produced a gradual rise in the heat capacity with a maximum at 103.5°C (Fig. 1). This behavior suggests the formation of several folding intermediates of the enzyme during exposure to increasing temperature. The asymmetry of the transition for all samples suggests a cooperative destabilization of the assembly, the small negative ΔCp suggests aggregation of the denatured protein. Re-scanning samples yielded a stable baseline indicating an irreversible process.

3. DISCUSSION AND CONCLUSIONS

In this study we have defined the minimal mutations necessary to elevate the thermostability of the GluDH from *T. litoralis*. Our working hypothesis was based on the proposed critical role of ionic interactions in the stabilization of enzymes at high temperatures (Yip et al., 1995; Chan et al., 1995; Korndörfer et al., 1995; Henning et al., 1995).

The replacement of a Thr by an ion-pairing Glu at position 138 resulted in a severe loss of thermostability. Although the single threonine to glutamic acid substitution at position 138 supposedly restored the six residues ion pair network in the thermococcal mutant enzyme, it also altered the net charge at the subunit interface, due to the presence in *T. litoralis* of the extra charge of Asp167. We speculated that the retainment of the local net charge across the two-fold axis of the molecule was a necessary structural requirement for

Table 1. Thermostability and T_m of native *P. furiosus* and recombinant wt and mutant *T. litoralis* GluDHs. Half-life measurements were conducted at 98°C and 104°C. The double mutant T138E/D167T is approximately 4-fold more stable than wild type

GluDH	$t_{1/2}$ at 98°C	$t_{1/2}$ at 104°C	T_m
P. furiosus native	6.7 h	4.6 h	114.5°C
T. litoralis recombinant T138-E/D167T	6.0 h	1.1 h	111.5°C
T. litoralis recombinant wt	6.1 h	0.3 h	109°C
T. litoralis recombinant T138-E	0.6 h	N.D.	103.5°C

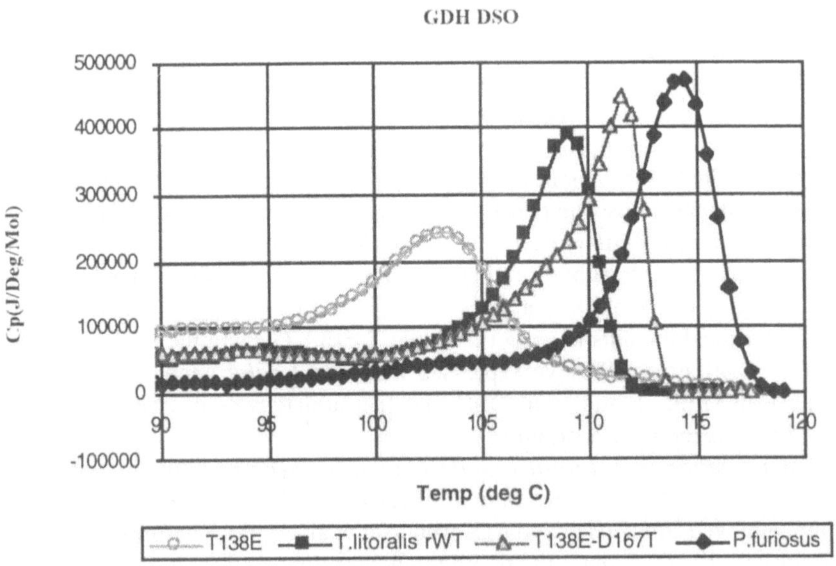

Figure 1. Relative heat capacity of GluDHs between 90°C and 120°C. The thermal stability of the double mutant T138E/D167T is elevated relative to *T. litoralis* wild type and approaches native *P. furiosus* enzyme. The single mutant T138E is relatively unstable.

the stability of the thermococcal GluDH. In the double Thr138–Glu/Asp167–Thr mutant we removed the extra charge, thus maintaining a local net charge identical to *P. furiosus* GluDH. The Thr138–Glu/Asp167–Thr mutant ($t_{1/2}$ = 1.1 h) showed a four-fold increase in the half-life at 104°C over the wt recombinant GluDH and an upward shift in the T_m from 109°C (wt) to 111.5°C (Thr138–Glu/Asp167–Thr mutant) (Table 1).

In this work we have shown experimentally that protein thermostability can be dramatically enhanced with minimal changes at key positions in the structure. Our findings are consistent with the observation that in a given protein the increase in stability commonly originates from small mutational changes, whose excess free energy of stabilization is equivalent to only a few weak bonds (DDG_{stab} = 50 kJ/mol) (Jaenicke, 1996) and that a single salt bridge can give a contribution of 3–5 kcal/mol to the free energy of folding (Anderson et al., 1990). This suggests a general strategy for elevation of thermostability by means of enhanced ion-pairing interactions combined with the removal of local repulsive electrostatic forces.

REFERENCES

Anderson, D.E., Becktel, W.J., & Dahlquist, F.W. , 1990, *Biochemistry* **29:** 2403–2408.

Britton, K.L., Baker, P.J., Borges, K.M.M., Engel, P.C., Pasquo, A., Rice, D.W., Robb, F.T., Scandurra, R., Stillman, T.J., & Yip, K.S.P. , 1995, *Eur. J. Biochem.* **229**: 688–695.

Chan, M.K., Mukund, S., Kletzin, Adams, M.W.W., & Rees, D.C. ,1995, *Science* **267:**1463–1469.

Di Ruggiero, J., & Robb, F.T. , 1995, *Appl. Environ. Microbiol.*, **61:**159–164.

Di Ruggiero, J., Tolliday, N., Borges, K.M., Veillerot, E., Vetriani, C., & Robb, F. T. Submitted to *Extremophiles*.

Henning, M., Darimont, B. Sterner, R., Kirshner, K., & Jansonius, J.N., 1995, *Structure* **3:**1295–1306.

Jaenicke, R. , 1996, *FEMS Microbiol. Rev.* **18:** 215–224.

Klump, H., Di Ruggiero, J., Kessel, M., Park, J.-B., Adams, M.W.W., & Robb, F.T. ,1992, *J. Biol. Chem.* **267:**22681–22685.

Korndörfer, I., Steipe, B., Huber, R., Tomschy, A., & Jaenicke, R. , 1995, *J. Mol. Biol.* **246:** 511–521.
Russel, R.J.M., Hough, D.W., Danson, M.J., & Garry, M.T. , 1994, *Structure* **2:**1157–1167.
Sedelinkova, S.E., Yip, K.S.P., Stillman, T.J., Ma, K., Adams, M.W.W., & Robb, F.T., 1996, *Acta Cryst.* **D52:**1185–1187.
Spassov, V.Z., Karshikoff, A.D., & Ladenstein, R. , 1995, *Protein Sci.* **4 :** 1516–1527.
Tanner, J.J., Hecht, R.M., & Krause, K.L. , 1996, *Biochemistry* **35:** 2597–2609.
Vihinen, M. , 1987, *Protein Eng.* **1:** 477–480.
Yip, K.S.P., Stillman, T. J., Britton, K.L., Artymiuk, P.J., Baker, P. J., Sedelnikova, S. E., Engel, P. C., Pasquo, A., Chiaraluce, R., Consalvi, V., Scandurra, R., & Rice, D. W. , 1995, *Structure* **3:**1147–1158.

48

LIGAND-ACTIVATED Ca^{2+} CHANNELS IN THE NUCLEAR ENVELOPE OF STARFISH OOCYTES

Luigia Santella[1] and Keiichiro Kyozuka[2]

[1]Laboratory of Cell Biology
Stazione Zoologica "A. Dohrn"
Naples, Italy
[2]Marine Biological Station of Asamushi
Tohoku University
Aomori, Japan

1. INTRODUCTION

The regulation of the calcium pool in the cell nucleus is a controversial issue. The presence in the envelope of large pores has often been taken as an indication for the free and unhindered traffic of Ca^{2+} between the cytoplasm and the nucles. Accordingly, experiments in a number of Laboratories, especially with nuclear targeted fluorescent indicators (Brini et al., 1993; Allbritton et al., 1994) have detected insignificant delays in the transmission of cytosolic Ca^{2+} waves to the cytoplasm. Others, however, have measured significant delays even using essentially the same methods (Badminton et al., 1995). In addition, patch-clamp work on nuclear envelopes have shown giga-ohm resistance, and single channel activity, despite the presence of dozens of pores in the patch (Mazzanti et al., 1990; Dale et al., 1994; Santella & Kyozuka, 1994; Santella, 1996). Very likely, then, the nuclear pores exist in both Ca^{2+} impermeant and Ca^{2+} permeant states: a mechanism for the gating on the pores based on the state of depletion of the envelope Ca^{2+} store has been recently proposed (Greber & Gerace, 1995; Stehno-Bittel et al., 1995).

Irrespective of the ease with which Ca^{2+} is transported through the pores, Ca^{2+} channels modulated by the ligands inositol 1,4,5-trisphosphate ($InsP_3$) (Nicotera et al., 1990) and cyclic ADP-ribose cADPr) (Gerasimenko et al., 1995) have been shown to be active in the envelope by experiments on isolated nuclei. Taken together with the demonstration that the machinery for the synthesis of $InsP_3$ exists in the nuclei (Divecha et al., 1991), and with the finding of a SERCA-type Ca^{2+} ATPase in the nuclear envelope (Nicotera et al., 1989; Lanini et al., 1992) these results argue in favour of the signal-mediated, independent regulation of the nuclear Ca^{2+} pool. One piece of information which is still missing is the location of the ligand-modulated Ca^{2+} channels in the envelope, i.e., are they activated by their ligands

New Developments in Marine Biotechnology, edited by Le Gal and Halvorson.
Plenum Press, New York, 1998.

from inside the nucleus, or do the ligands act on the external side to liberate Ca^{2+} in the spaces surrounding the nucleus, followed by its transport to the nucleoplasm across the pores? This succint report decribes experiments on starfish oocytes that strongly support the first alternative. More comprehensive discussion of the problem of Ca^{2+} regulation in the nucleus can be found in a number of recent reviews (e.g., Santella & Carafoli, 1997).

2. MATERIALS AND METHODS

2.1. Preparation of Gametes

The starfish *Astropecten auranciacus* used in most experiments were collected in the Bay of Naples in February-June and kept in running sea water. Prophase-I arrested oocytes, dissected from the ovaries in natural filtered sea water (NSW), were washed two to three times with NSW and kept in it for one hour prior to use. Oocytes in which the GVBD spontaneously occurred during this period were discarded; in the others the GVBD was promoted by the addition of 1-MA (10 μM).

2.2. Microinjection of Nuclei

$InsP_3$, cADPr (the agonist of the cADPr-sensitive channel), ryanodine or ruthenium red (which inhibits the same cADPr-sensitive channel) were injected into the germinal vesicles of prophase arrested oocytes to test their effect on the GVBD. The glass needles used for the injection (tip, 1 μm diameter) were filled with 100 mM sodium aspartate, 10 mM HEPES, pH 7.0, and either 250 μM $InsP_3$, 2 mM ryanodine, 250 μM cADPr, or 10 mM ruthenium red. The oocytes were put in a holder with a small amount of sea water and their nuclei (or cytoplasm) were impaled with an Eppendorf Transjector 5246 pressure injector (Eppendorf, Hamburg, Germany).

2.3. Measurement of Nuclear Calcium

A computer controlled photomultiplier system (CRS-400, Biorad, Hercules, CA) equipped with a 75W Xenon lamp with 490 nm excitation and 530 nm emission band pass filters was used. The beam of the lamp had a diameter of 50 μm and could thus be easily focused only on the nucleus, which in these oocytes has a diameter of 60 μm. The emission filter was removed automatically for 15 sec to irradiate the nucleus with ultraviolet light and photorelease the effectors, and the Ca^{2+} changes were then followed for the times shown in the Figure 1.

3. RESULTS AND DISCUSSION

3.1. Nuclear Ca^{2+} Changes Induced by the Photoreleasing of Caged $InsP_3$ or Caged cADPr

The possibility that uncaged $InsP_3$ diffused out of the nucleus had promoted the emptying of the cytoplasmic stores, followed by the entry of the released Ca^{2+} into the nucleus, was ruled out by injecting heparin, an inhibitor of the $InsP_3$ receptor, into the cytoplasm prior to the introduction of caged $InsP_3$ into the nucleus. The uncaging reaction

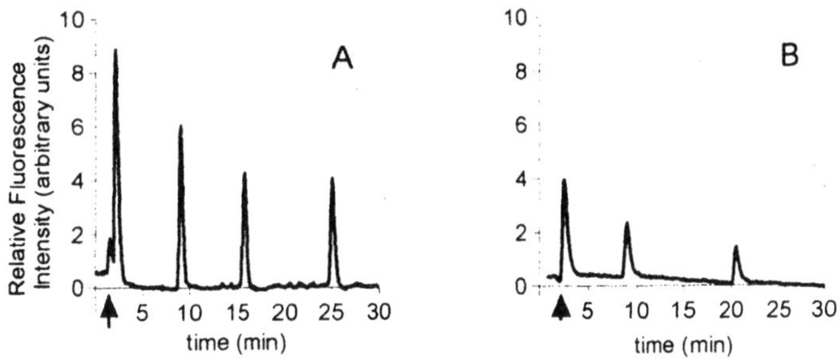

Figure 1. Ca^{2+} oscillations in the nucleus after photoreleasing cADPr. **A:** The height of the repetitive spikes evoked in the nucleus by the uncaging of cADPr (arrow) decreased progressively, possibly due to the time-dependent inactivation (degradation) of the uncaged cADPr. The injection of ruthenium red into the cytoplasm to block cADPr-sensitive channels before irradiating the nucleus **B:** had no effect on the repetitive character of the nuclear Ca^{2+} spiking.

induced nearly the same nuclear Ca^{2+} response, i.e., $InsP_3$ had acted on Ca^{2+} channels of the envelope that face the nucleoplasm. The introduction of heparin or of antibodies against $InsP_3$ receptors into the nucleus together with caged $InsP_3$ blocked the Ca^{2+} transient induced by the irradiation.

To establish whether cADPr receptors (channels) were also present and operative in the nucleus, caged cADPr was injected into the nucleus together with calcium green dextran. Uncaging of cADPr produced Ca^{2+} spike that reached a maximal height of about 14 arbitrary units in a few sec and decayed as rapidly as in the case of $InsP_3$. A control was performed by injecting the antagonist ruthenium red into the cytoplasm prior to introducing caged cADPr into the nucleus. The irradiation still induced a nuclear calcium transient, although in this case its height was lower, presumably, some ruthenium red (final concentration of 100 μM) diffused from the cytoplasm into the nucleus. Injecting the specific antagonist of cADPr, 8-amino-cADPr or injecting antibodies against the cADPr receptors into the nucleus eliminated the Ca^{2+} spike induced by the uncaging of cADPr.

3.2. Repetitive Nuclear Ca^{2+} Transients Induced by the Photoreleasing of Caged cADPr

Interestingly, in most of the experiments with cADPr, the spiking induced by its uncaging has repetitive character (Fig. *1A*), the amplitude of the spikes decreasing progressively with a period varying between 5 and 10 min. Repetitive transients were never observed by uncaging nuclear $InsP_3$.

In conclusion, the experiments shown in this contribution have demonstrated that $InsP_3$ and cADPr-sensitive Ca^{2+} channels are present on the inner membrane of the nuclear envelope of starfish oocytes and could be induced to liberate Ca^{2+} from the envelope store from within the nucleus. It will be now important to identify signaling routes that transmit the information delivered by first messengers to the plasma membrane receptors to the nuclear compartment.

REFERENCES

Allbritton, N., E. Oancea, M. A. Kuhn, and T. Meyer. 1994, *Proc Natl Acad Sci USA* **91:** 12458–12462.
Badminton, M. N, J. M. Kendall, G. Sala-Newby, and A. K. Campbell. 1995, *Exp Cell Res.* **216:** 236–243.

Brini, M., M. Murgia, L. Pasti, D. Picard, T. Pozzan, and R. Rizzuro.1993, *Embo J.* **12:** 4813–4819.
Dale, B, L. J. DeFelice, K. Kyozuka, L. Santella, and E. Tosti. 1994, *Proc . Soc London (B),* **255:** 119–124.
Divecha, N., H. Banfic, and R. F. Irvine. 1991, *Cell* 1991, **74:** 405–407.
Gerasimenko, O. V., J. V. Gerasimenko, A. V. Tepikin, and O. H. Petersen. 1995, *Cell* **80:** 439–444.
Greber, U. F., and L. Gerace. 1995, *J Cell Biol.* **128:** 5–14.
Lanini, L., O. Bachs, and E. Carafoli. 1992, *J Biol Chem.* **267:** 11548–11552.
Mazzanti, M., L. J. DeFelice, J. Cohen J, and H. Malter. H. 1990, *Natur,e* **343:** 764–767.
Nicotera, P., D.J. McConkey, D. P. Jones, and S. Orrenius. 1989, *Proc. Natl. Acad. Sci. USA* **86**: 453–457.
Nicotera, P., S. Orrenius, T. Nilsson, and P. O. Berggren. 1990, *Proc Nat Acad Sci USA* **87:** 6858–6862.
Santella, L., and K. Kyozuka.1994, B*iochem Biophys Res Commun.* **203:** 674–680.
Santella, L. 1996, *J Membr Biol.* **153:** 83–92.
Santella, L. and C. Carafoli. 1997, *FASEB J.* 11:1091–1109.
Stehno-Bittel, L., C. Perez-Terzic, and D. E. Clapham. 1995, *Science* **270:** 1835–1838.

49

INTRON AS A SOURCE OF GENETIC POLYMORPHISM FOR FISH POPULATION GENETICS

Seinen Chow*

National Research Institute of Far Seas Fisheries
Shimizu 424, Japan

By incorporating polymerase chain reaction (PCR), DNA-level assays on nuclear genome (nDNA) such as microsatellite detection are becoming conventional. Yet, polymorphism at PCR priming sites may result in differential amplification of alleles (Pemberton et al., 1995). Furthermore, designing species-specific primers may be costly. Intron-targeted PCR employing primers designed from the conservative exon sequences may be an alternative candidate, because intron may accumulate much higher genetic variation than exons and polymorphism at priming sites is likely to be very rare. Quantity of fish genome data deposited to GenBank is much less than that of higher vertebrates, and most of the data is of cDNA. However, exon-intron arrangement appears to be highly conserved, making determination of intron position in the cDNA data possible. GenBank data search was performed to find homologous genes from lower to higher vertebrates, within which conserved sequences in exon regions were determined to design semi-universal primers for amplifying homologous intron among fish species. Here, I present genetic polymorphism observed in intron regions of several genes, and using the variation population genetic survey was performed in the swordfish (*Xiphias gladius*) and some tuna (*Thunnus* spp.) species.

Two types of polymorphisms were observed; one was length polymorphism in amplified fragments and the other restriction site polymorphism detected by RFLP analysis. Length polymorphisms in the amplified fragments were observed in the 1st intron (RP1) of the S7 ribosomal protein gene (RP) and 6th intron (CK6) of the creatine kinase gene (CK) of the swordfish (Figs. 1 and 2). Nucleotide sequence analysis revealed that presence of TG repeat (microsatellite) was responsible for the highly polymorphic fragment patterns observed in the RP1 and presence or absence of 24bp nucleotide block for longer or shorter CK6. Whereas, RFLP was observed in the CK6 of the Atlantic northern bluefin

* Tel: +81-543-36-6045; Fax: +81-543-35-9642; E-mail: chow@enyo.affrc.go.jp

New Developments in Marine Biotechnology, edited by Le Gal and Halvorson.
Plenum Press, New York, 1998.

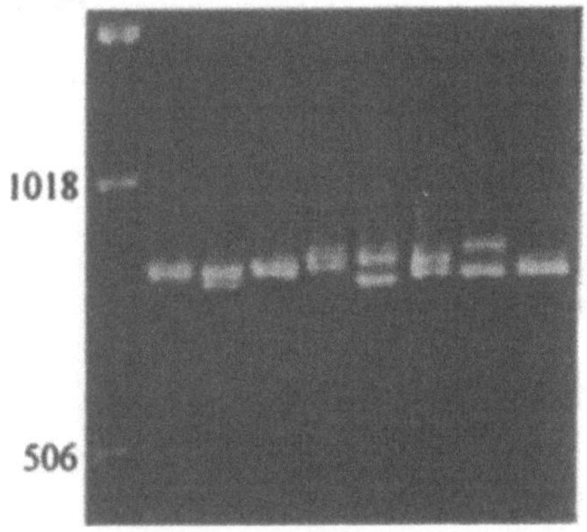

Figure 1. Fragment length polymorphism observed in the 1st intron of the S7 ribosomal protein gene of the swordfish (*Xiphias gladius*).

tuna (*Thunnus thynnus thynnus*). 2nd intron (RP2) of RP gene of the swordfish did not show any length variation nor RFLP. On the other hand, RFLP analysis on RP2 detected high variation in tuna species (*T. albacares*, *T. alalunga*, *T. obesus* and *T. t. thynnus*).

In the swordfish, allele interpretation in RP1 was not easy, while simple electrophoretic pattern in CK6 made allele interpretation possible. Two alleles (*A* and *B*) were determined in CK6, and genotype analysis in this locus was performed among 286 individuals of swordfish collected over six distant localities (Mediterranean Sea, North and South Atlantic Ocean, Indian Ocean, and North and South Pacific Ocean). Highly significant deviation from Hardy–Weinberg (H–W) equilibrium ($P<0.005$) caused by homozygote excess was observed only in the South Atlantic sample indicating that extensive population mixing with very little gene flow was occurring in this area. Previous mitochondrial DNA (mtDNA) analysis indicated that the the world-wide swordfish population is genetically structured not only among but also within ocean basins (Chow et al., 1997). RFLP observed in CK6 of the northern bluefin tuna also allowed genotype analysis on samples collected in the western and central Atlantic and Mediterranean Sea, in which no heterogeneity was observed within and between the samples. RP2 variation was used to compare other tuna samples of bigeye (*T. obesus*), yellowfin (*T. albacares*) and albacore (*T. alalunga*) collected in the Atlantic, Indian and Pacific Oceans. All samples were in accord with H–W equilibrium, while significant differences in the genotype frequencies were observed between local samples of all these three tuna species.

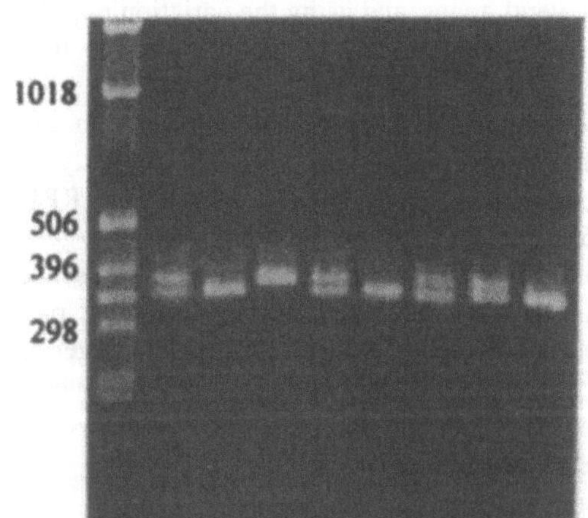

Figure 2. Fragment length polymorphism observed in the 6th intron of the creatine kinase gene of the swordfish (*Xiphias gladius*).

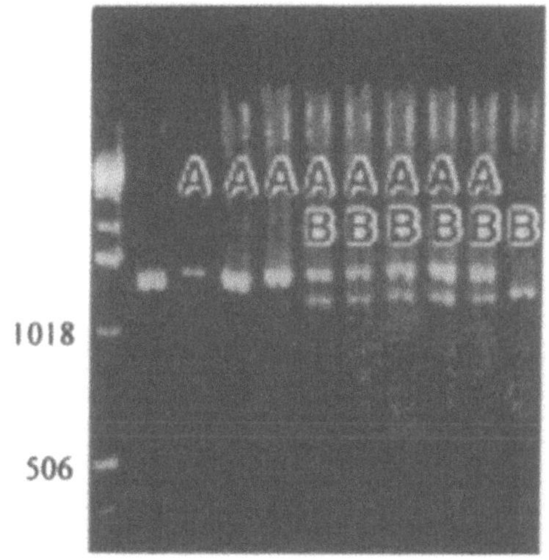

Figure 3. RFLP observed in the 2nd intron of the S7 ribosomal protein gene of the yellowfin tuna (*Thunnus albacares*).

The results of this investigation indicate that introns may be good source of intraspecific genetic variation for population genetic study and that the same set of primers can be used to amplify homologous intron region even among distant species. Further, the conserved exon primed PCR strategy may prevent artifacts due to priming site polymorphism.

REFERENCES

Chow, S. Okamoto, H., Uozumi, Y., Takeuchi, Y. and Takeyama, H., 1997, *Mar. Biol.* **127**:359–367.
Pemberton, J. M., Slate, J., Bancroft, D. R. and Barrett, J. A., 1995, *Mol. Ecol.* **4**:249–252.

50

POLYMORPHISM OF DIGESTIVE ENZYMES CODING SEQUENCES IN THE CRUSTACEA *PENAEUS VANNAMEI* (CRUSTACEA DECAPODA)

D. Sellos, C. Le Boulay, B. Klein, I. Cancre, and A. Van Wormhoudt

Station de Biologie Marine
Muséum National d'Histoire Naturelle et Collège de France
225, 29900 Concarneau, France

1. INTRODUCTION

Adaptation of digestive enzymes to the main constituant of food is well established in vertebrates (Howard and Yadkin, 1963; Puigserver *et al.*, 1985; Scheele, 1993). Grossman *et al.* (1943) early demonstrated the accretion of trypsin with a high protein diet. Later, Johnson *et al.* (1977) showed that casein was the best inducer of trypsin and demonstrated the different effects of the quality of food on induction. The increase of trypsin is modulated via transcription by casein-induced diets whilst other diets induce adaptation via post-translational events (Lhoste *et al.*, 1994). Adaptation to the main component of the diet is also described for chymotrypsin and amylase in the rat (Reboud *et al.*, 1961; Giorgi *et al.*, 1984). The molecular regulation of these adaptations appears to be highly species dependent and each enzyme is regulated, for a given species, by different pathways, i.e. transcription for chymotrypsin or translation for amylase in the rat (Lhoste *et al.*, 1993). In Palaemon serratus, convergent results are obtained between the enzyme activities, the level of dietary substrates in food and the growth (Van Wormhoudt *et al.*, 1980). The maximal specific protease activity was occuring when the level of dietary protein reached 45%. The shrimp *Penaeus vannamei* was studied as it is considered to be one of the most important penaid species for commercial production. The study of digestive enzymes is an essential step toward our understanding of the mechanism of digestion and a better adjustment of nutritional needs to decrease pollution impact of aquaculture production farms. To specify the existence and the level of control of digestive enzyme expression by the constituants of the food, molecular biology technics were applied to the characterization of the nucleotidic precursors of these enzymes for animals fed with two different diets containing 25 and 40% of casein.

New Developments in Marine Biotechnology, edited by Le Gal and Halvorson.
Plenum Press, New York, 1998.

2. MATERIALS AND METHODS

Living animals were obtained from the "Centre Océanologique du Pacifique" (IFREMER), Tahiti, and reared for 20 days in 100 liter tanks in recirculated water. Crude extracts were obtained from freeze-dried hepatopancreas from individual shrimp. Proteins were separated by native electrophoresis (Davis, 1964), and electrotransferred on PVDF membrane (Immobilon). ADN and ARN were extracted following conventional methods or the instruction of the Micro-scale total RNA separator kit (Clontech) respectively. PCR experiments were conducted using Pfu DNA polymerase (Stratagene) and two oligonucleotides based on invariable sequences of the α-amylase cDNAs (Van Wormhoudt and Sellos, 1996).

3. RESULTS AND DISCUSSION

Electrophoretic characterization of the major enzymatic activities of the digestive gland shows a high degree of polymorphism (Fig. 1). If chymotrypsin appears only as two

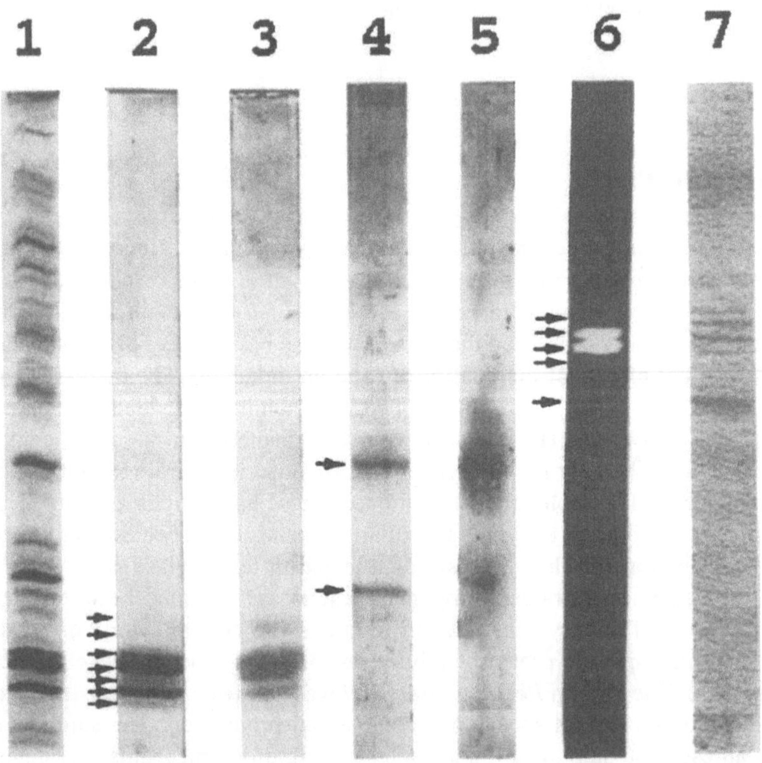

Figure 1. Electrophoretic characterization of the major digestive enzymes. Lane 1: Crude extract from the digestive gland revealed by silver staining. Lane 2: 1 µg of *Penaeus vannamei* pure trypsin revealed by silver staining. Lane 3: Trypsin revealed by BANA (N-L-arginine-naphtylamide) as a specific substrate (Zwilling et al., 1969). Lane 4: 1 µg of *Penaeus vannamei* pure chymotrypsin. Lane 5: Revelation of chymotrypsin antigens in crude extract with specific antibodies (1/500). Lane 6: Revelation of α-amylase activity on amylodextrine plates (Van Wormhoudt and Favrel, 1988). Lane 7: Revelation of amylase antigens in crude extract with antibodies against *Penaeus japonicus* amylase (1/500).

very distinct bands, trypsin is composed of at least seven sub-fractions, all of them being recognized with a specific antibody. Amylase is formed of two main fractions with at least three faint subfractions. The enzymatic profiles have been shown to be modified during development in the case of chymotrypsin or as an adaptative response to different alimentary diets in the case of amylase (Le Moullac *et al.*, 1996).

To demonstrate that these enzyme variants are synthetized from different mRNA, we established a shrimp digestive gland cDNA library, and used it to determine nucleotidic sequences of the two mRNA coding the two chymotrypsic variants (Sellos and Van Wormhoudt, 1992), of three mRNA coding three trypsic subfractions (Klein *et al.*, 1996) and also three amylase mRNA coding the two major forms and one less expressed sub-fraction (Van Wormhoudt and Sellos, 1996). This study allow us to demonstrate that in crustacea as in vertebrates, proteolytic enzymes are synthetized as a preproenzyme with an hydrophobic peptide (allowing translocation into the endoplasmic reticulum) and a zymogen peptide (allowing the storage of the protein under an inactive form).

In order to prove the existence of the adaptative response at the level of the transcription of specific genes coding the different amylase subfractions under different alimentary diets (25 and 40% of casein), we conducted RT-PCR experiments with individual mRNA extracts from animal reared with these conditions. This confirmed that when animals are fed with a 25% casein diet, and when we observe two majors amylase subfraction, we also show the presence of the two mRNA encoding these proteins. On the other hand, for animals fed with the 40% casein diet, where only one major amylase form is present, we could obtained only nucleotidic sequences representing this protein.

To study polymorphism at the gene level, we conducted PCR experiments using genomic DNA and a set of specific primers deduced from the chymotrypsin cDNA sequence. The coding part of the gene is 1522 base pairs long and is composed, as in mammals, of seven exons interupted with six introns (Fig. 2). The AG-GT junction rule is always conserved. Establishing the sequences of the genes encoding the two chymotrypsin fractions of the digestive gland and one another subfraction, we have shown that for each of the genes, the introns are different in size and in nucleotidic sequence. These specific sequences

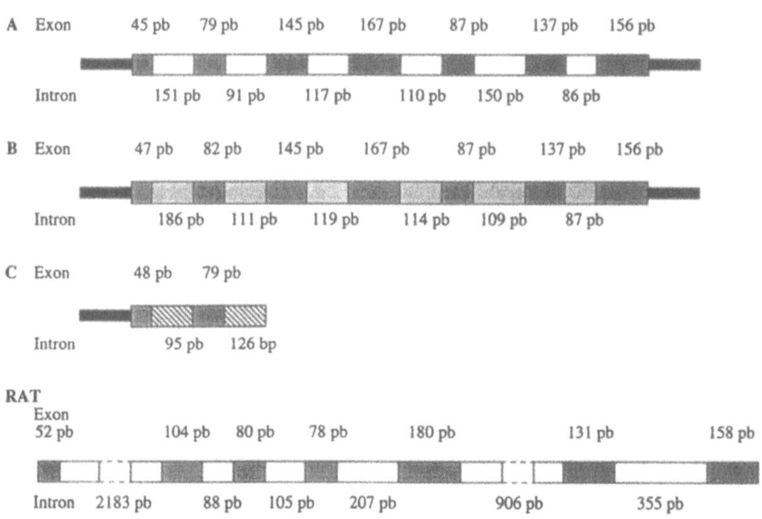

Figure 2. Schematic representation of chymotrypsin gene structures. The sizes of exons (in grey) and introns (in white or light grey) are given for the three crustacean chymotrypsin genes and for the rat counterpart.

could represent signals allowing differential expression during the developpment or for an adaptative response to environmental conditions.

We are currently studying the organization of other protease genes to establish structural comparisons with vertebrate counterparts. Characterization of promotor sequences of these genes is in progress.

4. SUMMARY

A high degree of polymorphism was observed by electrophoretic technics when studying digestive enzymes (trypsin, chymotrypsin and amylase), in crustacea. For all of these proteins, several variants were observed. In the same manner, screening a crustacean digestive gland cDNA library, we found a high level of polymorphism in digestive enzyme coding cDNA sequences. The existence of several variants for one enzymatic activity could be a response to adaptation needs to internal or external constraints. For an example, studying amylase electrophoretic profiles, we could observed that the two main variants are present when the animals are fed with an alimentary diet containing 25% of protein and that only one of the main amylase component is observed with a 40% protein diet. Isolation of cDNAs confirmed the expression of corresponding messengers. Taking another example, (chymotrypsin), investigation on organization and nucleotidic sequences of digestive enzyme genes have shown a complex gene structure with the existence of seven exons and six introns. For all the genes coding for this enzyme, the same exon/intron organization was observed, a high degree of polymorphism in nucleotidic sequences is seen in exons reflecting the messenger complexity, but also in introns.

REFERENCES

Davis, B.J., 1964, Disc electrophoresis-II: method and application to human serum proteins. *Ann. N.Y. Acad. Sci.*, **121**:404–427.

Giorgi, D., Bernard, R., Lapointe, R. and Dagorn, J.C., 1984, Regulation of amylase messenger RNA concentrations in rat pancreas by food content. *The Embo J.*, **3** (7):1521–1524.

Grossman M.I.,Greengard, H. and Ivy, A.C., 1943, The effect of dietary composition on pancreatic enzymes. *Am. J. Physiol.*, **138**:676–682.

Howard, F.and Yadkin, J.,1963, Effect of dietary change upon the amylase and trypsin activities of the rat pancreas. *Brit. J. Nutr.*, **17**:281–294.

Johnson, A., Hurwitz, R. and Kretchner, N., 1977, Adaptation of rat pancreas amylase and chymotrypsinogen to changes in diet. *J. Nutr.*, **107**:87–96.

Klein, B., Le Moullac, G., Sellos, D. and Van Wormhoudt, 1996, Molecular cloning and sequencing of trypsin cDNAs from *Penaeus vannamei* (Crustacea Decapoda): Use in assessing gene expression during the moult cycle. *Int. J. Biochem. Cell. Biol.*, **28**:551–563.

Le Moullac,G., Klein, B., Sellos, D. and Van Wormhoudt, A., 1996, Adaptation of trypsin, chymotrypsin and α-amylase to casein level and to protein source in *Penaeus vannamei* (Crustacea Decapoda). *J. Exp. Mar. Biol. Ecol.*, **208**:107–125.

Lhoste, E.F., Fiszlewicz, M., Gueugneau, A.M., Wicker-Planquart, C., Puigserver, A. and Corring, T., 1993, Effect of dietary proteins on some pancreatic mRNA encoding digestive enzymes in the pig. J. Nut. Biochem., **4**:143–152.

Lhoste, E.F., Fiszlewicz, M., Gueugneau, A.M. and Corring, T., 1994. Adaptation of exocrine pancreas to dietary proteins: effect of the nature of protein and rat strain on enzyme induction and messenger mRNA levels. *J. Nut. Biochem.*, **5**:84–94.

Puigserver, A., Wicher, C. and Gaucher, C., 1985, Aspects moléculaires de l'adaptation des enzymes pancréatiques et intestinales au régime alimentaire. *Reprod. Nutr. Develop.*, **25**:787–802.

Reboud, J.P., Ben Abdeljlil, A. and Desnuelle, P., 1961, Variation de la teneur en enzymes du pancréas de rat en fonction de la composition des régimes. *Bioch. Biophys. Acta*, **58**:326–337.

Scheele, G.A., 1993, Regulation of pancreatic gene expression in response to hormones and nutritional substrates. IN The pancreas: Biology, Pathology and disease, II edition, edited by W. Vay Liang, W. Go, *et al.*, Raven Press N.Y., 103–119.

Sellos,D. and Van Wormhoudt, A., 1992, Molecular cloning of a cDNA that encodes a serine protease with chymotrypsic and collagenolytic activities in the hepatopancreas of the shrimp *Penaeus vannamei* (Crustacea Decapoda). *FEBS LETTERS*, **309**:219–224.

Van Wormhoudt A., Ceccaldi, H.G. and Martin, B., 1980, Adaptation de la teneur en enzymes digestives de l'hépatopancréas de *Palaemon serratus* (Crustacea Decapoda) à la composition d'aliments expérimentaux. *Aquaculture*, **21**:63–78.

Van Wormhoudt, A. and P. Favrel, 1988, Electrophoretic characterization of *Penaeus elegans* (Crustacea Decapoda) α-amylase system: study of amylase polymorphism during the intermolt cycle. *Comp. Biochem. Physiol.*, **98B**:201–207.

Van Wormhoudt,A. and Sellos, D., 1996, Cloning and sequencing analysis of three cDNAs in the shrimp Penaeus vannamei: Evolutionary aspects. *J. Mol. Evol.*, **42**:543–551.

Zwilling, R.; Pfeiderer, G.; Sonneborn, H.H.; Kraft, V. and Stucky, I.1969, The evolution of endopeptidases. V: common and different traits of bovine and crayfish trypsin. *Comp. Biochem. Physiol.* **28**:1275–1287.

51

MATING DYNAMICS OF THE SNOW CRAB (*CHIONOECETES OPILIO*, BRACHYURA: MAJIDAE)

An Analysis Using DNA Microsatellite Markers

N. Urbani,[1] B. Sainte-Marie,[2] J.-M. Sévigny,[2] D. Zadworny,[1] and U. Kuhnlein[1]

[1]Department of Animal Science
McGill University
21 111 Lakeshore Road
Sainte Anne de Bellevue, Québec H9X 3V9, Canada
[2]Division des Invertebrés et de la Biologie Expérimentale
Institut Maurice-Lamontagne
Ministère des Pêches et des Océans
850 Route de la Mer, C.P. 1000
Mont Joli, Québec G5H 3Z4, Canada

1. INTRODUCTION

Snow crab *Chionecetes opilio* is a common inhabitant of the deep bottoms of the Northwest Atlantic, the Bering Sea and the Sea of Japan. Because of its size and abundance it supports important fisheries around the world that are dependent on the fluctuations of the populations (Davidson et al. 1985). Despite its economic importance, there are still gaps in the knowledge of the biology of *C. opilio*. In particular, patterns of sperm storage and paternity remain unclear in a species where females can mate with several males and have the capacity to store their sperm in paired spermathecae.

The objective of this study was to study patterns of sperm competition and expression of paternity in primiparous females (first-time spawners) mated in the laboratory with tagged males, and in wild-caught primiparous females. Highly polymorphic microsatellite loci were isolated from a DNA genomic library, and two markers were used to genotype by PCR adults, spermathecae and embryos.

New Developments in Marine Biotechnology, edited by Le Gal and Halvorson.
Plenum Press, New York, 1998.

2. MATERIALS AND METHODS

2.1. Genetic Analysis

Microsatellite loci were identified from a size-selected genomic library of *C. opilio* (Urbani et al. In press), and loci Cop3-4 and Cop24-3 were chosen for having numerous alleles (22 and 14 in 20 unrelated individuals, respectively) and high levels of heterozygosity (95% and 90%, respectively) giving an average probability of 4.5×10^{-5} for two individuals to share a particular allelic combination by chance. DNA was extracted from male mandibular palps, female pereopods, spermathecal contents and egg clutches (approx. 10 000 eggs).

2.2. Dissection of Females

One pereopod and the egg clutch were removed from each female and preserved in 100% ethanol for genetic analyses. Because sperm delivery to paired spermathecae is usually balanced (Sainte-Marie and Lovrich 1994), the two spermathecae of an individual female were considered replicates and were used to assess complementary aspects of the reproductive process. Left and right spermathecae were used for genetic and histological analysis, respectively. Left spermathecae were preserved in 100% ethanol while right spermathecae were fixed in Bouin for 48 h and then kept in 70% ethanol at 4°C for 4 months or fixed in Bouin added with 1.5% acetic and 1.5% trichloroacetic acids for 48h and then kept in 70% ethanol at 4°C for 4 months. They were then embedded in paraffin and cut in medio-sagittal sections 7 µm thick. Sections were stained using the technique of Dominici and analysed by light microscopy.

2.3. Mating Experiments

Crabs were held in laboratory tanks supplied with fresh flowing seawater at –0.5° to 0.3°C and photoperiod was controlled to reflect natural light rhythms. Mating experiments were conducted following either one of two methods. In the first, a molting female was placed in a 500-l tank containing 1, 2, 3 or 4 tagged males and monitored for 24 h using a ceiling-mounted low-light video camera and time-lapse VCR recorder at one image per second. Video recordings were analysed to determine the copulating sequence, identity of mates and duration of copulations. In the second, a recently molted female was placed in a 50-l tank containing a single tagged male until copulation occurred and post-copulatory guarding was initiated by the male. She was then separated from her mate and transferred to another tagged male so that she might copulate anew. A total of seven females and 10 different males were used.

2.4. Collection of Wild-Caught Females

A total of 7 females carrying recently extruded eggs were collected near Sept-Iles (Québec, Canada), and were immediately processed for histological and genetic analyses. As spermathecae contained more or less distinct masses (putatively, different ejaculates), using a scalpel, we attempted to isolate for genetic analyses these masses from the left spermathecae by cross-sectioning the contents more or less perpendicular to the sagittal plane. Resulting cross-sections were numbered and preserved individually in 100% ethanol.

3. RESULTS AND DISCUSSION

3.1. Controlled Matings

Females copulated between 2 and 4 times with up to 3 different males. Total copulatory time for individual females varied from 25 to 63 min and correlated positively with spermathecal dry weight. Histological examination of the right spermathecae revealed the presence of 1 to 2 rather discrete structural units, putative ejaculates, arranged sequentially along the longitudinal axis of the organs. Obviously, some smaller ejaculates were not resolved by histology. Each unit was usually composed of a concentration of spermatophores dorsally capped by a relatively large amorphous mass. The genetic analysis revealed that the genotypes of all the males involved in the matings with the 7 females were present in their spermathecae, but their progeny had been fathered by a single male in each case. Two patterns of paternity were apparent. In one female, it was the first of two successive mates that sired the progeny. In that case there was a long time lapse between matings (20.5 h) and extended female guarding by the first mate before copulation with the second male occurred. In the other females, it was the last of successive mates that fathered the progeny, coincident with a shorter time lapse of 4 h between matings. It appeared that the last male to inseminate a female before oviposition gained paternity of the offsprings. The most parsimonious mechanism to explain last-male sperm precedence was sperm stratification, i.e. displacement of rival sperm towards the blind-end of the spermatheca.

3.2. Wild-Caught Females

Histological examination of the right spermathecae revealed the presence of 2 to 6 (mean=4.2) discrete structural units arranged sequentially along the longitudinal axis of the organs. Each unit was composed of an aggregate of spermatophores usually capped by a larger volume of amorphous mass and presumably represented individual ejaculates. Since each male contributes one or two alleles, the least number of males which must have contributed to the sperm present in a particular spermatheca is given by half the number of alleles (rounded to the next whole number in the case of an uneven number of alleles). Since two loci were assayed, the least number of males genotypes per spermatheca was determined from the locus revealing the highest number of alleles. The number of genotypes per organ ranged from 3 to 5 (mean=3.9) and roughly agreed with the number of structural units determined from the histological examination. Different combinations of alleles occurred in successive cross-sections of the spermathecae indicating that sperm deposits from different males were stratified within the organs. The analysis of the progeny revealed that each brood had been sired by a single male. The paternal alleles were mostly present in the cross-sections towards the blind-end of the receptacles, and indicated that the females must have re-mated after extruding their eggs. Thus a male may assist a female in molting, promptly inseminate her, and then guard her until she extrudes her eggs. Once released, the female may remain receptive to males for several days after oviposition and may mate again.

Our results confirmed the polyandrous nature of female snow crabs during their first breeding period and their capacity to store sperm of several males. Sainte-Marie and Carriere (1995) demonstrated that females effectively fertilize a second egg clutch from sperm acquired at the molt to maturity, indicating that sperm stored in spermathecae may remain viable for several months. Several lines of evidence indicated that spermathecae may be important for females as an "insurance policy", since they may not always find a male with which to copulate (Beninger et al. 1993). Sperm storage could be particularly important when the number of males available for reproduction is low as population size have been shown to fluctuate (Sainte-Marie et al. 1996).

4. SUMMARY

Male snow crabs are subject to an extensive and lucrative fishery. Concerns that the current management regimes do not maintain the reproductive potential and genetic integrity of the populations generated considerable interest in the reproductive biology of snow crabs. In order to study sperm competition and mating dynamics, we established a genomic DNA library and isolated a series of highly polymorphic microsatellite loci. Two markers were used to determine expression of paternity and number of male genotypes present in female spermathecae. The two loci were amplified by PCR from DNA of adults, spermathecae and embryos of primiparous females from the wild and from multiple mating experiments under laboratory settings. Single paternity was observed for the progeny of all females. The analysis of wild females revealed that their spermathecae contained the sperm of at least three different males. The analysis of laboratory females showed that sperm displacement was the mechanism by which single paternity was obtained by the last males to mate. Microsatellite DNA markers are powerful tools for investigating mating dynamics in the snow crab and could be applied to other crab species with sperm storage organs.

REFERENCES

Beninger, P.G., Lanteigne, C, and R.W.Elner, 1993, *J. Crustacean Biol.* **13**: 1–16.

Davidson, K., Roff, J.C., and R.W. Elne, 1985, *Can. J. Fish. Aquat. Sci.* **42**: 474–482.

Sainte-Marie, B., and G. A. Lovrich, 1994, *J Crustacean Biol.* **14**: 508–521.

Sainte-Marie, B., and C. Carriere, 1995, *Fish. Bull. U.S.* **93**: 758–763.

Sainte-Marie, B., Sévigny, J.-M., Smith, B.D., and G.A. Lovrich, 1996, Alaska sea grant Coll. Prog. Rep. 96–02, pp.451–478.

Urbani, N., Sévigny, J.-M., Sainte-Marie, B., Zadworny, D., and U. Kuhnlein, *Mol. Ecol.* (In Press)

52

DENATURATION OF ALGAL PHYCOBILIPROTEINS CAN BE USED AS A THERMAL PROCESS INDICATOR

A. Orta-Ramirez,[1] D. M. Smith,[1] and J. E. Merrill[2]

[1]Department of Food Science and Human Nutrition
[2]Department of Microbiology
Michigan State University
East Lansing, Michigan 48824

1. INTRODUCTION

Thermal processing is a very common method of food preservation. The thermal treatment consists of applying heat to a food product at a specific temperature for a determined interval of time. The process time is calculated to achieve sufficient microbial destruction to comply with public health standards and prevent spoilage of the food product. The thermal processing kinetics can be defined in terms of D, z and F values (Teixeira, 1992; Singh and Heldman, 1993; Hendrickx et al., 1995). D value is the time in minutes required to decrease a quality attribute by 90% at a constant temperature. The z value is the temperature increase necessary to reduce the D value by 90%. The F value is the time required to achieve a stated reduction in a population of microorganisms or spores. This time is usually expressed as a multiple of the D value. Many thermal processes have been standardized, and published recommendations for specific conditions are available (National Canners Association, 1968). Even with reliable thermal processing schedules and equipment, though, there is a need to determine process compliance to ensure food safety. The evaluation of thermal processes in foods can be done using *in situ* methods, physical-mathematical models or time-temperature integrators (TTIs). These TTIs are devices capable of predicting the time-temperature response of a quality or safety index in a food product without the need for extensive mathematical modeling of the thermal process. This occurs when the thermal destruction of both TTI and target index are identical ($Z_{TTI} = Z_{target}$) (Hendrickx et al., 1995; Van Loey et al., 1996).

The United States Department of Agriculture Food Safety Inspection Service (USDA-FSIS) has recently proposed a modification of the current thermal processing regulations for meat products based on destruction of the pathogenic microorganism *Salmonella*. The USDA is recommending that manufacturers use a thermal process sufficient to achieve a

New Developments in Marine Biotechnology, edited by Le Gal and Halvorson.
Plenum Press, New York, 1998.

7-D reduction in *Salmonella* in cooked/roasted beef, cooked corned beef and poultry products, and a 5-D reduction in cooked uncured meat patties (USDA-FSIS, 1996). Reported z values for *Salmonella* were 5.6°C in ground beef (Orta-Ramirez et al., 1996) and 6.24°C in turkey (Veeramuthu, 1996). Therefore, a TTI with the same z value as *Salmonella* could be used in beef and poultry products to determine the adequacy of any 7-D or 5-D process.

The overall goal of this research is to develop a time-temperature integrator (TTI) to monitor thermal processing of foods with special emphasis on meat products. The main component of this TTI is the naturally-fluorescent pigment protein *phycoerythrin* (PE). The unique spectral properties of PE are given by the specific chromophores (CHRs) and their interaction with the apoprotein. Manipulation of the microenvironment of the CHRs will result in alteration of the fluorescence properties. The studies reported here were undertaken to determine the thermal inactivation kinetics of PE in a range of solution compositions. By characterizing the factors or conditions that alter the kinetics of PE thermal destruction, we can evaluate the usefulness of the protein in relation to different thermal food processes.

2. MATERIALS AND METHODS

Phycoerythrin was obtained from fresh cultures of *Porphyra yezoensis* by ammonium sulfate (AS) fractionation and gel filtration. Fractions from gel filtration with A_{565}/A_{280} >5 were pooled and stored as a stock solution in 60% AS at 4°C. Purity was confirmed by presence of a single band on native polyacrylamide gel electrophoresis (PAGE) (Davis, 1964).

Immediately prior to use, PE stock solution was dialyzed against water at 4°C overnight, diluted in working buffer to achieve an absorbance of 0.1 at 565 nm, and 2-mL transferred to 10 × 75 mm thermal death time (TDT) tubes. The tubes were sealed, placed in a wire rack and immersed in a circulating controlled temperature bath. Temperature was monitored using a thermocouple inserted in a control TDT tube containing 2 mL of the same solution. Tubes were removed at predetermined intervals of time, cooled in an ice bath, then kept at 4°C until analyzed within 12–24 h. Each study was done in triplicate.

To test the effect of pH, dialyzed PE was diluted in 50 mM NaCl, 20 mM buffers ranging from pH 3 to 10. Buffers employed were: citrate (pH 3.0, 4.0 and 5.0), phosphate (pH 6.0 and 7.0), trizma (pH 8.0 and 9.0), and glycine (pH 10.0).

The effects of sucrose, glycerol, detergent, and urea on PE stability and thermal inactivation were also studied. Each compound was dissolved in working buffer to achieve specified concentrations and pH adjusted if necessary. PE from dialyzed stock solution was added to achieve 0.1 absorbance at 565 nm. Compounds tested and their concentrations were: glycerol (20, 40 and 60%), sucrose (20, 40 and 60%), urea (0.5, 1, 2, 4, 6 and 8 M for pH 7.0; and 0.5, 1, 2 and 4 M for pH 4.0), and SDS (0.1, 0.5, 1, 2.5 and 5%). Each experiment was done in triplicate.

Fluorescence in arbitrary units (a.u.) for each sample was determined using a computer enhanced SLM-4000 fluorometer (SLM Instruments). Maximum excitation ranged from 493 to 498 nm and maximum emission between 578 and 580 nm. Changes in fluorescence are expressed in relation to fluorescence in the unheated control sample under same conditions (F0), where the control sample contained the same concentration of PE in buffer solution without presence of any other substance.

Fluorescence counts were converted to logarithms. D values (in minutes) were calculated according to the Laboratory Manual for Food Canners and Processors (1968) except that survivor curves at each temperature were calculated by linear regression. TDT curves were determined by plotting log of D values *vs* temperature. The z values were determined as the reciprocal of the slope of the TDT curve by linear regression.

3. RESULTS AND DISCUSSION

PE was stable at pH range 5.0–9.0 and fluorescence decreased at pH 4.0 and 10.0. PE completely lost fluorescence at pHs below 4.0 and above 10.0. The z values ranged from 3.89°C at pH 10.0 to 6.26°C at pH 8.0 (Table 1).

Thermal inactivation of PE was studied in different concentrations of glycerol and sucrose at pH 7.2 (Table 2a and 2b). PE was more stable in presence of sucrose than of glycerol. The D values increased with increasing concentrations of sucrose and decreased with increasing concentrations of glycerol. The z values ranged from 5.03°C (40%) to 6.56°C (60%) for glycerol, and from 5.85°C (40%) to 6.37°C (60%) for sucrose (Table 2).

Thermal inactivation of PE in different concentrations of sodium-dodecyl sulfate (SDS) and urea at pH 7.0 was also studied (Table 2c and 2d). Overall, PE was more stable in urea than in SDS as reflected by the lower D values for SDS at all temperatures. The D values decreased with increasing concentrations of SDS or urea. The D values differed slightly among the three concentrations of SDS for all the temperatures tested. The D values of PE were similar in 1 and 2M urea, while 3M had a much greater effect on PE at all temperatures. The z values ranged from 7.45 (2%) to 8.55°C (3%) for SDS, and from 8.34 (2M) to 11.17°C (3M) for urea. All z values were higher than that for PE in buffer without SDS or urea (Table 2).

The thermal inactivation behavior of PE was successfully modified by using different agents. The modified z values ranged from 3.89 to 11.17°X. By immobilization and manipulation of the PE environment, the z value might be shifted to approximate the z values of *Salmonella* in beef and poultry products (Orta-Ramirez et al., 1996; Veeramuthu, 1996). Current research includes the utilization of some of the denaturing agents in combination with extreme pHs (4.0 and 10.0) to study the PE thermal inactivation under such conditions.

4. ABSTRACT

The objective of this study was to evaluate the applicability of phycoerythrin as the detecting component of a time temperature integrating (TTI) device for use in food process and other industrial thermal process monitoring. The unique spectral properties of PE

Table 1. D- and z-values for PE thermal destruction at pH 4.0–10.0*

	pH						
	4.0	5.0	6.0	7.0	8.0	9.0	10.0
D60	304.6						475
D65	91.15						5.87
D70	2.01	9273	12245	2195	2028	93.73	1.28
D80		168.4	91.66	132.3	3.23	5.87	
D90		0.40	1.43	1.03	1.30	0.61	
z-value	4.59	4.58	5.09	6.01	6.26	5.05	3.89

*Buffers as specified in the text. D-values are decade reduction in fluorescence (see text) at constant temperature indicated (°C). z-value calculations as per text.

Table 2. D- and z-values for PE thermal destruction analyses in sucrose (a), glycerol (b), sodium dodecyl sulfate (c), and urea (d) solutions*

a. Sucrose	Control	20%	40%	60%
D80	19	46	99	291
D83	10	14	84	202
D85	9	4	16	54
D87	2.5	4	8	26
z-value	8.75	6.28	5.85	6.37
b. Glycerol	Control	20%	40%	60%
D70	1612	1428	971	474
D73	578	498	376	149
D77	84	70	52	28
D80	18	16	11	16
z-value	5.10	5.05	5.03	6.56
c. SDS	Control	1%	2%	3%
D50	2702	146	152	152
D53	1613	52	61	60
D57	135	17	18	18
D60	131	10	7	7
z-value	6.65	8.55	7.45	7.50
d. Urea	Control	1M	2M	3M
D50	2702	6250	6666	588
D53	1613	3846	3846	291
D57	135	1428	961	148
D60	131	408	472	70
z-value	6.64	8.54	8.34	11.17

*Working buffer was 50 mM phosphate, pH 7.0, with other components added as indicated to final PE concentration of 0.1 absorbance at 565 nm. D-values are decade reduction in fluorescence (see text) at constant temperature indicated (°C). z-value calculations as per text.

are conferred by specific interaction of the chromophores with the apoproteins. Variations in the chemical composition of the suspending aqueous solution can significantly alter the denaturation kinetics of PE. In this study we have determined inactivation parameters (D and z values) of PE, as determined by fluorescence measurements, in a variety of conditions. PE was dissolved in solutions of varying pH (4.0–10.0), as well as in varying concentrations of glycerol (20–60%), sucrose (20–60%), urea (1–3 M), and sodium-dodecyl sulfate (SDS) (1–3%), all at pH 7.0. These solutions were placed in thermal death time tubes and held at 60, 65, 70, 80, or 90°C for predetermined time intervals. D values (decade activity decrease) ranged from 12245 min at pH 6.0 (70°C) to 0.40 min at pH 5.0 (90°C). The z values spanned a range of values from 3.89°C (pH 10.0) to 11.17°C (3% SDS, pH 7.0). These data demonstrate that by modifying the environmental conditions, the inactivation kinetics of PE can be adjusted to match those of several common thermal food processes. The use of colorimetric or protein-based tests is a current approach for monitoring thermal food processes, and PE appears to have considerable promise in this potential application.

REFERENCES

Davis, B.J. 1964, *Ann. N.Y. Acad. Sci.* **121**:404–427.

Hendrickx, M., Maesmans, G., De Cordt, S., Noronha, J., Van Loey, A., and Tobback, P. 1995, *Rev. Food Sci. Nut.* **35**:231–262.

National Canners Association. 1968, Laboratory Manual for Food Canners and Processors. Vol. 1. AVI Publishing Company, Inc. Westport, CT.

Orta-Ramirez, A., Price, J.F., Hsu, Y-.C., Veeramuthu, G.J., Cherry-Merritt, J.S. and Smith, D.M. 1996, *J. Food Prot.* **60**:471–475.

Singh, R.P. and Heldman, D.R. 1993, Introduction to Food Engineering. 2nd ed. Academic Press Inc. New York, NY.

Teixeira, A. 1992, Thermal process calculations. Ch.11, In *Handbook of Food Engineering*. D.R. Heldman and D.B. Lund (Ed.), 563–620.Marcel Dekker, Inc., New York.

USDA-FSIS. 1996, Performance standards for the production of certain meat and poultry products. Federal Register. 61(86), 19564. US Department of Agriculture, Food Safety Inspection Service, Washington, DC.

Van Loey, A., Hendrickx, M., De Cordt, S., Haentjens, T. and Tobback, P. 1996, *Trends Food. Sci. Tech.* **7**(1):1–33.

Veeramuthu, G.J. 1996, Personal communication.

STRESS RESPONSIVE GENE FOR UV-A IN MARINE CYANOBACTERIUM *OSCILLATORIA* sp.

Tadashi Matsunaga and Akira Yamazawa

Department of Biotechnology
Tokyo University of Agriculture and Technology
2-24-16 Naka-cho, Koganei, Tokyo 184, Japan

Recently, we have isolated a new type of UV-A resistant marine cyanobacteria *Oscillatoria* sp. which grow under UV A irradiation and produce bioputerin as an UV-A absorbing pigment (Matsunaga et al., 1993). This pigment was related to the pterine pigments found in butterfly wings UV-A sensitive isolates did not produce significant levels of this chromophore. UV-A radiation was very effective in eliciting biopterin glucoside synthesis, in which intracellular level of biopterin glucoside was correlated with the intensity of UV-A. These results suggested that biopterin glucoside may protect the cyanobacterium from adverse effects of UV-A irradiation. Furthermore, the effect of an UV-A light intensity on cell growth, biopterin glucoside production and photosynthetic activity was investigated (Wachi et al., 1995). The marine *Oscillatoria* sp. NKPB 091600 could grow at UV-A intensities of up to 800 $\mu W \cdot cm^{-2}$ and could grow at 300 $\mu W \cdot cm^{-2}$ as well as in the absence of UV-A irradiation. The photosynthetic activity of pre-cultured cells with UV-A was protected well from UV-A induced inhibition. It was demonstrated that a rapid increase in bioterin glucoside content was observed after 10 hr under UV-A radiation.

In order to elucidate the UV-A resistant characteristics in the marine *Oscillatoria* sp., specific protein induced by UV-A irradiation was isolated (Yamazawa et al., submitted). Protein profile under normal light conditions (cool-white fluorescent light irradiation; 500 $\mu W \cdot cm^{-2}$) was compared with that under UV-A conditions (UV-A irradiation; 900 $\mu W \cdot cm^{-2}$ and cool-white fluorescent light irradiation; 500 $\mu W \cdot cm^{-2}$). Cell suspension of the marine *Oscillatoria* sp. grown under normal light conditions was distributed into several petri dishes (20 ml). These were exposed to UV-A or normal light irradiation for 15 hr. Protein profiles in water-soluble cell fractions were analyzed by SDS-PAGE in both samples. A 60 kDa protein was specifically induced under UV-A irradiation. Furthermore, induction of this protein by UV-A irradiation was correlated with that of the UV-A absorbing compound, biopterin glucoside (Fig. 1). This 60 kDa protein purified by electroelution was subjected to N-terminal amino acid sequencing. The result indicated that this UV-A responsive protein was homologue to GroEL well known as a bacterial molecular chaperonin. Especially, this

New Developments in Marine Biotechnology, edited by Le Gal and Halvorson.
Plenum Press, New York, 1998.

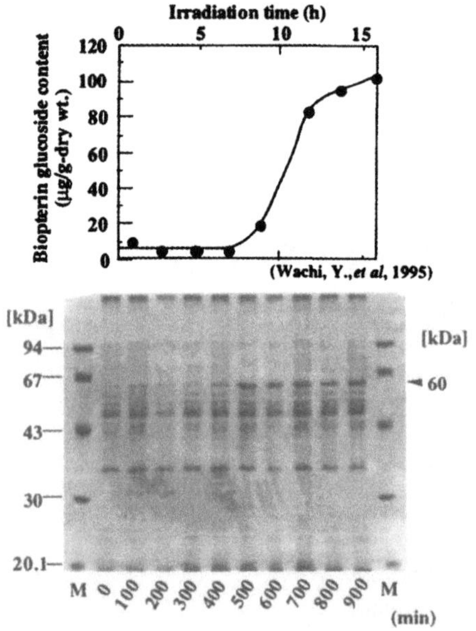

Figure 1. SDS-PAGE of induced protein under UV-A irradiation (900 μWcm^{-2}) at 0~900 min. Lane M shows molecular weight marker. Gel was stained by CBB.

protein had high similarity with the GroEL of *Synechococcus* sp. PCC7942 and *Synechocystis* sp. PCC6803. This 60 kDa of protein was confirmed to be a GroEL by immunological analysis using anti-heat shock protein 60 (anti-cyanobacterial GroEL).

The transcriptional regulation of *groEL* under UV-A irradiation was investigated in the marine *Oscillatoria* sp. Genomic DNA fragment of containing *groESL* genes was cloned in pUC18 and sequenced. This sequence had high similarity with *groESL* of cyanobacterium, *Synechococcus* sp. PCC 7942. Northern hybridization using DIG-labeled 0.9 kb fragment containing 5′ region of *groEL* gene as a probe was carried out (Fig. 2). The levels of mRNA of *groEL* started to increase after 300 min of UV-A irradiation, and kept

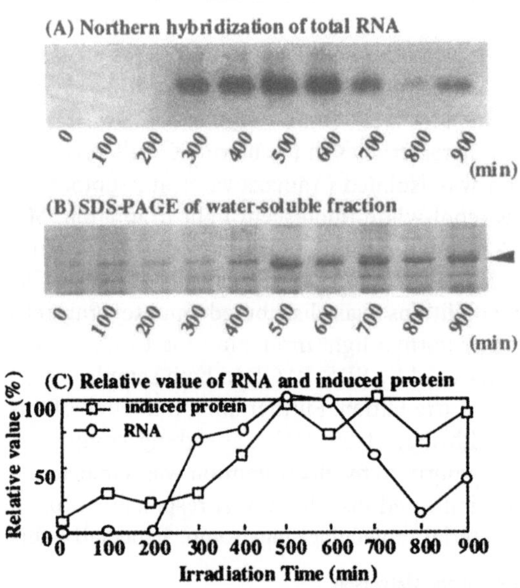

Figure 2. Northern hybridization of total RNA and SDS-PAGE of water-soluble fraction form *Oscillatoria* sp. Incubated under UV-A irradiation (900 μWcm^{-2}). Arrow shows an induced protein in water-soluble fraction.

at high level for 300 min. Certain amount of GroEL protein was observed to be expressed in early time, while drastic increase of expression was observed after start of high transcription of mRNA. UV-A responsive regulation of GroEL is obviously controlled by transcriptional level, and this regulation pathway may link to biopterin glucoside synthesis.

Since GroEL is well known as a heat shock protein, the effect of heat shock (45°C) on GroEL expression were investigated in this *Oscillatoria* sp. Cell suspension was incubated at 45°C under a cool-white fluorescent light for 15 hr. Although the induction of GroEL was observed, amount of GroEL protein produced was much smaller than that by UV-A irradiation. Furthermore, the expression was gradually induced not like drastic increase in UV-A response. These results suggested that regulation mechanisms for *groEL* expression under UV-A irradiation may differ from that under heat shock conditions. The promoter activation of GroEL by UV-A in the marine *Oscillatoria* sp. may be different.

In order to investigate the relationship between biopterin glucoside synthesis and GroEL expression, UV-A non-resistant mutants of the marine *Oscillatoria* sp.have been constructed and their molecular characterization has been carried out. The information obtained from these mutants may contribute to understanding of the UV-A resistance mechanisms in the marine *Oscillatoria* sp.

REFERENCES

Matsunaga, T., et al., 1993, *Applied Microbiology and Biotechnology* **39**:250–253
Wachi, Y., et al., 1995, *Biochimica et Biophysica Acta* **1244**: 165–168
Yamazawa, A., et al., submitted

54

ANALYSIS OF STRESS RESPONSIVE GENE FOR SALINITY IN A MARINE CYANOBACTERIUM *SYNECHOCOCCUS* sp.

Haruko Takeyama and Hideki Nakayama

Department of Biotechnology
Tokyo University of Agriculture and Technology
2-24-16 Naka-cho, Koganei, Tokyo 184, Japan

There are many stress conditions in natural environment where organisms have been developed their physiological functions to adapt to these conditions. Marine organisms have several strategies for adapting osmotic stress. We have screened many useful marine photosynthetic microorganisms from several oceanic area and found a unique marine cyanobacterium *Synechococcus* sp. NKBG042902 (Matsunaga et al., 1990) which grows under wide salinity range of 0–5% NaCl and quickly adapts to different salinity conditions. The NKBG042902 have an interesting endogenous plasmid, pSY10, which copy number is controlled by salinity conditions (Takeyama et al., 1991). This marine plasmid pSY10 is maintained at high copy number under seawater conditions (including 3% NaCl), whereas its copy number is depressed under fresh water conditions (without NaCl) (Figure 1). The size of this plasmid was determined 2561 bp and *repA* gene sequence encoding replication protein was found (Kawaguchi et al., 1994). The replication of this plasmid was presumed to be mediated by RepA protein. At upstream region of the *repA* gene, two promoter regions (Prep1, Prep2) having high similarity to strong promoter of *E. coli* sigma 70 factor were found. The replication mechanism of pSY10 mediated by RepA protein was elucidated under different salinity conditions.

By measuring the copy number of recombinant plasmids including the *repA* gene and several region of pSY10, it was demonstrated that downstream region of the *repA* gene have negative effect for replication of pSY10 under fresh waster conditions in NKBG042902. However, this regulation of pSY10 replication was not observed under seawater conditions and in the other cyanobacterial host, and the plasmids were maintained at high copy number. Since RepA protein seemed to be responsible for regulating the copy number, binding of RepA protein was analyzed under seawater and fresh water conditions by gel shift assay. When the water-soluble fraction prepared from NKBG042902 harboring native pSY10 grown under seawater conditions (in the BG

New Developments in Marine Biotechnology, edited by Le Gal and Halvorson.
Plenum Press, New York, 1998.

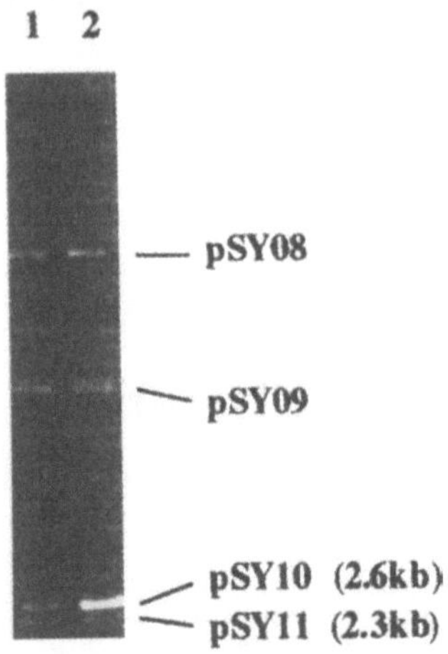

Figure 1. Agarose gel electrophoresis of plasmid DNA extracted from a marine *Synechococcus* sp. NKBG042902. Lane 1: Cultured in BG11 medium without NaCl. Lane 2: Cultured n BG11 medium containing 3% NaCl.

medium including 3% NaCl) was used, band shift of downstream region of the *repA* gene (minus replication region) was observed. On the other hand, no binding was observed when the extract prepared from the NKBG042902 cell grown in the medium without NaCl was used. Furthermore, the cell extracts prepared from the other marine cyanobacterial cells which contained a recombinant plasmid including pSY10 showed the band shift of minus replication region of pSY10 under both salinity conditions. These results suggested that under seawater conditions the negative effect of minus replication region was blocked by binding of the RepA protein to it and pSY10 was maintained at high copy number. On the other hand, transcription of the *repA* gene was depressed under fresh water conditions and only small amount of RepA protein was available to bind, resulting low copy number by effect of minus replication region.

The amount of expressed RepA protein was demonstrated to affect on the copy number of pSY10 directly. It was speculated that transcription of the *repA* gene was controlled by salinity conditions. The band shift analysis was carried out to elucidate the regulation system in upstream of the *repA* gene including its promoter region. Only the extract from the NKBG042902 cells grown under fresh water conditions showed the existence of certain protein(s) which bound *repA* gene promoter region. This phenomenon was observed only in a specific host *Synechococcus* sp. NKBG042902 under fresh water conditions. These results suggested that the protein(s) regulating *repA* gene transcription was expressed from the host chromosome under fresh water condition. Furthermore, the promoter activity of the *repA* gene was directly measured under both salinity conditions using CAT (chloramphenicol acetyltransferase) gene as a reporter gene. Higher CAT activity was obtained under seawater conditions than that under fresh water conditions. When the cells cultured under seawater conditions were transferred to fresh water conditions, the CAT activity decreased rapidly to the level of that in the cells under fresh water conditions. On the other hand, CAT activity increased to the level of seawater conditions after 6 hrs when the cells cultured under fresh water conditions were transferred to seawater conditions. These results showed that stress-response protein(s) was expressed from chromosome under fresh water conditions in a marine *Synechococcus* sp. NKBG042902 and depressed transcription of the *repA* gene following low copy number of pSY10.

Plasmids independently replicates without chromosomal regulation. However, the replication of a marine plasmid pSY10 is obviously controlled by chromosomal functions and therefore this is a novel plasmid replication mechanisms. We have tried to purify the host specific protein(s) expressed through chromosome under fresh water conditions. It may reveal detail regulation mechanisms of salinity stress response gene(s) and the function of this cryptic marine plasmid pSY10.

REFERENCES

Kawaguchi, R., Nagaoka, T., Burgess, J.G., Takeyama, H., and Matsunaga,.T., 1994, *Plasmid* **32**:245–253
Matsunaga,T., Takeyama, and H., Nakamura, N. 1990, *Appl. Biochem. Biotechnol.*, **24/25**:151–160
Takeyama, H., Burgess, J.G., Sode, K., and Matsunaga, T., 1991, *FEMS Microbiol. Lett.* **90**:95–98

55

MUSSELS *MYTILUS* AS MODEL ORGANISMS IN MARINE BIOTECHNOLOGY

Polypeptide Markers of Development and Sexual Differentiation of the Reproductive System

Alexander T. Mikhailov,[1] Mario Torrado,[1] and Josefina Méndez[2]

[1]Institute of Health Sciences
[2]Department of Cell and Molecular Biology
University of La Coruña
Campus de Zapateira
La Coruña 15071, Spain

Bivalve molluscs belonging to the genus *Mytilus* are distributed throughout the world and are extensively used as model organisms in many farming studies: over 20 countries report regular harvests of farmed mussels but world production is dominated by two countries, China and Spain (Hichman, 1992). New information coming from molecular and genetic studies of mussel reproductive biology and ecology (see Rawson and Hilbish, 1995; Saavedra et al., 1996) is characterized by possible implications for the assessment of stocks, for their conservation, maintaining, and restocking. At the same time, because mussels *Myilus,* more than other group of bivalve molluscs, are now used in developmental biology (Mikhailov et al., 1995, 1996a; Togo and Morisawa, 1997), the information presented in our report will serve as a starting point for those wishing to extend such studies to marine biotechnology.

Our experiments were performed on the gonad/mantle tissue of *Mytilus galloprovincialis* Lmk. of Galicia (NW Spain).

MYTILUS GONAD/MANTLE AS A MODEL FOR STUDYING INTO REPRODUCTIVE BIOLOGY

Reproduction involves growth, differentiation, and maturation of reproductive organs, gametogenesis, and spawning. In *M. galloprovincialis*, gonad is not a permanent, differentiated organ, but consists of gonoducts that invade the mantle somatic tissue dur-

New Developments in Marine Biotechnology, edited by Le Gal and Halvorson.
Plenum Press, New York, 1998.

ing each annual reproductive cycle (i.e., gonad and mantle tissues are morphologically fused in this species). The animals undergo one period of reproduction per year, and such repeated reproduction is associated with annual breakdown, resorption and subsequent redevelopment of the gonad and reproductive tract. In mussels *Mytilus*, which exhibit a significant loss of economically important conditions following gonad development and spawning, the use of molecular markers for predicting the gonad/mantle proportions can indicate when the population may be harvested or studied most effectively. We showed for the first time (Mikhailov et el., 1996 b) that *M. galloprovincialis* mantle, of both males and females contains polypeptides (with an apparent MW of 45 to 53 kDa) specific for connective tissue: "mantle connective tissue polypeptide" (MCTPs). Electrophoretic, immunoblotting and immunofluorescent experiments shown that MCTPs are primary localized in the adipogranular (ADG) mantle cells, and their expression in the mantle/gonad is seasonally and morphologically regulated. MCTP expression directly correlates with the volume of mantle connective tissue, but inversely correlates with gonad duct/follicle concentration. MCTPs are overexpressed during a period of sexual rest, when the mantle consists of connective tissue cells, in which reserves of protein, fat and glycogen are accumulated. In contrast, mature gonads contain only trace amounts of MCTPs. Moreover, there is a temporal correlation between the onset and decrease of MCTP expression and the appearance and disappearance of the ADG cells in the mantle. In addition to the ADG cells, positive MCTP immunofluorescence was also detected in the membrane of gonad follicles/ducts (but not in oocytes or spermatozoa) and superficial mantle epithelium. Three types of MCTP distribution in the mantle can annually be distinguished: (1) the cytoplasmic type; (2) the membrane-associated type; and (3) the labeling associated with extracellular granules. It could be speculated that MCTPs are synthesized by ADG cells, excreted as protein-rich granules which can be used to support the processes of gonad morphogenesis and germ cell differentiation. Although, it is unclear whether MCTPs have any functional significance, our data indicate to the possibility to use them as a markers of mantle/gonad conditions. Our preliminary results shown that, using SDS-PAGE and Western blot, up to 100 gonad/mantle samples (obtained by mantle biopsy of living mussels) could be investigated within 2–3 days. Moreover, during the comparative analysis of the sex-specific polypeptides in male/female mantles of *M. galloprovincialis* (see below), we used MCTPs as electrophoretical markers to control the stage of the gonad development/involution directly on the corresponding proteinograms.

MYTILUS GONAD/MANTLE AS A MODEL FOR STUDYING SEX DIFFERENTIATION

A low incidence of hermaphroditism indicates that *M. galloprovincialis* is a strictly gonochoric species. Meanwhile, we could not detect the sex chromosomes or any phenotypic sex-specific characteristics. As in the majority of bivalves, the reproductive system in *Mytilus* is very simple, and there is no sexual dimorphism apart from color of the gonad. This difference is in most cases subtle and is of little use in distinguishing the sexes (Mikhailov et al., 1996b). Our study focused on sexual differentiation of the reproductive system in bivalve molluscs, led to identification of sex-associated protein expression in the gonad/mantle tissue of *M. galloprovincialis*. One of these proteins named "male-associated polypeptide" (MAP) was biochemically and immunochemically characterized (Mikhailov et al., 1995, 1997; Torrado et al., 1995, 1997). Using 2D-electrophoresis followed by Western blot, MAP was identified as a single spot having an apparent MW about of 40 kDa and p*I* of about 6.0. MAP is a monomeric nonglycosylated protein; the molecule contains no cysteines and tryptophans. At the amino acid level, mussel MAP is similar to mammalian carboxylesterases. Levels of MAP expression depended on the stage of male gonad development/involution: its concentration in mature gonads achieved up to 10% of the total soluble protein while in rest gonads only traces of the protein could be detected. However, at each stage of gonad/mantle annual morphogenesis, MAP concentration in males was much higher than that in females. In the females, its concentration either did not exceed 0.5%, or the corresponding fraction could not be detected. We suggested that in female gonad/mantle MAP synthesis is a constitutive process, and activation of its expression is unrelated to differentiation of female gonad phenotype. Such interpretation was reinforced by the results of studying of MAP enzymatic activities. Basing on the immunochemical similarity between mussel MAP, *Drosophila* male-specific esterase S, and rabbit liver carboxylesterase and using the optimized protocol, we could recover MAP carboxylesterase activity after SDS-PAGE of mussel male gonad (Mikhailov et al., 1997; Torrado et al., 1997). Applying the same protocol, we did not detect MAP esterase activity in female gonads, although the same female gonad extracts revealed a weak positive MAP immunoreactivity on Western blots. These data highlight MAP detected in male gonad of *M. galloprovincialis* as a new member of male-predominant carboxylesterases which have been identified in the reproductive system in insects and mammals (see Mikhailov et al., 1997). In mussels, MAP is expressed in male mantle connective tissue (mainly around the gonad ducts), membranes of gonad follicles, as well as in intra-follicular liquid and duct fluid, but not in sperm cells. Nevertheless, MAP concentration depends on the presence or absence of mature spermatozoa in the gonad follicles/ducts. In ripe gonads before spawning, MAP was expressed at high levels, while in spent gonads (after complete spawning) only traces of this polypeptide could be detected (in spent or resting male gonads, the only cells currently detected by us to secrete MAP are the fibroblast-like cells). Moreover, we found MAP in spermatozoa and cell-free fluid obtained from white stream suspensions followed from spawning mussels. We suggest that MAP can be implicated in the process of sperm maturation and emission. Concerning MAP esterase activity, rat male-specific carboxylesterase is also over-expressed in interstitial cells of the testis, and it can bind to several organophosphorous compounds (Yan et al., 1995). The latter suggests that a male-specific polypeptides with carboxylesterase activity may play a role in detoxifying process in male reproductive system. The overall levels of mussel MAP expression in male mature gonad may be determined by tissue- and/or cell-specific transcription factors. This interpretation is supported by our results of MAP over-expression in male morphological compartment of common, hermaphroditic gonad of the scallop, *Pecten maximus.*

REFERENCES

Hickman, R.W. 1992, Mussel cultivatiuon. In: The Mussel *Mytilus:* Ecology, Physiology, Genetics and Culture. E. Gosling (ed.). Elsevier:Amsterdam, pp.465–510.

Mikhailov, A.T., Torrado, M., Méndez, J. 1995, Sexual differentiation of reproductive tissue in bivalve molluscs: identification of male associated polypeptide in the mantle of the *Mytilus galloprovincialis*. *Int. J. Dev. Biol.* **39**:545–548.

Mikhailov, A.T., Torrado, M., Méndez, J. 1996a, Molecular studies of soma-germ cell interactions using mussel gonad as an experimental model. *Int. J. Dev. Biol.* Suppl.**1**:203–204.

Mikhailov, A.T., Torrado, M., Méndez, J., López, M.J. 1996b, Annual cycle of expression of the connective tissue polypeptide markers in the mantle of the mussel, *Mytilus galloprovincialis* Lmk. *Mar. Biol.*, **126**:77–90.

Mikhailov, A.T., Torrado, M., Korochkin, L.I., Kopantzeva, M.A., Méndez, J. 1997, Male-predominant carboxylesterase expression in the reproductive system of molluscs and insects: immunochemical and biochemical similarity between *Mytilus* male associated polypeptide (MAP) and *Drosophila* sex-specific esterase S. *Comp. Biochem. Physiol.* **117B**(4): in press.

Rawson, P.D., Hilbish, T.J. 1995, Distribution of male and female mtDNA lineages in populations of blue mussels, *Mytilus trossulus* and *M. galloprovincialis*, along the Pacific coast of North America. *Mar. Biol.* **124**:245–250.

Saavedra, C., Stewart, D.T., Stanwood, R.R., Zouros, E.,1996, Species-specific segregation of gender-associated mitochondrial DNA types in an area where two mussel species (*M. edulis* and *M. trossulus*) hybridize. *Genetics* **143**:1359–1367.

Togo, T., Morisawa, M. 1997, Aminopeptidase-like protease released from oocytes affects oocyte surfaces and suppresses the acrosome reaction in establishment of polyspermy block in oocytes of the mussel *Mytilus edulis*. *Dev. Biol.* **182**:219–227.

Torrado, M., Méndez, J., Mikhailov, A.T. 1995, Polypeptide markers of the sexual differentiation of reproductive tissue in Mytilus galloprovincialis Lmk. In: Proc. 5th Natl. Congr. on aquaculture. St.Carles de la Rápita, Spain, pp. 198–203.

Torrado, M., Mikhailov, A.T., Korochkin, L.I., Méndez, J. 1997, Sex-dependent carboxylesterase expression in the reproductive system of bivalve molluscs: an approach to substrate-specific detection of male associated polypeptide (MAP) after SDS-electrophoretic separation of crude gonad extracts. Invertebrate Reprod. Devel. in press.

Yan, B., Yang, D., Brady, M., Parkinson, A. 1995, Rat testicular carboxylesterase: cloning, cellular localization, and relationship to liver hydrolase A. *Arch. Biochem. Biophys.* **316**:899–908.

56

MOLECULAR ECOLOGY OF MARINE INVASIONS

Two Case Studies

Jonathan B. Geller

Department of Biological Sciences
University of North Carolina at Wilmington
Wilmington, North Carolina 28403

1. INTRODUCTION

Biological invasions are the establishment of populations of species in regions where they were formerly absent. Recent studies have shown that marine invasions have dramatically increased in the past several decades (Carlton and Geller 1993, Cohen and Carlton 1995). However, only in the past few years have the ecological consequences of invasions been carefully studied (e.g., Brenchley and Carlton 1983, Bertness 1984, Carlton 1989, Carlton et al. 1990, Shushkina and Musayeva 1990, Carlton and Geller 1993, Nalepa and Schloesser 1993, Grosholz and Ruiz 1995). Some examples of invasions which have had profound ecolgical impacts include the zebra mussel *Dreissena polymorpha* in the Great Lakes (Hebert *et al.* 1989), the Asian clam *Potomocorbula amurensis* (Carlton *et al.* 1990), the ctenophore *Mnemiopsis leidyi* in the Black Sea (Vinogradov *et al.* 1989) among others. The U.S. National Research Council has recognized marine invasions as one of five major threats to biodiversity (NRC 1995).

The primary mechanism for modern marine invasions is the use of ballast water in ocean-going ships (Carlton 1985, Carlton and Geller 1993). Ships take up seawater, without treatment or effective filtration against plankton, for stability and maneuverability. Planktonic organisms in one port are captured in ballast tanks and, after a voyage of only 10–14 days across an ocean, are released into another bay or estuary that is environmentally similar to their native home. Carlton and Geller (1993) sampled plankton from 150 ships arriving in Coos Bay, Oregon (USA) from various ports in Japan and found a minimum of 367 species from most major phyla and plant divisions. In short, it is likely that thousands of exotic species are released each day into ports around the world, producing a pattern of global homogenization of marine biotas in urbanized areas (Carlton 1989). For some questions, molecular methods assist in the study of marine invasions, and I have discussed these in detail elsewhere (Geller 1996).

New Developments in Marine Biotechnology, edited by Le Gal and Halvorson.
Plenum Press, New York, 1998.

In this short report, I review two case studies. One involves the invasion of the European mussel *Mytilus galloprovincialis* to southern California (Geller et al. 1994, Geller 1998, Suchanek et al. 1997), and the other focuses on the invasion of Green crabs, *Carcinus maenas*, to five regions globally (Geller et al. 1997).

2. *MYTILUS GALLOPROVINCIALIS*

Mytilus galloprovincialis, the Mediterranean Blue mussel, is an invader in California and many other shores outside its native range in southern Europe (Koehn 1991). This mussel belongs to a species complex which also includes *M. edulis*, native to the North Atlantic, and *M. trossulus*, native to the North Pacific (Koehn 1991). These morphologically similar species can be distinguished genetically using allozymes, mitochondrial DNA (mtDNA) and nuclear DNA markers (McDonald and Koehn 1988, Sarver and Foltz 1993, Geller et al. 1993, 1994, Inoue and Odo 1994, Rawson and Hilbish 1995). Before 1988, all modern Californian populations of Blue mussels were thought to be *M. edulis*, but subsequent genetic analyses revealed that all populations south of San Francisco Bay are *M. galloprovincialis* (McDonald and Koehn 1988, Sarver and Foltz 1993, Geller et al. 1994, Rawson and Hilbish 1995). For example, Suchanek et al. (1997) reported the geographic distribution of diagnostic alleles of a gene encoding a byssal adhesive protein, showing this pattern.

The dominance of an introduced species of mussel in southern California raises several questions: 1) which mussel species was present before the invasion occurred? 2) if mussels prior to invasion were native, did the invader actively displace that native species?

To answer the first of these questions, Geller (in preparation) sequenced a portion of the mitochondrial 16S rRNA gene DNA from mussels collected in southern and central California in the period 1870–1900 and compared them to sequences from all three species of Mytilus . Phylogenetic analyses clearly showed that mussels collected in Monterey Bay and Catalina Island, sites which are now dominated by *M. galloprovincialis*, were *M. trossulus* (Figure 1).

Mussel populations have been continuously present in California since the time these museum specimens were collected. Therefore, the decline of *M. trossulus* did not precede the invasion of *M. galloprovincialis* and the two species co-occurred for some period of time. These observations do not directly support or refute the hypothesis that *M. galloprovincialis* displaced native mussels through competitive interactions, but the temporal overlap between the two species is a requisite condition for that hypothesis. Studies of *M. galloprovincialis* in other areas of the world indicate that it is a strong competitor with other mytilid species (Coustau et al. 1991, Gardner and Skibinski 1991, Skibinski and Roderick 1991, Willis and Skibinski 1992, Hockey and van Erkom Schurink 1992, Gardner 1994), suggesting that competitive displacement is plausible. This historical reconstruction suggests that transplantation experiments might shed light on the fate of *M. trossulus* in southern California.

3. *CARCINUS MAENAS*

European green crabs in the genus *Carcinus* have invaded five regions around the globe in the past century: both coasts of North America, South Africa, Australia, and Japan. (Cohen *et al.* 1995, Grosholz and Ruiz 1996) The genus *Carcinus* is monotypic or

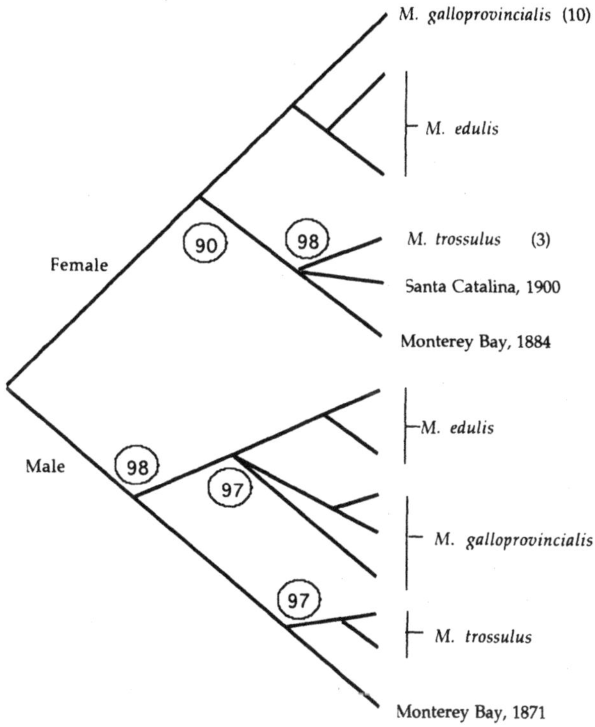

Figure 1. Phylogenetic placement of sequences obtained from 19th century mussels collected in southern and central California in clades with recently collected *Mytilus trossulus*. Polytomies are represented by single branches with the number of sequences indicated in parentheses and relevant bootstrap values are circled.

may consist of a sibling species pair, with *C. maenas* on Atlantic shores and *C. aestuarii* (= *C. mediterraneus)* in the Mediterranean Sea differing in degree of setation of male pleopods. However, neither rigorous morphometrics, genetic, nor mating studies have been conducted, leading to confusion in systematics and identification, especially in sites of invasion. Geller et al. (1997) examined mitochondrial 16S ribosomal mtDNA in order to clarify systematic relationships and identify sources of invasion by Green crabs.

Two mitochondrial 16S rDNA haplotypes, corresponding to Atlantic (Dutch, English, Spanish) and Mediterranean (French and Italian) populations, were seen in the European populations of *Carcinus*. No sequence variation was seen within populations or among sites except between the Atlantic and Mediterranean coasts. Over 395 bp, the two *Carcinus* haplotypes differed by 2.5%. This degree of sequence divergence is consistent with species differences seen in both sibling species and morphologically distinct congeners of brachyuran crabs (Geller et al. 1997). This supports the recognition of Mediterranean populations as a separate species, *Carcinus aestuarii*.

Assigment of crabs from all five invaded regions indicated that populations in California, New England, and Tasmania were solely composed of individuals with the Atlantic haplotype (Table 1). However, our samples from Japan and South Africa contained both haplotypes (Table 1). This contrasts the view that Japanese populations were solely *C. aestuarii* (Ikeda 1989) and that South African populations were solely *C. maenas* (Le Roux et al. 1990).

We have subsequently examined one microsatellite locus (CMCA6) from *Carcinus* (Geller and Bagley, in prep). The Japanese population at this locus is quite differentiated from all European populations that we have examined, so a source cannot be specified. However, there is no relationship between microsatellite genotypes and mitochondrial

Table 1. Frequency of Mediterranean (*Carcinus aestuarii*) and Atlantic (*C. maenas*) mitochondrial 16S rDNA haplotypes in native and introduced populations[a]

Site	Altantic haplotype	Mediterranean haplotype
Native		
Banyul-Sur-Mer, France	0	15(10)
Naples, Italy	0	3(3)
Rio de Betanzos, Spain	14(5)	0
Cadiz, Spain	34	0
Texel Isl. and Den Helder, Netherlands	45(19)	0
Isle of Wight, England	4(3)	0
Introduced		
Bodega Harbor, California, USA	20(5)	0
Mystic, Connecticut, USA	18(9)	0
Capetown, South Africa	48(6)	4(2)
Falmouth, Tasmania, Australia	30(6)	0
Tokyo Bay, (Shinhama), Japan	2(1)	13(6)
Tokyo Bay, (Tokyo University of Fisheries), Japan	8(3)	7(5)

[a]Assignments were based on direct sequencing of PCR products or the presence of an *Alu* I restriction site uniquely in Mediterranean haplotypes. Total sample size is shown, with number of individuals sequenced in parentheses. Data from Geller et al. (1997).

Table 2. Expected heterozygosity at microsatellite locus CMCA6 in native and introduced *Carcinus* populations[a]

Region	n	Heterozygosity
Netherlands	9	0.90
France	9	0.91
Spain	15	0.92
California	23	0.80
South Africa	14	0.89
Japan	32	0.77

[a]Data from Geller and Bagley, in prep. Site information included in legend to Table 1.

haplotypes. It is plausible that Japanese populations are *C. aestuarii* with introgressed *C. maenas* mitochondrial DNA. In general, heterozygosity was lower in introduced populations, with Japan<California<South Africa (Table 2). The loss of heterozygosity, especially in Japan, is suggestive of a genetic bottleneck. If so, this might suggest that cohorts of larvae released by ships are not a broad sample of source populations, but instead represent the products of a subset of the source populations.

In conclusion, molecular approaches to the study of marine invasions are useful when species identifications are in doubt (see Geller 1996). Population genetic studies potentially add information about source regions, secondary spread following invasion, loss of genetic variability, and differentiation of introduced populations relative to sources.

REFERENCES

Bertness, M.D. 1984, *Ecology* **65**:370–381.
Brenchley, G.A., and J.T. Carlton, 1983, *Biological Bulletin* **165**:543–558.

Carlton, J.T. 1985, *Oceanography and Marine Biology Annual Reviews* **23**:313–371
Carlton, J.T. 1989, *Conservation Biology* **3**:265–273.
Carlton, J.T., and J.B. Geller. 1993, *Science* **261**:78–82.
Carlton, J.T., J.K. Thompson, L.E. Schemel, and F.H. Nichols. 1990, *Marine Ecology Progress Series* **66:**81–94.
Cohen, A.N, and J.T. Carlton, 1995, *Report to the U.S. Fish and Wildlife Service,* Washington, D.C.
Cohen, A.N., Carlton, J.T., and M.C. Fountain. 1995, *Marine Biology* **122**:225–237.
Coustau, C., Renaud, F., Maillard, C., Pasteur, N., and B. Delay. 1991, *Genetical Research* **57**:207–212.
Gardner, J.P.A. 1994, *Hydrobiologia* (suppl.) **99**:37–71.
Gardner, J.P.A., and D.O.F. Skibinski. 1991, *Marine Ecology Progress Series* **71**:235–243.
Geller, J. B., Carlton, J.T. , and D.A. Powers. 1993, *Molecular Marine Biology and Biotechnology* **2**:44–50.
Geller, J.B. 1996, *Marine Biology* **122**:239–247.
Geller, J.B. 1998, *Conservation Biology*, in press.
Geller, J.B., and D.A. Powers. 1994. Nautilus **108:** *Suppl.* **2**, 141–4.
Geller, J.B., Carlton, J.T., and Powers, D.A. 1994, *Marine Biology* **119**:243–249.
Geller, J.B., Walton, E.D., Grosholz, E.D., and G.M. Ruiz. 1997, *Molecular Ecology*, **6**:901–906
Grosholz, E.D., and G.M. Ruiz. 1995, *Marine Biology* **122**:239–247.
Hebert, P.D.N., B.W. Muncaster, and G.L. Mackie. 1989, *Canadian Journal of Fisheries and Aquatic Science* **46**:1587–1591.
Hockey, P.A.R., and C. van Erkom Schurink. 1992, *Transactions of the Royal Society of South Africa.* **48**:123–140.
Ikeda, H. 1989,. *Kanagawa Shizenshi Shiryo* **10**:83–85.
Inoue K., and S. Odo 1994, *Biological Bulletin* **186**:349–355.
Koehn, R.K. 1991,. *Aquaculture* **94**:125–145.
Le Roux, P.J., G.M. Branch, and M.A. Joska. 1990,. *South African Journal of Marine Science* **9:**85–92.
McDonald, J.H., and R.T. Koehn 1988, *Marine Biology* **99**:111–118.
Nalepa, T.F., and D.W. Schloesser 1993, Zebra mussels, biology, impacts, and control, (CRC Press, Boca Raton, FL).
National Research Council, 1995, Understanding marine biodiversity, a research agenda for the nation. (National Academy Press, Washington, D.C.).
Rawson, P.D., and T.J. Hilbish 1995, *Molecular Biology and Evolution* **12**:893–901.
Sarver, S.K., and D.W. Foltz 1993, *Marine Biology* **117**:105–112.
Shushkina, E.A., and E.I. Musayeva. 1990, *Oceanology* **30**:225–228.
Skibinski, D.O.F., and E.E. Roderick. 1991, *Biological Journal of the Linnean Society* **42**:351–366.
Suchanek, T., Geller, J.B., Kreiser, B., and J.B. Mitton. 1997, *Biological Bulletin*, in press.
Vinogradov, M. Ye., E.A. Shuskina, E.I. Musayeva, and P. Yu. Sorokin. 1989, *Oceanology* **29**:220–224.
Willis, G.L., and D.O.F. Skibinski. 1992, *Marine Biology* **112**:403–408.

57

A SUPER HEAT-STABLE EXTRACELLULAR PROTEINASE FROM THE HYPERTHERMOPHILIC ARCHAEON *AEROPYRUM PERNIX* K1

P. Chavez C., Y. Sako, and A. Uchida

Laboratory of Marine Microbiology
Division of Applied Bioscience
Graduate School of Agriculture
Kyoto University
Kyoto 606, Japan

1. INTRODUCTION

Many proteases have been isolated and characterized from thermophilic microorganisms with the expectation that these enzymes would have higher intrinsic stability than mesophilic proteases. This expectation is based on the fact that there is a positive correlation between the growth temperature of the microorganisms and its enzyme stability (Cowan et al., 1985). In addition, the properties which allow thermostables enzymes to withstand high temperatures also confer resistance to other factors such as, detergents and denaturing agents, thereby offering great potential advantages for a range of biotechnological processes (Adams et al., 1995). The most detailed studies have been on Thermolysin, a metalloprotease isolated from *B. thermoproteolyticus* (Endo, 1962). It is the only thermophilic protease currently used on an industrial scale. Recently, the search for more thermostable proteases has been focused on hyperthermophilic archaea. Most of the proteases isolated correspond to intracellular serine enzymes. However, although proteases from hyperthermophiles are certainly more stables, 100°C appears to be the upper limit for long enzyme half lives.

Only few proteases have been reported from hyperthermophilic microorganisms (Cowan et al. 1985). All purified archaeal proteases show a high thermostability and thermoactivity. One of the most thermostable protease reported is pyrolysin (Eggen et al. 1990) from *Pyrococcus furiosus* (Fiala and Stetter, 1986) with a half-live value of 4 h at normal boiling point of water. The half life of pyrolysin at 110°C is 3 min. *Aeropyrum*

New Developments in Marine Biotechnology, edited by Le Gal and Halvorson.
Plenum Press, New York, 1998.

Table 1. Summary of purification of an extracellular proteinase (Aeropyrolysin) from *Aeropyrum pernix* K1

Preparation	Volume (ml)	Total Protein (mg)	Total Activity (U)	Yield (%)	Specific Activity (U/mg)	Purification fold
Supernatant	20,000	N.D.	N.D.	N.D.	N.D.	N.D.
$(NH_4)_2SO_4$	80	58	130,000	100	1,961	1
Q Sepharose	40	25	70,400	54	2,843	1.5
DEAE-cellulose	18	10	219,600	169	21,960	11.2
Mono Q	5	3.6	128,500	99	35,694	18.2
Superose 12	2.5	1.3	73,750	57	57,617	29.3

pernix K1 is the first strictly aerobic hyperthermophilic archaeon that has been described to date (Sako et al. 1996). It grows optimally above 90°C with a doubling time of 200 min. Large scale culture can be achieved with high growth and high cell yield. These characteristics make this archaeon a potential biotechnological resource for the purification of the most stable enzymes. Recently, we have purified and biochemically characterized an extraordinarily hyperthermostable proteinase (aeropyrolysin) from the new isolated archaeon *Aeropyrum pernix* K1 (Sako et al. 1997). The characterization of this enzyme is showed in this study.

2. MATERIALS AND METHODS

The enzyme was purified from free-cell supernatant of *Aeropyrum pernix* K1 culture (Sako et al. 1996). Purification (Table 1) was carried out by amonium sulfate precipitation (80%), anionic exchange chromatography (Q-Sepharose and Mono Q, Pharmacia columns) and gel filtration chromatography (Superose 12, Pharmacia). The purity of the enzyme was analyzed by 10% SDS-PAGE.

The catalytic activity of the purified enzyme with regard to temperature, pH, inhibitors, substrate specificity and denaturing agents was examined as follows: enzyme solutions were incubated in sealed glass capillary tubes in a temperature-controlled oil bath at intervals. Protease activity was determined by the hydrolysis of azocasein at 90°C for 20 min. One unit of proteinase activity was defined as giving an absorbance change of 0.1/min at 340 nm.

3. RESULTS AND DISCUSSION

Azocasein hydrolysis activity was observed at extracellular culture of *Aeropyrum pernix* K1, with a maximal activity at the early stationary phase. From supernatant cultures a single monomeric proteinase of 52 kDa was purified (Table 1).

Biochemical characterization of the enzyme was examined. The optimum temperature for azocasein hydrolysis was 100°C for native enzyme. The addition of 1 mM $CaCl_2$ to the enzymatic reaction mixture increased the activity at temperatures above 100°C. The optimum activity of the enzyme in the presence of calcium ion was 110°C (Fig. 1).

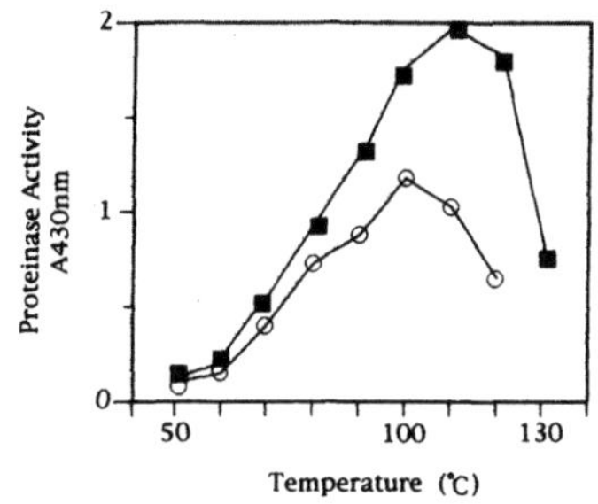

Figure 1. Effect of temperature on Aeropyrolysin proteolytic activity. Proteolytic activity was assayed at the indicated temperature in the presence (■) and absence (O) of 1 mM $CaCl_2$ for 20 min.

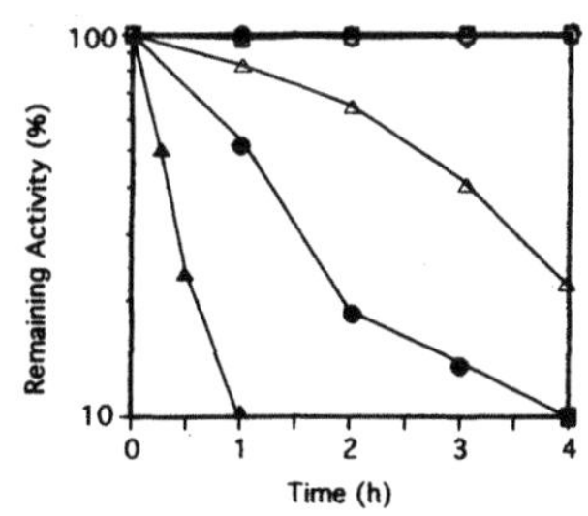

Figure 2 Thermostability of Aeropyrolysin at temperatures between 90°C and 125°C. Purified enzyme (2.5 μg/ml in 50 mM Tris-HCl (pH 8) containing 1 mM $CaCl_2$) was incubated at 90°C (O), 100°C (□), 110°C (◆), 120°C (△) and 125°C (●). Native enzyme (▲) was incubated at 100°C in the absence of 1 mM $CaCl_2$. Samples of enzyme solution were removed at intervals, cooled, and the residual proteolytic activity was determinated as described in Materials and Methods.

The activity of the enzyme at 90°C at different pH solutions was examined. The enzyme showed an apparent maximal activity at pH 6–8 for the hydrolysis of azocasein. Removal of calcium ion by the addition of EDTA or EGTA inhibited proteolytic activity indicating that it corresponds to a metalloproteinase. No significant effect was observed for other protease inhibitors such as, PMSF, DFP, N-ethylmaleimide, iodoacetic acid and p(chloromercuri)benzoic acid. In addition, the protease showed a high activity with the metalloproteinase specific substrate MOCAc–Pro–Leu–Gly–Leu A2pr(Dnp)–Ala–Arg–NH_2. The activity of the enzyme was not greatly affected by the presence of chaotropic agents, reducing agents or detergent. On the contrary, the presence of 1%, 2%, 3% and higher concentrations of SDS increased the proteolytic activity at 90°C. Thermostability experiments showed half lives of 2.5 h at 120°C and 1.2 h at 125°C (Fig. 2). The enzyme presented in this study showed the highest thermostability values to date reported. A comparison of this enzyme with other proteases from hyperthermophilic microorganisms is shown in Table 2. The characteristics of aeropyrolysin suggested that this enzyme could be an attractive candidate for some biotechnological applications.

Table 2. Comparison of *A. pernix* proteases and other thermostable proteases from archaea

Protease	Mol. mass (kDa)	Optimum pH	Optimum temp. (°C)	Active Site	Half-life	Origin
Aeropyrolysin	52	10.4	100	Metallo	120°C - 2.5h 125°C - 1.2h	*Aeropyrum pernix*
Pyrolysin	140, 130, 115, 100, 65	6.5-10.5	115	Serine	95°C - 9h	*Pyrococcus furiosus*
TT protease	44	7.0	110	Cysteine	100°C - 1h	*Pyrococcus sp.*
S66	66	7.0	105	Serine	98°C - 33h	*Pyrococcus furiosus*
Archaelysin	52	7.0	98	Serine	105°C - 8-9min	*Desulfurococcus sp.*
Thermopsin	45	2.0	90	Aspartate	90°C - 90min	*Sulfulobus acidocaldarius*

REFERENCES

Adams, M.W., Perler, F.B., Kelly, R.M. 1995, *Biotechnology*. **13**:662–668.
Cowan, D.A., Daniel, R.M., Morgan, H.W. 1985, *Trends Biotechnol.* **3**:68–72.
Eggen R., Geerling A., Watts J., M. de Vos W. 1990, *FEMS Microbiology Letters* **71**:17–20.
Endo, S. 1962, *Hakko Kogaku Zasshi*. **40**:34~353.
Fiala, G., and Stetter, K.O., 1986, *Arch. Microbiol.* **145**:56–60.
Sako, Y., Chavez-Croocker, P., and Ishida, Y., 1997, *FEBS Lett.* **415**:329–334.
Sako, Y., Nomura, N., Uchida, A., Ishida, Y., Morii, H., Koga, Y., Hoaki, T., Maruyama, T., 1996, *Int. J. Syst. Bacterol.* **46**: 1070–1077.

58

MANNOSE ADHESIN–GLYCAN INTERACTIONS IN THE *EUPRYMNA SCOLOPES–VIBRIO FISCHERI* SYMBIOSIS

M. McFall-Ngai,[1,*] C. Brennan,[3] V. Weis,[2] and L. Lamarcq[1]

[1]Pacific Biomedical Research Center
University of Hawaii
[2]Department of Zoology
Oregon State University
Corvallis, Oregon 97331-2914
[3]Department of Biological Sciences
University of Southern California
Los Angeles, California 90089-0371

The recognition of and adherence to animal epithelia by bacteria is often mediated by bacterial surface lectins, called adhesins, that bind to sugars on the brush borders of host cells (Giampapa *et al.*, 1987; Abraham *et al.*, 1988; Sharon *et al.*, 1989). We report here results of experiments that provide evidence for the role of adhesin-glycan interactions in the initiation of the model symbiosis (Ruby, 1996) between the sepiolid squid *Euprymna scolopes* and its light organ symbiont *Vibrio fischeri*.

1. MATERIALS AND METHODS

Animals were collected and maintained as previously described (McFall-Ngai *et al.*, 1991). To investigate whether bacterial adhesin-animal glycan interactions are involved in the successful colonization of the *E. scolopes* light organ by *V. fischeri*, we determined: i) the type of adhesin(s) present on the surfaces of *V. fischeri*; ii) the nature of the glycans on light organ epithelia that interface with the symbionts; and, iii) the ability of sugars and sugar analogs to inhibit the experimental inoculation of the symbiotic organ.

Standard hemagglutination assays were employed for the detection of adhesins on *V. fischeri* (Duguid *et al.*, 1980). A hemagglutination deficient mutant of *E. coli* (pilE$^-$, strain

* Corresponding author: PBRC, U of Hawaii, 41 Ahui St., Honolulu, HI 96813. Telephone: (808) 539-7310; Fax: (808) 599-4817; e-mail: mcfallng@hawaii.edu

New Developments in Marine Biotechnology, edited by Le Gal and Halvorson.
Plenum Press, New York, 1998.

ORN133) and its parent strain (Maurer *et al.*, 1987) were used as negative and positive controls, respectively. Sugars on the surfaces of light organ host cells were detected by light microscopy on paraffin-embedded tissue sections using specific lectins conjugated to colorimetric reporters (Bry *et al.*, 1996). In colonization experiments, a light organ isolate, *V. fischeri* strain ES114 (Lee *et al.*, 1994), was grown in seawater-based media, washed and inoculated into seawater containing sugars or sugar analogs; specific sugars that are metabolizable by *V. fischeri* were substituted with nonmetabolizable analogs. After 15 min, the seawater was dispensed as 5 ml aliquots into scintillation vials and a single newly hatched squid was introduced to each vial. Host tissue colonization was detected both by monitoring *V. fischeri* luminescence with a photometer and by enumerating colony-forming units (CFU) of *V. fischeri* that resulted from the plating of light organ homogenates.

2. RESULTS

In analyses of *V. fischeri* for the presence of specific adhesins, both the symbiosis-competent strain ES114 and the symbiosis-noncompetent strain MJ1 were found to agglutinate guinea pig red blood cells (RBCs). These data suggest that *V. fischeri* has mannose adhesins either on fimbriae or on the cell wall.

Because these results suggested that *V. fischeri* has mannose-recognizing adhesins on their surfaces, we analyzed host tissues for the presence of mannose residues that might behave as receptors for these adhesins. Histological sections of the juvenile light organ stained positively for the presence of mannose along the apical surfaces of the host epithelial cells, whereas there was no detectable staining in samples incubated with the fucose-recognizing lectin, which was used as a control for the possible non-specific binding of lectins to host tissue. In addition, we analyzed host tissue extracts for the presence of 5 common glycans (Glycan Differentiation Kit, Boehringer Mannheim, Inc., Indianapolis, IN, USA). Whereas control extracts from mantle tissue had all 5 of the glycans detected by this method, only mannose was abundant enough in light organ extracts to be detected.

In determinations of the ability of several sugars and sugar analogs to block infection (Fig. 1), only the mannose analog, methyl-a-D-mannopyranoside (Sigma Chemical Co., St. Louis, MO, USA), was capable of blocking infection; i.e., only the animals exposed to this sugar under prescribed conditions (see below) showed neither luminescence nor detectable CFUs in light organ homogenates. In the presence of other sugars or sugar

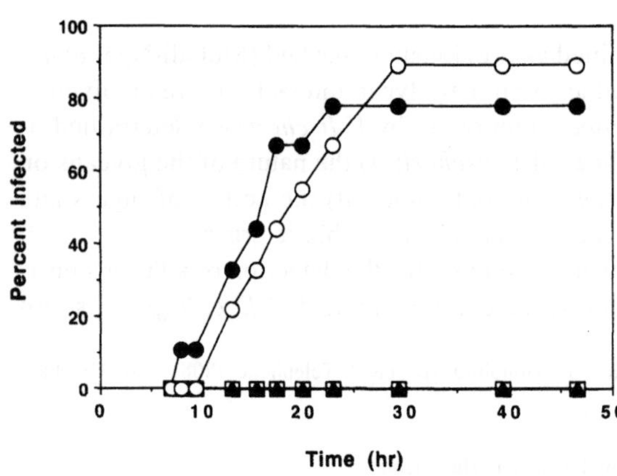

Figure 1. The effect of the presence of sugars on colonization of the *E. scolopes* light organ. Exposure to sugars (0.25%) such as a galactose analog (○), fucose, or a glucose analog resulted in similar colonization levels of the light organ as control animals not exposed to sugars during the colonization period (●). Exposure to the mannose analog (0.25%) inhibited infection fully (□). The control animals not exposed to added *V. fischeri* (▲) also showed no light organ colonization.

analogs, under all conditions, the juveniles exhibited luminescence and averaged 3.0×10^5 CFU/animal, similar to the colonization level of control animals not exposed to sugars. The growth conditions of the bacteria, i.e., grown in either a rich or a minimal medium, had no discernible effect on the inhibition patterns of the sugars.

We investigated the conditions under which mannose analogs were capable of inhibiting inoculation of the light organ. With strain ES114 at a concentration of 700 cells/ml, the mannose analog at concentrations from 2% down to 0.25% was capable of blocking infection in 100% of the juveniles. [Animals were healthy for up to 6 d when continuously exposed to 2% sugar in seawater; however, at higher percentages of sugar, they showed signs of stress, and 5% sugar was lethal within a few hours.] At slightly higher inoculation densities of the bacteria, approximately 1000 cells/ml, the mannose analogs were less effective at blocking infection. However, in those animals that did become infected, the onset of infection was always delayed by several hours relative to control animals, and the extent of the infection was always lower in comparison with controls exposed to no sugar or sugars other than the mannose analog. When bacteria were present at a concentration of 5000 cells/ml, full colonization was not inhibited in juvenile animals by the mannose analog, but was delayed up to 10 hr. These data suggest that inhibition by mannose analogs is only effective at concentrations of bacteria similar to those the animal experiences in its natural environment (Lee *et al.*, 1994).

We also found that the continued presence of the mannose analog is required to block infection by the bacteria. After the animals were exposed to bacteria in the presence of the mannose analog for 4 hr, the animals were placed in seawater without the sugar. These juveniles became infected, although the infection was delayed a few hours in comparison with the infection kinetics of the control groups. Also, when added after colonization, mannose analogs were not capable of eliminating the infection or changing the colonization level, even when added within a few hours of exposure to *V. fischeri* cells. Finally, animals that had been exposed to the mannose analog for 48 hr, rinsed and placed in seawater without analog, but containing 700 *V. fischeri* cells/ml, were colonized within 20 hr following exposure to symbionts. These data provide evidence that long periods of exposure to the sugar did not permanently alter the animal's ability to become colonized by its bacterial partner.

3. DISCUSSION

These data provide evidence for the presence of mannose-recognizing molecules on the surface of *V. fischeri* cells and mannose glycans on the apical surfaces of the host epithelia that interact with the bacteria during colonization. Colonization experiments with mannose analogs showed that this specific sugar is capable of inhibiting infection of the light organ, presumably by binding to the adhesins on *V. fischeri*, thereby blocking those sites from interaction with host cells. The final proof of the involvement of adhesin-glycan interactions in this symbiosis awaits the production of adhesin-defective mutants of *V. fischeri*.

ACKNOWLEDGMENTS

We thank Professor S. Clegg, University of Iowa, for assistance with the study of *V. fischeri* adhesins. This work was funded by NSF grant #IBN 96-01155 and ONR grant #N00019-J-1357.

REFERENCES

Abraham, S. N., D. Sun, J. B. Dale, and E. H. Beachey. 1988, *Nature* **336**:682–684.

Bry, L., P. R. Falk, T. Midtvedt, and J. I. Gordon. 1996, *Science* **273**:1380–1383.

Duguid, J. P., and D. C. Old. 1980, pp. 184–217 in Beachey, E. H. (ed.), *Bacterial Adherence.* Chapman & Hall, New York.

Giampapa, C. S., S. N. Abraham, T. M Chiang, and E. H. Beachey. 1987, *J. Biol. Chem.* **263**:5362–5367.

Lee, K.-H. and E. G. Ruby. 1994, *Appl. Environ. Microbiol.* **60**:1565–1571.

Maurer, L., and P. E. Orndorff. 1987, *J. Bacteriol.* **169**:640–645.

McFall-Ngai, M. J. and E. G. Ruby. 1991, *Science* **254**:1491–1494.

Ruby, E. G. 1996, *Annu. Rev. Microbiol.* **50**:591–624.

Sharon, N. and H. Lis. 1989, *Science* **246**:227–234.

59

TEMPORAL CONTROL OF *LUX* GENE EXPRESSION IN THE SYMBIOSIS BETWEEN *VIBRIO FISCHERI* AND ITS SQUID HOST

Karen L. Visick and Edward G. Ruby*

Pacific Biomedical Research Center
University of Hawaii
Ahui Street, Honolulu, Hawaii 96813

A long-standing question in the study of luminous bacterial symbionts has been: What is the importance of light-emission to the bacteria? We have used the symbiosis between *Vibrio fischeri* and the Hawaiian sepiolid squid *Euprymna scolopes* as a model system (McFall-Ngai & Ruby, 1991; Ruby, 1996) with which to discover the answer to this question. Within hours of hatching, juvenile squids become inoculated by a few cells of *V. fischeri* present in the ambient seawater (Ruby & Asato, 1993). These bacteria rapidly colonize the epithelium-lined crypts of the nascent light-emitting organ, attaining a population size of about one million cells. Their light emission is used by the host in its nocturnal behavior (McFall-Ngai, 1997), and evidence is accumulating that suggests that the host regulates the bacterial luminescence on multiple levels (Boetttcher & Ruby, 1995; Boettcher et al., 1996).

In culture, luminescence is under the control of the LuxR activator protein complexed with autoinducer, a homoserine lactone derivative (*N*-(3-oxo-hexanoyl)-L-homoserine lactone) synthesized by LuxI. The LuxR-autoinducer complex induces the transcription of the *lux* operon, which includes the *luxI* gene, the genes for the luciferase enzyme (*luxAB*), and the genes for the synthesis of the aldehyde substrate (*luxCDE*). Strains of bacteria that are defective for either structural (*luxA*) or regulatory (*luxR*, or *luxI*) luminescence genes can initiate the colonization of the light organ just as well as the wildtype, but are unable to maintain a normal level of colonization beyond 24 to 48 hours (Visick et al., 1996; Visick & Ruby, 1996). This result suggests that the ability to express the luminescence reaction is itself a requirement for symbiotic competency.

To further elucidate the role of luminescence in the symbiosis, we sought to control the expression of the lux genes independently of their known regulatory genes, *luxR* and *luxI*. This was accomplished by constructing a plasmid, pKV44, that carries mutations in

* Phone: (808) 539-7309, Fax: (808) 599-4817, E-mail: eruby@hawaii.edu

New Developments in Marine Biotechnology, edited by Le Gal and Halvorson.
Plenum Press, New York, 1998.

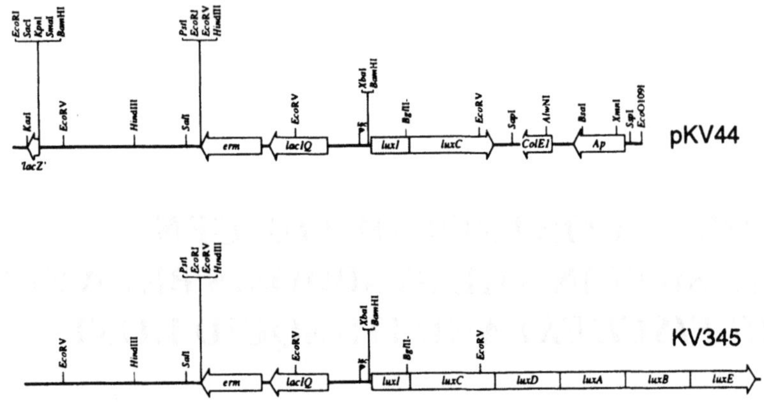

Figure 1. Plasmid pKV44 and its use to construct a chromosomal *lux* operon under the control of the *tac* promoter. The *luxR* gene was replaced with a gene for erythromycin resistance, the *tac* promoter and the regulatory gene, *lacIQ*. The *luxI* defect was constructed by the insertion of a 2-bp frameshift mutation at the BglII restriction site.

both the *luxR* and *luxI* genes, as well as the *tac* promoter (P*tac*) upstream of the *lux* operon (Fig. 1). The LacI^Q repressor, which negatively regulates expression from P*tac*, was also inserted upstream of the promoter, and can be manipulated by the addition of isopropyl thiogalactopyranoside (IPTG) to the culture medium to induce expression of P*tac*-regulated genes. Plasmid pKV44 was electroporated into *V. fischeri* strain ESR1, and a recombinant, KV345, that had incorporated the mutations (D*luxR*::*erm* P*tac* *luxI*$^-$) into the chromosome was isolated (Fig. 1).

The luminescence of KV345 was examined to determine whether the expression of the *lux* genes was up-regulated in response to IPTG added to the culture medium. When growing in culture, KV345 cells produced a moderate level of luminescence, and upon the addition of IPTG, this luminescence was increased over 100-fold (Fig. 2). These data

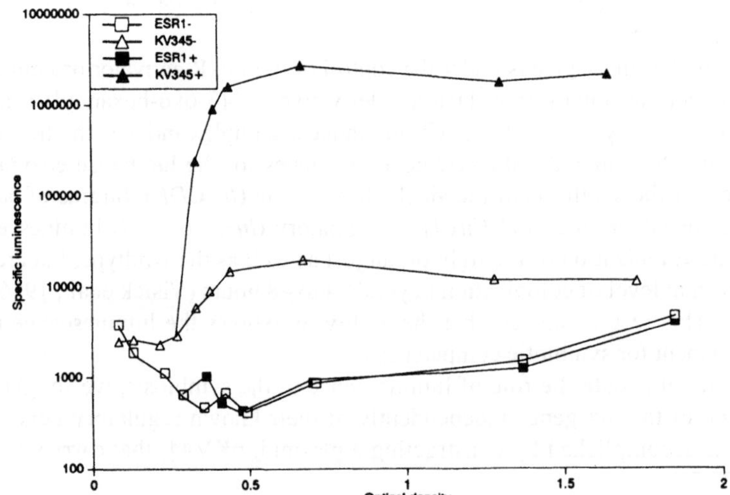

Figure 2. Specific luminescence of the *lux* mutant strain KV345 and its parent, ESR1, in the presence (+) or absence (–) of 235 ug of IPTG per ml in SWT broth.

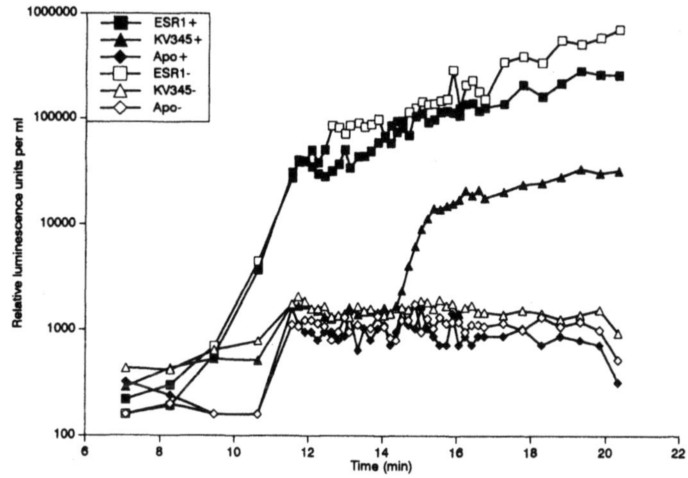

Figure 3. Luminescence of juvenile squids infected with either the parent strain (ESR1) or with the *lux* mutant strain (KV345) plotted versus time. IPTG was added to some vials (+) 14 h after initiation of the association.

demonstrate that KV345 is in fact dependent upon IPTG-mediated induction for maximal luminescence in culture.

It was not clear whether the basal level of luminescence produced by this strain would be sufficiently high for detection in juveniles infected with this strain. In addition, although *V. fischeri* inside the light organs of juvenile or adult animals are known to be susceptible to certain antibiotics when added to the seawater, IPTG had never before been tested for its ability to penetrate the light organ matrix to regulate bacterial genes. To test whether KV345 would produce luminescence in the association, juveniles of *E. scolopes* were incubated in vials of seawater containing either *V. fischeri* strain KV345 or its parent, ESR1. Measurements over the first 14 h revealed that, although juveniles infected with ESR1 cells had begun to luminesce, no light was detected in squid infected with KV345. However, KV345-infected squid became detectably luminous within 30 min after the addition of IPTG (Fig. 3). Thus, it is possible to manipulate the expression of the *lux* genes through the use of the tac promoter and the addition of IPTG to the seawater surrounding juveniles of *E. scolopes*. Use of IPTG to change the temporal expression of luminescence will facilitate our experiments to determine why *lux* mutants are unable to maintain a normal colonization in the symbiosis. Finally, this procedure can be generally applied to control the expression of other bacterial genes that are important in the symbiosis.

REFERENCES

Boettcher, K. J. and E. G. Ruby, 1995, *J. Bacteriol.* **177**:1053–1058.

Boettcher, K. J., E. G. Ruby and M. J. McFall-Ngai,1996, *J. Comp. Physiol.* **179**:65–73.

McFall-Ngai, M. J., 1997, *Amer. Zool.* (in press).

McFall-Ngai, M. J. and E. G. Ruby, 1991, *Science* **254**:1491–1494.

Ruby, E. G. 1996, *Ann. Rev. Microbiol.* **50**:591–624.

Ruby, E. G. and L. M. Asato, 1993, *Arch. Microbiol.* **159**:160–167.

Visick, K. L., V. O. Orlando, and E. G. Ruby, 1996, *Abst. Gen. Meet. Am. Soc. Microbiol.* **96**:509.

Visick, K. L. and E. G. Ruby, 1996, *Gene* **175**:89–94.

60

BACTERIAL SYMBIONTS OF THE BRYOSTATIN-PRODUCING BRYOZOAN *BUGULA NERITINA*

Margo G. Haygood and Seana K. Davidson

Marine Biology Research Division and Center for Marine Biotechnology and Biomedicine
Scripps Institution of Oceanography
University of California, San Diego
La Jolla, California 92093-0202

Bugula neritina (Cheilostomata, Cellularoidea) is an arborescent bryozoan with a cosmopolitan distribution in temperate marine waters. *Bugula neritina* is a subject of current research interest because it is the source of the bryostatins (Pettit, 1991) a family of cytotoxic macrocyclic lactones, one of which, bryostatin 1, is currently in clinical trials for treatment of cancer (Jayson et al., 1995; Philip et al., 1993; Prendiville et al., 1993). Many bryozoans produce alkaloids (Anthoni et al., 1990) presumably as a chemical defense, but *B. neritina* is the only bryozoan currently known to produce complex polyketides. Polyketides are compounds typically produced by actinomycetes and fungi, and in at least one case by a proteobacterium (Schupp et al., 1995).

Bugula neritina harbors bacteria in the funicular cords (Woollacott & Zimmer, 1975) and larvae (Woollacott, 1981) . We have initiated investigations of *B. neritina*-associated microorganisms to determine their role in the relationship, and whether they influence production of bryostatins. The subject of this report is the symbionts found in *B. neritina* larvae. In *B. neritina*, the larvae are coronate and anenteric. They are brooded in the adult colony, then released into the seawater where they settle on suitable substrate within ten hours and metamorphose (Woollacott, 1971). Bacteria in the larvae of *B. neritina*, *B. simplex* and *B. pacifica* were described by R. M. Woollacott (1981) as extracellular rod-shaped bacteria located in the pallial sinus.

The goals of this study were 1) to obtain small subunit (SSU, or 16S) rRNA gene sequence from the larval symbiont of *B. neritina*, 2) to verify the authenticity of the sequence obtained by *in situ* hybridization with specific oligonucleotide probes based on the SSU sequence, 3) to analyze the phylogenetic affiliations of the symbiont, 4) to determine whether any of the bacteria that can be readily cultured from the larvae correspond

New Developments in Marine Biotechnology, edited by Le Gal and Halvorson.
Plenum Press, New York, 1998.

to the larval symbiont and 5) to explore whether the larval symbiont can be reduced or eliminated from colonies in order to enable us to investigate whether the symbionts are required for bryostatin synthesis.

SEQUENCING AND OLIGONUCLEOTIDE DESIGN

Larvae were collected, washed three times in sterile seawater and DNA extracted. SSU rRNA genes were amplifed with universal primers and sequenced. Sequence was obtained for the region corresponding to *E. coli* positions 38–1265. Phylogenetic analysis of the sequence using maximum parsimony and neighbor-joining showed that the larval symbionts are gamma proteobacteria not closely related to any sequences in the Genbank or RDP databases (Figure 1). The sequences from six populations of *B. neritina* from California and one from North Carolina are very closely related.

Oligonucleotides for in situ hybridization and PCR were designed based on the sequences. A negative control probe with the same base composition as Bn1253r, but a scrambled sequence was also designed. Database searches found no sequences that matched the oligonucleotides perfectly. A specific oligonucleotide, Bn1253r, and the negative control probe were tested by in situ hybridization with strains with the most similar sequence in the probe region that we could obtain. Test strains were hybridized at temperatures ranging from 42–55°C. The eubacterial probe bound but Bn1253 did not, confirming its specificity. The negative control probe produced no signal.

IN SITU HYBRIDIZATION

Whole larvae were separately hybridized with the eubacterial probe (positive control) the specific probe Bn1253r and the negative control probe. Both the eubacterial probe and the specific probe bound to the bacteria in the pallial sinus, the negative control probe did not. With the universal probe, bacteria elsewhere were only rarely observed. Binding of the specific probe confirms the authenticity of the sequence obtained.

CULTIVATION OF BACTERIA FROM *B. NERITINA* LARVAE

Larvae were plated on complex media but relatively few bacteria were recovered. Typically a variety of colony types was observed at about 1–5 colony forming units per larva. A specific PCR assay for the larval symbiont was developed using two specific oligonucleotides as primers was developed. DNA was extracted and tested from 15 representative isolates, none of which was positive.

CURING OF NEWLY SETTLED COLONIES

Larvae were settled in seawater containing various antibiotics. Treatments extended for six days, with daily change of antbiotics. Colonies were then raised for three months in flowing seawater and fed a mixture of phytoplankton. Colonies were then sacrificed and assayed for the presence of the larval symbiont with the specific PCR assay. Treatment with gentamicin was most effective, resulting in a 50–100-fold reduction in symbiont number.

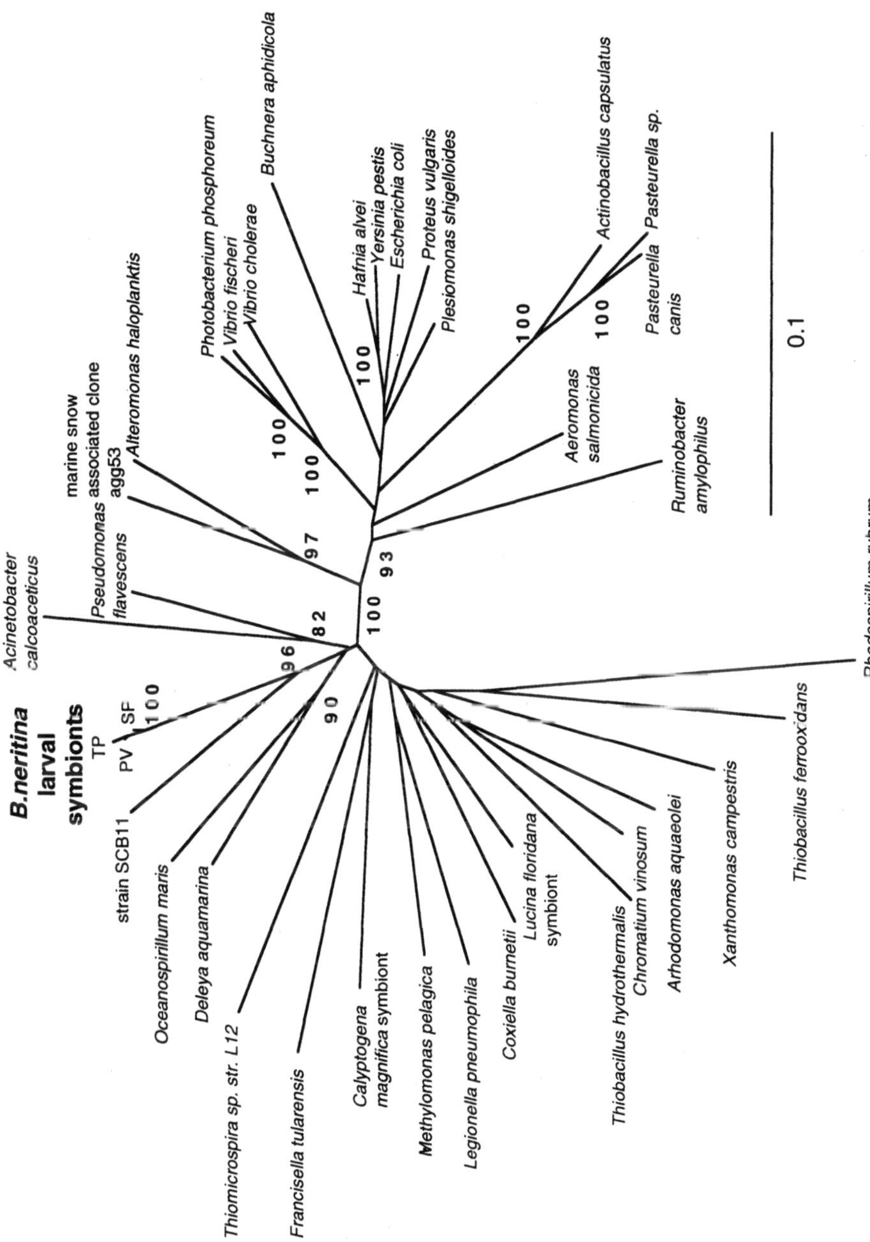

Figure 1. Gamma proteobacteria neighbor joining phylogenetic tree.

The larval symbionts of *B. neritina* are a novel type of gamma proteobacteria. Because they cannot readily be isolated we believe they are previously unknown and not simply unrepresented in sequence databases. We have proposed the taxonomic status "*Candidatus* Endobugula sertula" for this organism. Candidatus status is used for incompletely described prokaryotes in cases when a type strain is not available.

ACKNOWLEDGMENTS

This work was funded by a grant from the National Sea Grant College Program, National Oceanic and Atmospheric Administration, U.S. Department of Commerce, under grant number NA36RG0537, project number RMP-61 through the California Sea Grant College System, and in part by the California State Resources Agency. The views expressed herein are those of the authors and do not necessarily reflect the views of NOAA or any of its sub-agencies. The U.S. Government is authorized to reproduce and distribute for governmental purposes.

We are grateful to CalBioMarine Technologies, Inc. for advice, technical assistance and the donation of *B. neritina* samples. We thank Linda Z. Holland for advice on *in situ* hybridization methods, Edward F. DeLong for the gift of oligonucleotides, and Ronald McConnaughey for expert assistance with collecting. Meredith Leong assisted with bacterial isolations and PCR evaluation of isolates.

REFERENCES

Anthoni, U., P. H. Nielsen, M. Perieira, and C. Christophersen. 1990,. *Comp. Biochem. Physiol.* **96B:**431–437.

Jayson, G. C., D. Crowther, J. Prendiville, A. T. McGown, C. Scheid, P. Stern, R. Young, P. Brenchley, J. Chang, S. Owens, and et al. 1995,*British Journal of Cancer.* **72**(2):461–8.

Pettit, G. R. 1991, *Prog. Chem. Org. Nat. Prod.* **57:**153–195.

Philip, P. A., D. Rea, P. Thavasu, J. Carmichael, N. S. Stuart, H. Rockett, D. C. Talbot, T. Ganesan, G. R. Pettit, F. Balkwill, and et al. 1993, *Journal of the National Cancer Institute.* **85**(22):1812–8.

Prendiville, J., D. Crowther, N. Thatcher, P. J. Woll, B. W. Fox, A. McGown, N. Testa, P. Stern, R. McDermott, M. Potter, and et al. 1993, *British Journal of Cancer.* **68**(2):418–24.

Schupp, T., C. Toupet, B. Cluzel, S. Neff, S. Hill, J. J. Beck, and J. M. Ligon. 1995, *Journal of Bacteriology.* **177**(13):3673–3679.

Woollacott, R. M. 1981, *Mar. Biol.* **65:**155–158.

Woollacott, R. M., and R. L. Zimmer. 1975, *J. Morph.* **147:**355–378.

Woollacott, R. M. a. R. L. Z. 1971, .*J. Morph.* **134:**351–382.

ARE GAMMA *PROTEOBACTERIA* THE PREDOMINANT SYMBIONTS IN THE SQUID *LOLIGO PEALEI*?

Elena Barbieri,[1,*] Deborah Hughes,[2] Rebecca Ericson,[3] and Andreas Teske[4]

[1]Marine Biological Laboratory
Woods Hole, Massachusetts
[2]Scripps Institution of Oceanography
La Jolla, California
[3]Forsyth Dental Center
Boston, Massachusetts
[4]Woods Hole Oceanographic Institution
Woods Hole, Massachusetts

1. INTRODUCTION

Like many cephalopods, females of the squid species *Loligo pealei* have a pair of accessory nidamental glands (ANG) that harbor a dense bacterial community (Bloodgood, 1977). The ANG is closely associated with the nidamental gland, which produces capsule-like membranes surrounding the eggs (Fig. 1). Because of its close association with the nidamental gland and the oviduct, the ANG probably plays a role in protecting the eggs by coating them with bacteria (Barbieri et al., 1997; Biggs & Epel., 1991). Recently, identical or very similar bacteria have been found in the ANG and in the egg cases of *Loligo pealei* (Barbieri et al., unpublished results).

The ANG, which is colorless in immature animals, becomes red-orange in reproductively mature females. The color has been attributed to symbiotic bacteria (Bloodgood, 1977; Dunlap, 1990) which were tentatively identified as *Pseudomonas* and *Vibrio* species

* *Corresponding author. Mailing address: Istituto di Scienze Tossicologiche Igienistiche ed Ambientali, University of Urbino, Via S. Chiara 27, 61029 Urbino, Italy. Phone 39 (0) 722 350585. Fax 39 (0) 722 4714. E-mail: barbieri@fis.uniurb.it

New Developments in Marine Biotechnology, edited by Le Gal and Halvorson.
Plenum Press, New York, 1998.

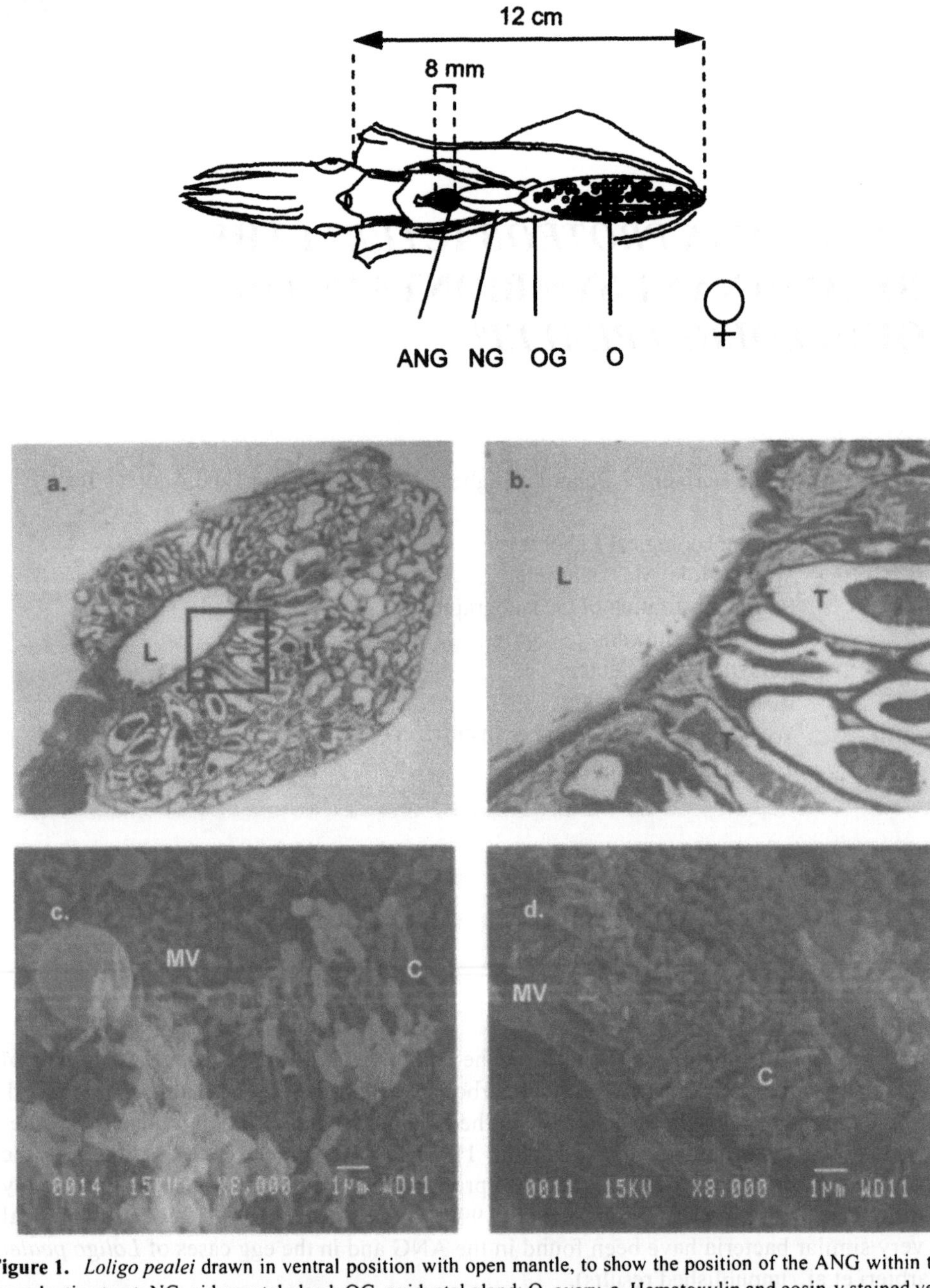

Figure 1. *Loligo pealei* drawn in ventral position with open mantle, to show the position of the ANG within the reproductive tract. NG, nidamental gland; OG, oviductal gland; O, ovary. **a.** Hematoxylin and eosin-y stained ventral thin section (6 µm) of ANG section in paraffin. L, lumen, T, tubule. 5×. **b.** Detail from **a**, 40×. **c** and **d.** SEM micrograph showing ANG tissue with associated bacteria. C, cilia; MV, microvilli. 8000×.

(Lum-Kong & Hastings, 1992). In the phylogenetic framework of 16S ribosomal RNA, these genera belong to the gamma-*Proteobacteria* group (Woese, 1987). Recent 16S rRNA-based studies have also identified members of the alpha-*Proteobacteria*, a different group, in the ANG of *Loligo pealei* (Barbieri et al., 1996).

Here we investigate the ANG, the egg case membranes and their bacterial populations by histological and ultrastructural approaches.

2. MATERIALS AND METHODS

ANG samples were obtained by aseptic dissection from a mature female *Loligo pealei*. Egg membranes came from a freshly deposited egg case. Both samples were provided by the Marine Resources Center, Marine Biological Laboratory, Woods Hole. For histological staining, ANG and egg case tissues were fixed in 4% Glutaraldehyde/filtered seawater for 24 h, kept in 50% ethanol for 2 h and 70% ethanol for 8 h, followed by paraffin embedding, thin sectioning (6 μm) and hematoxylin and eosin-y stain (McCafferty, 1997). For Scanning Electron Microscopy (SEM), ANG tissue and interior egg case membranes were fixed with glutaraldehyde and postfixed with osmium tetroxide, dehydrated in an ethanol series and gold-palladium coated before viewing with a JEOL-ISM-840 Scanning Electron Microscope (Watson et al., 1984).

3. RESULTS AND DISCUSSION

The ANG mid ventral section showed the gland tubular tissue with the gland lumen in the center (Fig. 1.a,b). The lumen surface was formed by a ciliated epithelium which continued into the connecting ducts towards the ANG tubuli, and was covered with a continuos layer of approx. 1 μm-long microvilli. The cilia were bent in the same direction. By their coordinated movement they probably flushed materials and liquids towards the tubuli, to facilitate bacterial colonization and accumulation of secreted products and bacteria in the tubuli. Extracellular fibrous and granulous material was visible free or in association with bacteria (Fig. 1.c,d).

Two distinct bacterial morphologies were observed in different tubuli of the gland. Large rod-shaped bacteria (1.5 × 0.4 μm) were found near the lumen and in only one tubule of the gland, suggesting a more peripheric association with the ANG (Fig. 1.c). In contrast, small coccoid bacteria (0.5 × 0.3 μm) formed homogeneous clumps in the majority of the tubuli examined (Fig. 1.d).

The sheath covering the egg case was an acellular matrix, 1–2 mm thick, and consisted of approximately 100 membrane layers. Most internal layers surrounded the corion of each egg (Fig. 2.a). Between the membrane layers, large rod-shaped bacteria (1.5 × 0.4 μm) occurred together with a higher number of small coccoid bacteria (0.5 × 0.3 μm) (Fig. 2.c,d). Rod-shaped cells growing in filaments were also present (10 × 0.4 μm), and resembled in that respect a new *Shewanella* species isolated from *Loligo pealei* ANG (Leonardo et al., unpublished results).

The bacteria within the gland and the egg case were stable, dividing populations that benefit from their stable and protected environment, probably including a steady nutrient supply within the ANG. Since morphologically homogeneous bacterial cell clusters are found in specific localities of the ANG, we suggest that bacterial colonization of the ANG is selective with respect to bacterial species and corresponding host tissue of the ANG. We are working on the identification of these bacterial populations. So far it seems doubtful to regard the small coccoid bacteria, the principal inhabitants of the ANG tubuli, as previously cultured *Vibrio*, *Pseudoalteromonas* or *Shewanella* species of the gamma-*Proteobacteria*. We are evaluating the hypothesis that these organisms are alpha-*Proteobacteria* which have been detected repeatedly by 16S rRNA cloning and sequencing (Barbieri et al., 1996).

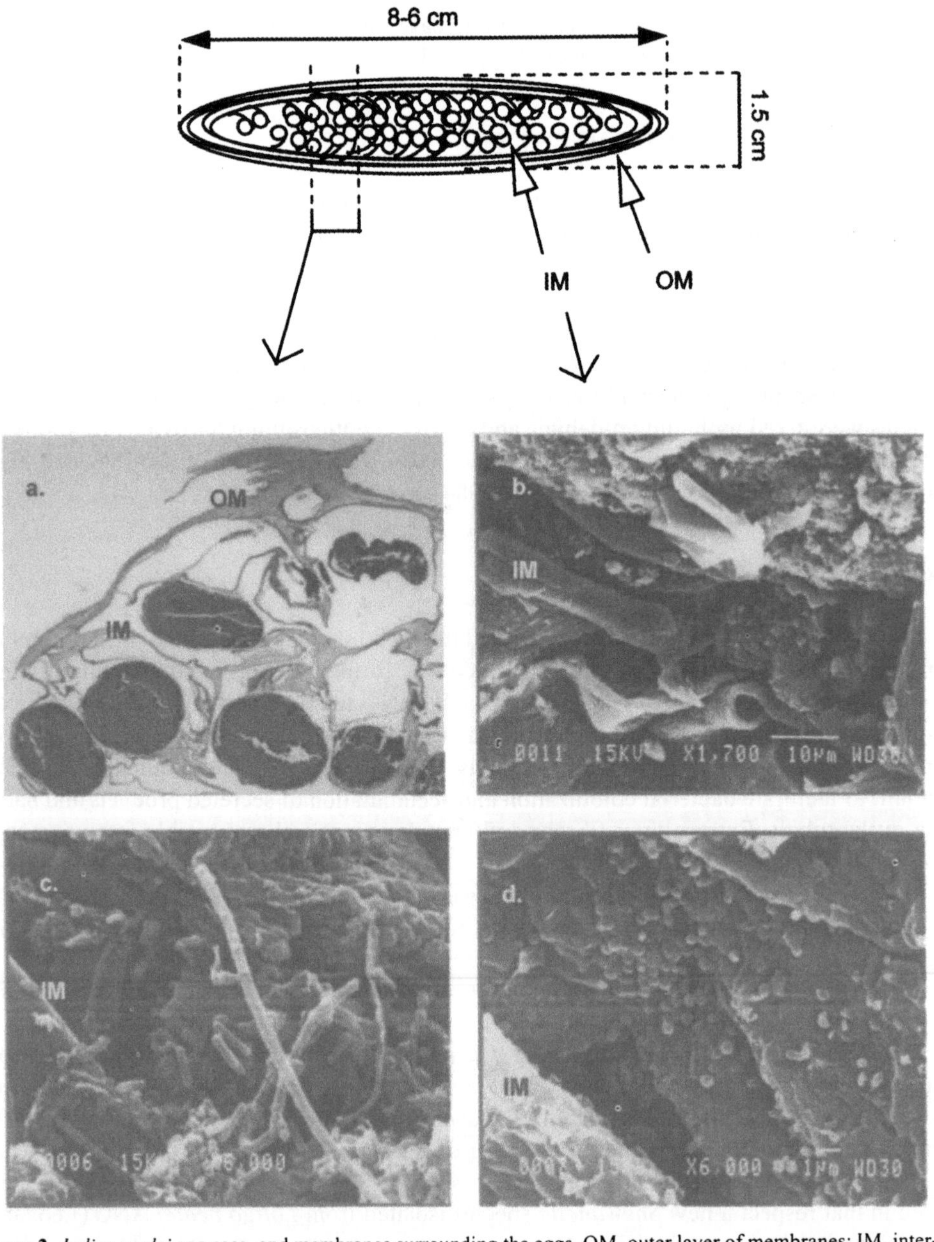

Figure 2. *Loligo pealei* egg case, and membranes surrounding the eggs. OM, outer layer of membranes; IM, internal membranes. Dotted lines in the drawing indicate the sampling region, 1 cm within the central egg case. **a.** Hematoxylin and eosin-y stained thin section (6 μm) of eggs and egg case membranes. 5×. **b.** SEM micrograph of egg case membranes. 1700×. **c** and **d.** SEM micrograph of egg case membranes with associated bacteria. 8000×.

ACKNOWLEDGMENTS

We thank instructors and students of the MBL Microbial Diversity summer course, where this study was initiated and the Marine Resources Center for technical assistance.

This project was supported by the MBL Bernard Davis summer research fellowship, and work during the Microbial Diversity Course '97 was supported by grants from the US Department of Energy and the Office of Naval Research.

REFERENCES

Barbieri, E., B.J. Paster, D. Hughes, L. Zurek, D. P. Moser, A. Teske and M. L. Sogin. Unpublished results.

Barbieri, E., J. Gulledge, D. Moser, and C.C. Chien. 1996, *Biol. Bull.* **191**:316–317.

Barbieri, E., K. Barry, A. Child, and N. Wainwright, 1997, *Biol. Bull.*,**193:**275–276..

Biggs, J., and D. Epel, 1991, *J. Exp. Zool.* **259**:314–318

Bloodgood, A. R, 1977, *Tissue & Cell* **9(2)**:197–208

Dunlap., P. 1990, 56th Annual Meeting of the American Malacological Union. Abst. p. 36.

Leonardo, M.R., D.P. Moser, E. Barbieri, C.A. Brantner, and K.H. Nealson. Unpublished results.

Lum-Kong, A., and T. S. Hastings, 1992, *J. Zool. Lond.* **228**:395–403.

McCafferty, M. 1997, Pers. Comm. Laboratory for Aquatic Animal Medicine and Pathology (LAAMP), Marine Biological Laboratory, Woods Hole, MA, USA..

Watson, L.P., A.E. McKee, and B.R. Merell, 1984, Preparation of Microbiological Specimens for Scanning Electron Microscopy. pages 45–56. In: J.A. Murphy and G.M. Roomans (eds.), Preparation of Biological Specimens for Scanning Electron Microscopy. Published by Scanning Electron Microscopy Inc., AMF O'Hare, IL 60666, USA.

Woese, C. R., 1987, *Bacteriol. Rev.* **51**:221–271.

62

MOLECULAR DETECTION OF AQUATIC BIRNAVIRUSES FROM MARINE FISH AND SHELLFISH

Satoru Suzuki*

Department of Aquaculture
Kochi University
Nankoku, Kochi 783, Japan

1. INTRODUCTION

Birnaviruses have been isolated from several kinds of marine fin fish (Kusuda et al., 1993) and shellfish (Rivas et al., 1993). The isolates are distinguishable from infectious pancreatic necrosis virus (IPNV) in terms of antigenicity and genogrouping by using VP2-NS junction region (Hosono et al., 1996). Hosono et al., 1996 proposed the tentative term, marine birnavirus (MABV), which includes yellowtail ascites virus (YAV) (Sorimachi & Hara, 1985), viral deformity virus (VDV) (Nakajima et al., 1993) and other birnaviruses isolated from marine fishes (Kusuda et al., 1989; Kusuda et al., 1994). A molecular detection technique was needed to determine the distribution of MABV and to determine the variation in the viral genome from different MABV isolates. This information is important in understanding the spread of MABV between susceptible hosts and reservoirs of infection in natural environment. We recently developed a 2-step PCR method specific for birnaviruses (Suzuki et al., 1997). In this paper, the MABV genome was surveyed by using the 2-step PCR on fish and shellfish samples.

2. MATERIALS AND METHODS

2.1. Fish and Shellfish Samples

For the detection of the MABV genome, cultured fish were collected in the summer of 1994 in Kochi prefecture, Japan. Shellfish samples included Agemaki (jack knife clam) *Sinonovacula constricta* imported from Korea to Saga prefecture, Japan in 1993 and 1994,

* Fax: +81-888-64-5197; e-mail: ssuzuki@cc.kochi-u.ac.jp

New Developments in Marine Biotechnology, edited by Le Gal and Halvorson.
Plenum Press, New York, 1998.

Table 1. Primers used for RT-PCR (P1–P2) and nested-PCR (P3–P4)

Primer code	Primer sequence	Primer orientation	Product length (bp)	Map position
P1	5'AGAGATCACTGACTTCACAAGTFAC	Sense	359	1403–1761 (IPNV
P2	5'TGTGCACCACAGGAAAGATGACTC	Antisense		Jasper/segment A)
P3	5'CAACACTCTTCCCCATG	Sense	168	74–241 (Y-6/VP2/NS
P4	5'AGAACCTCCCAGTGTCT	Antisense		junction part)

and Japanese pearl oyster *Pinctada fucata* collected in Ehime prefecture, Japan in 1996. Spleen was tested from fish samples and intestine was used for shellfish.

2.2. 2-Step Polymerase Chain Reaction (PCR)

Extraction of nucleic acids from fish and shellfish was performed by the method of Davis and Boyle (Davis & Boyle, 1990) Primers for reverse transcription (RT)- and nested-PCR are shown in Table 1. Primers P1 and P2 were added to the purified nucleic acids; the solution was then denatured by heating it at 100°C for 5 min followed by cooling on ice. The reverse transcription mixture was then added. Final composition in 20 μl was 50 mM Tris-HCl pH 8.3, 74 mM KCl, 3 mM $MgCl_2$, 1 mM each dNTP, 20 U RNase inhibitor (HPRI), 2 mM DTT and 40 U MMLV reverse transcriptase. RT was performed at 37°C for 1 h followed by heating at 100°C for 5 min. The product was allowed to cool to 37°C; a 5 μl portion was then taken and put into a PCR mixture containing 100 mM Tris-HCl pH 8.3, 500 mM KCl, 15 mM $MgCl_2$, 0.1% gelatin, and 20 mM each dNTP. This gave a final volume of 100 μl. The reaction was started by adding of 2.5 U of *Taq* polymerase (TaKaRa) involved 30 cycles: 1 min at 95°C for denaturation; 1 min at 48°C to allow annealing; 1 min at 72°C for extention. A 3 μl product of the RT-PCR was taken and employed as a template for the nested-PCR. Primer set P3–P4 was used for the reaction. Amplification conditions of nested-PCR were the same as RT-PCR as described above. Specificity and sensitivity of this 2-step PCR were confirmed as previously reported (Suzuki et al., 1997).

3. RESULTS

The results of the detection of MABV genome in fish and shellfish are summarized in Tables 2 and 3. Details of the samples and results in each case were reported elsewhere (Suzuki et al., 1997; Suzuki et al., 1997; Sukuzi et al., 1998). Virus isolation using the CHSE-214 cell line was also performed on all fish and shellfish samples. All samples that were found to be virus positive were positive in either RT- or nested-PCR. Moreover, even

Table 2. Detection of MABV genome by PCR in cultured fish and virus isolation in cell culture

Sample	PCR positive/ sample number	Virus isolation positive/ sample number
Yellowtail (*Seriola quinqueradiata*)	9/9	2/9
Amberjack (*Seriola dumerili*)	10/11	2/11
Red seabream (*Pagrus major*)	2/6	1/6
Sea bass (*Lateolabrax japonicus*)	3/5	1/5

Table 3. Detection of MABV genome by PCR in shellfish and virus isolation in cell culture

Sample	PCR positive/ sample number	Virus isolation positive/ sample number
Agemaki (*Sinonovacura consticta*)	23/30	1/30
Japanese pearl oyster (*Pinctada fucata*)	6/6	4/6

if the virus isolation was not successful, some samples in fish and most samples in shellfish were positive by PCR.

4. DISCUSSION

In recent years, PCR assays for IPNV have been reported by some researchers (Heppel et al., 1992; Lopes-Lastra, 1994). We modified these techniques for detection of MABV and established the 2-step PCR (Suzuki et al., 1997). This technique was used on clinical samples of fish and shellfish. The fish samples used in this study were considered to be infected with an iridovirus, because they had congestion of the liver and enlarged cells in the spleen, although an iridovirus was not isolated. The positive results in virus isolation and PCR suggest that these fish had a double infection of an iridovirus and MABV. It is known that birnaviruses become chronic infections in survivors, after acute infection, for a long time (Yamamoto, 1975). Speciments examined in this study may have been persistently infected, however, became activated when the fish were stressed by high temperature and/or other infections.

In the shellfish samples, many PCR positive results were also obtained. However, virus isolation in Agemaki was only successful 1 of the 30 samples. It has been reported that birnaviruses are distributed widely in the environment and infect molluscs without causing disease (Mortensen et al., 1992). The Agemaki case in this study suggests the bivalve molluscs could be the host and/or reservoir of MABV. In the Japanese pearl oyster, virus isolation correlated with PCR results. To answer whether the MABV infection is important in mortality in Japanese pearl oyster, further study is needed, since the pathogenicity of MABV for shellfish has not been clear so far. The results of this study that compared different detection techniques suggests that MABV not only replicates but also persists in various marine hosts.

5. SUMMARY

The genome of the marine birnavirus (MABV) was surveyed in fish and shellfish samples by using the 2-step PCR technique targeting the VP2-NS junction region of genome segment A. The MABV genome was detected in a high ratio in fish and shellfish. This indicates that MABV is distributed widely in both cultured and in wild fish and shellfish.

ACKNOWLEDGMENT

We thank Dr. S. LaPatra for critically reading this manuscript.

REFERENCES

Davis, V. S. and Boyle, J. A., 1990, *Anal. Biochem.*, **189**:30–34.

Heppell, J., Berthiaume, L., Tarrab, E., Lecomte, J. and Arella, M. , 1992, *J. Gen. Virol.*, **73** :2863–2870.

Hosono, N., Suzuki, S. and Kusuda, R. 1996,*J. Fish Dis.*, **19**:295–302.

Kusuda, R., Kado, K., Takeuchi, Y. and Kawai, K., 1989, *Suisan Zoushoku*, **37**:115–120.

Kudusa, R., Nagato, K. and Kawai, K., 1994, *Suisan Zoushoku*, **42**:145–149.

Kusuda, R., Nishi, Y., Hosono, N. and Suzuki, S. 1993, *Fish Pathol.*, **28:**91–92.

Lopes-Lastra, M., Gonzalez, M., Jashes, M. and Sandino, A. M., 1994, *J. Fish Dis.*, 17: 269–282.

Mortensen, S. H., Bachere, E., LaGall, G. and Mialhe, E. , 1992,*Dis. Aquat. Org.*, **12**: 221–227.

Nakajima, K., Maeno, Y., Arimoto, M., Inoue, K. and Sorimachi, M. 1993, *Fish Pathol.*, **28**:125–129.

Rivas, C., Cepeda, C., Dopazo, C. P., Novoa, B., Noya, M. and Barja, J. L. 1993, *Aquaculture*, **115**:183–194.

Sorimachi, M. and Hara, T. 1985, *Fish Pathol.*, **19:** 231–238.

Suzuki, S., Hosono, N. and Kusuda, R., 1997, *J. Mar. Biotechnol.*, **5**:205–209.

Suzuki, S., Nakata, T., Kamakura, M., Yoshimoto, M., Furukawa, Y., Yamashita, Y. and Kusuda, R., 1997, *Fisheries Sci.*, **63**:563–566.

Suzuki, S., Kamakura, M. and Kusuda, R.,1998, *Fisheries Sci.*, **64**: in press.

Yamamoto, T., 1975, *J. Fish. Res. Bd Can.*, **32**:568–570.

63

A SDS/PAGE/WESTERN BLOT/EIA PROTOCOL FOR THE SPECIFIC DETECTION OF SHRIMP VIRAL PATHOGENS

Philip C. Loh, Lourdes M. Tapay, E. Cesar, B. Nadala, Jr., and Yuanan Lu

Department of Microbiology
University of Hawaii
Honolulu, Hawaii 96822

1. INTRODUCTION

Shrimp viral diseases have seriously impacted the sustainability and economic success of the shrimp aquaculture industry worldwide. Among several of the recent viral pathogens which caused massive mortalities in cultured shrimp were the yellow-head virus (YHV) and the white spot baculovirus (WSBV) (also called the Chinese baculovirus [CBV] or the systemic ectodermal and mesodermal baculovirus [SEMBV]). Both YHV and CBV have been reported by our laboratory to be highly pathogenic for *Penaeus stylirostris* (blue shrimp) and *P. vannamei* (white shrimp), the two principal penaeid species commercially cultured in Hawaii and the Western Hemisphere (Lu et al. 1997). They are thus potentially a serious problem particularly to the brood stock industry.

A major problem to the control and prevention of shrimp viral diseases is the lack of relatively simple and cost-effective technologies for the early detection and diagnosis of viral infections, particularly asymptomatic infections. In this paper, we report the successful development of a less invasive, combined Western blot and enzyme immunoassay protocol for the early detection of both YHV and CBV in experimentally infected animals before the appearance of clinical symptoms. Detection of both viruses were also examined in infected primary shrimp cell cultures.

2. MATERIALS AND METHODS

2.1. Preparation of Anti-YHV and Anti-CBV Antibodies

Polyclonal antibody against YHV and CBV were prepared in New Zealand white rabbits (7–8 lbs) using purified virus as antigen. Immunoglobulin G was purified from the

New Developments in Marine Biotechnology, edited by Le Gal and Halvorson.
Plenum Press, New York, 1998.

antisera using recombinant bacterial protein-G columns (Gamma-Bind™). In order to remove antibodies that cross-react with normal shrimp antigens, the IgG was adsorbed onto acetone-dried, ground shrimp muscle tissue and also with shrimp hemolymph.

2.2. SDS-Polyacrylamide Gel Electrophoresis (SDS-PAGE)

SDS-PAGE was done according to the method of Laemmli (1970). Briefly, samples were suspended in loading buffer (0.5 mM Tris⁻Cl, pH 6.8, 2.5 ml; 10% sodium dodecyl sulfate, 4 ml; glycerol, 2 ml; β-mercaptoethanol, 1 ml; and deionized distilled water, 0.5 ml), boiled at 95°C for 5 min, loaded into the wells of 5% (YHV) and 10% (CBV) SDS-PAGE, and electrophoresed at 200 V.

2.3. Western Blotting

The electrophoresed gel was blotted onto a nitrocellulose membrane (pore size, 0.1 µm) in blotting buffer (3.03 g Tris base, 14.4 g glycine, and 200 ml methanol per liter) at 100 V for 1 hr. The membrane was then rinsed in PBS (pH 7.4), soaked in 5% skim milk (in PBS) for 1 hr, and rinsed in PBS for 5 min. The membrane was treated with 1:1000 dilution of the primary antibody (either a polyclonal hyperimmune anti-YHV IgG or anti-CBV IgG) for 1 hr, rinsed 3× with PBS for 5 min, and then treated with 1:2500 dilution of the secondary antibody (goat anti-rabbit IgG-horseradish peroxidase conjugate [Kirkegaard and Perry Laboratories, Inc., Gaithersburg, MD]) for 1 hr. The membrane was again rinsed 3× with PBS for 5 min and then treated with the substrate (3,3,5,5-tetramethylbenzidine or TMB [Kirkegaard and Perry Laboratories, Inc., Gaithersburg, MD]) until the bluish purple color developed. The reaction was stopped by soaking the membrane in distilled water. All of the incubations were done at room temperature (r.t., 25 ± 2°C) unless specified otherwise.

2.4. Time-Course Infectivity Experiments in Whole Animals

Ten *Penaeus vannamei* (40–60 gram) were injected intra-muscularly in the second segment of the abdomen with either 5% w/v of head soft tissues from CBV-infected shrimp or 10% w/v head soft tissues from YHV-infected shrimp at 0.2 ml per shrimp. At specific periods after injection, 0.1 ml hemolymph was collected from the hemocoel of at least 2 randomly selected shrimps using a 1 ml syringe with 26G needle. The hemolymph were diluted in equal volume of 20% citrate buffer and then stored at −80°C until needed.

For Western blotting, 200 µl aliquots of hemolymph samples were clarified at 8,000 × g for 5 min and then pelleted at 140,000 × g for 5 min. The pellets were resuspended in 100 µl 2x loading buffer and heated to 95°C for 5 min. Only 10 µl of the treated samples were loaded per well of the SDS-PAGE gel.

2.5. Time-Course Infectivity Experiments in Primary Lymphoid Cell Cultures

Forty-eight hr primary lymphoid cell cultures in 25 cm^2 Primaria™ flasks were inoculated with 0.5 ml of 100-fold dilution of CBV or YHV filtrate (clarified and filtered 10% w/v virus-infected head soft tissues). Adsorption was carried out at r.t. for 1.5 hr after which the excess filtrate was removed and replaced with 5 ml of growth medium (2X Leibovitz Medium-15 supplemented with 8% shrimp head extract, 5% fetal bovine serum, 6 ml salt solution and 100 I.U./100 µg/ml penicillin/streptomycin). The flasks were incubated at r.t. and were sampled at various days post-infection (p.i.) by freezing at −80°C. An uninoculated flask was frozen at 7 days and used as control. The flasks were then thawed and collected into each of sterile conical tubes for centrifugation at 1800 × g for 15 min at 5°C. The supernatant was stored at −80°C until needed.

For Western blotting, the samples were clarified by centrifuging 200 l aliquots at 8,000 × g for 5 min and then the pelleted at 140,000 × g for 5 min. The pellets were resuspended in 20 μl 2x loading buffer and heated to 95°C for 5 min. All 20 μl of the treated samples were loaded in the well of the SDS-PAGE gel.

3. RESULTS

3.1. Detection of Viral Proteins by NC-EIA

The nitrocellulose-enzyme immunoassay (NC-EIA) protocol when used for the detection of viral proteins was found to have a sensitivity of 0.4 ng for YHV and 1 ng for CBV. The polyclonal antisera used also exhibited strong cross-reactions with hemolymph and normal shrimp tissues. Such cross-reactivities however, could be significantly reduced

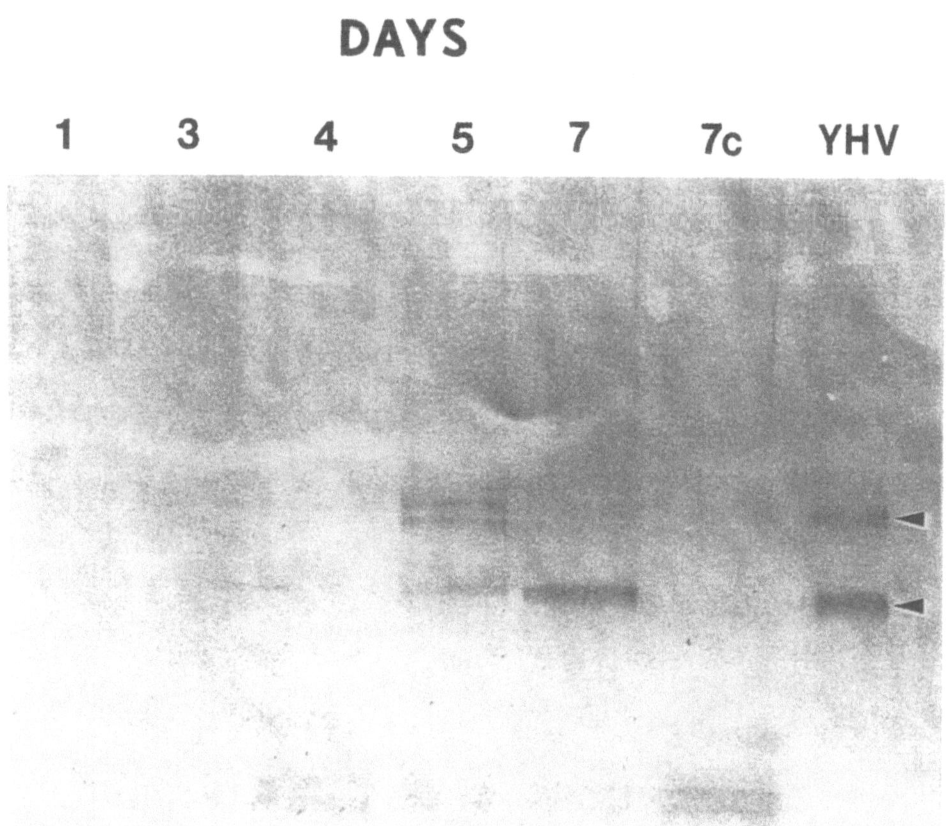

Figure 1. Western blot of YHV proteins (135 and 170 kDa, arrowheads) in hemolymph samples collected at various periods from shrimp (*P. vannamei)* experimentally infected with YHV.

by prior extensive adsorption with hemolymph. Since residual cross-reactivity could not be completely eliminated, the adsorbed antisera for both YHV and CBV were then employed in the Western blot format where virus-specific proteins were clearly dilineated from normal hemolymph and tissue proteins.

3.2. Detection of YHV Proteins in the Hemolymph and in Primary Cell Cultures

Protein analysis of purified virus preparations by SDS-PAGE had previously revealed four major bands with the following estimated molecular sizes (kDa): 175, 135, 67 and 22. (Tapay et al. 1997). These bands probably corresponded to the large (L), glycoprotein (G), nucleocapsid (N) and matrix (M) proteins of rhabdoviruses. Western blot analysis of these bands using polyclonal anti-YHV IgG showed strong reactivity of the antibody with the putative G protein band, as well as the L and N bands.

Western blot analysis of YHV-containing hemolymph samples obtained from experimentally-infected shrimp showed that the putative G protein (135 kDa) as well as the L protein could be specifically distinguished. Figure 1 shows that the viral protein first became detectable in the hemolymph 56 hours p.i. Overt symptoms in infected animals were not observed until 72 hours p.i.

Western blot analysis of YHV-infected primary lymphoid cell cultures had similar results. The putative G protein of the virus was easily distinguished from background staining, and was detected as early as 4 days p.i. (Figure 2).

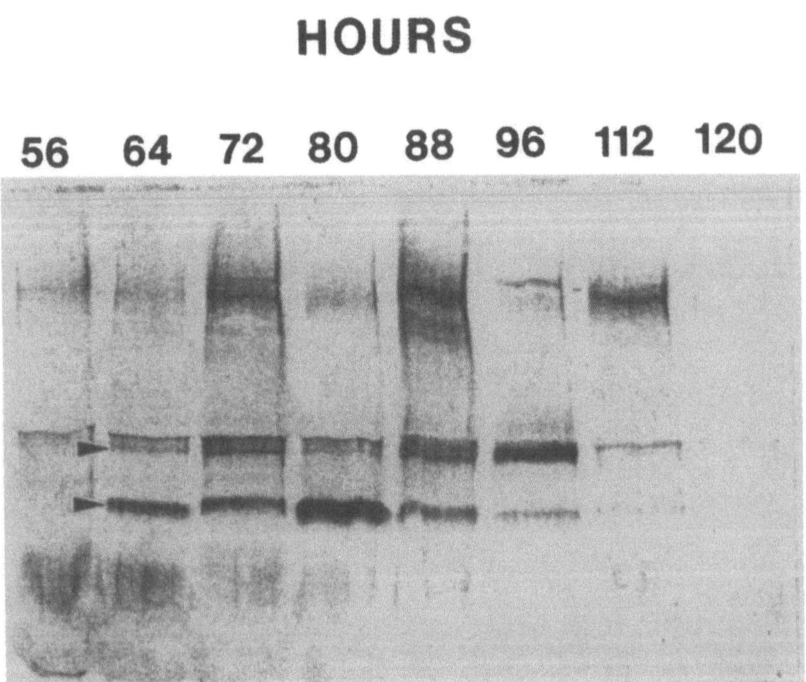

Figure 2. Western blot of YHV proteins (135 and 170 kDa, arrowheads) in cell culture lysates collected at various periods from primary shrimp lymphoid cell cultures infected with YHV. 7c = control uninfected cell culture lysate after 7 days.

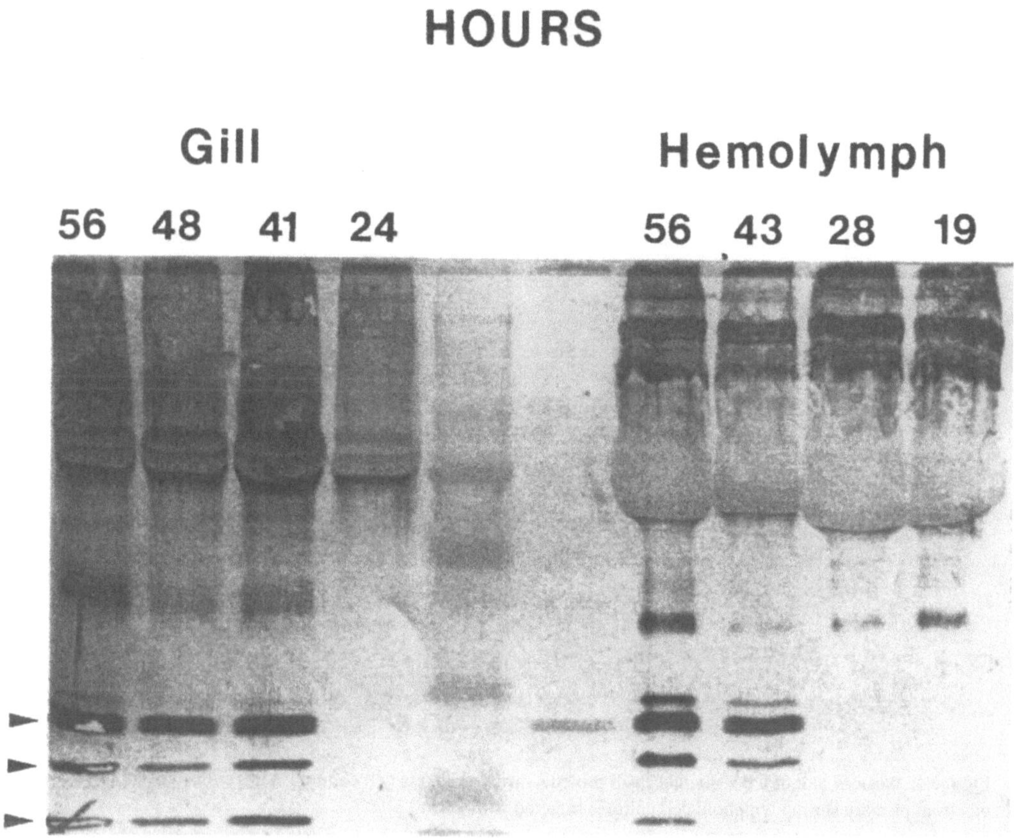

Figure 3. Western blot of CBV nucleocapsid proteins (arrowheads) in gill and hemolymph samples collected at various periods from shrimp (*P. vannamei*) experimentally infected with CBV.

3.3. Detection of CBV Proteins in the Hemolymph and in Primary Cell Cultures

SDS-PAGE analysis of purified CBV revealed among several protein bands three prominent bands with the following molecular sizes (kDa): 19, 23.5 and 27.5 (doublet band). The same bands were observed in Western blots of purified CBV using polyclonal anti-CBV IgG. Preliminary studies indicated that the 27.5 and 19 kDa protein bands are associated with the envelope component of the virus. The 23.5 kDa protein is identified with the capsid structure (Nadala and Loh, unpublished).

Western blot analysis of CBV-containing hemolymph and gill samples obtained from time-course experiments using experimentally-infected shrimps showed the appearance of the three prominent CBV-nucleocapsid bands 43 hours p.i. and 41 hours p.i., respectively (Figure 3). The infected animals did not show overt symptoms of disease until 72 hours p.i.

Western blot analysis of CBV-infected primary lymphoid cell cultures showed similar results. The three prominent bands of the virus were easily visible as early as 4 days post-infection (Figure 4).

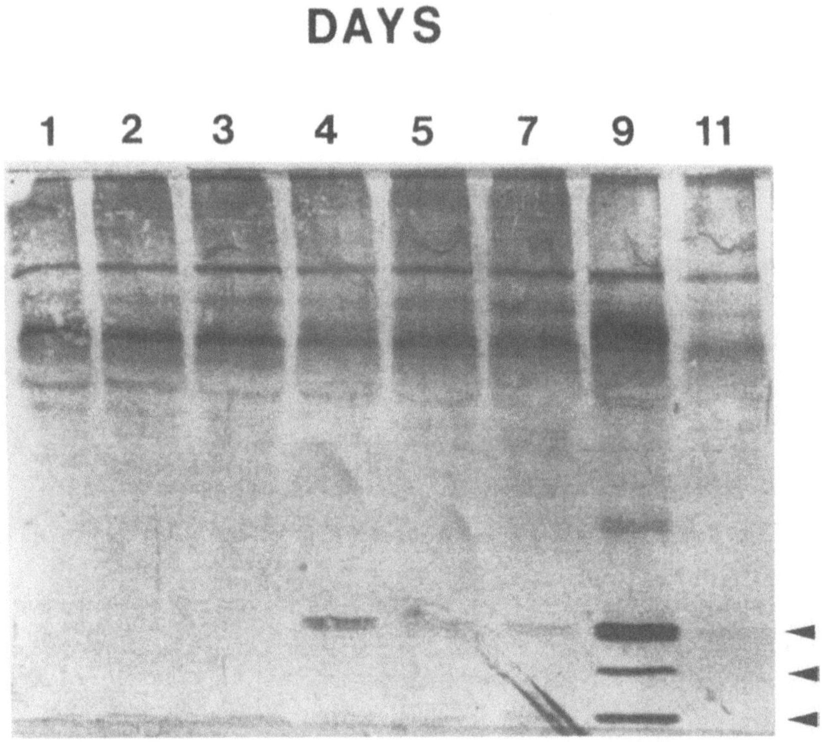

Figure 4. Western blot of CBV nucleocapsid proteins (arrowheads) in cell culture lysates collected at various periods from primary shrimp lymphoid cell cultures infected with CBV.

3.4. Field Studies

Table 1 shows some of the early results obtained when random samples of cultured shrimp from various hatcheries and farms in the Pacific, S.E. Asia and the U.S. were examined for CBV and YHV by the WB and NC-EIA methodologies.

The limited samples from the Pacific examined thus far were negative for both viral pathogens. On two occasions samples from Indonesia and the U.S. were positive for CBV (or WSBV). The recent occurrence of CBV (or WSBV) in the U.S. has aroused concern particularly since this is the second incident reported and also involved a new species of

Table 1. YHV and CBV detection results for different shrimp farms

Farm location	Sampling date	YHCV	CBV
US	9/23/96	–	–
New Caledonia	11/15/96	–	–
Tahiti	11/23/96	–	–
Indonesia	11/8/96	–	–
	12/4/96	–	–
	1/25/97	–	+
US	3/25/97	–	–
	5/6/97	–	+
	5/23/97	+	–

penaeid shrimp indigenous to the east coast of north and central Americas (*P. setiferus*). On one occasion the animals were positive by WB for YHV.

4. DISCUSSION

The production of high titered specific anti-CBV and anti-YHV polyclonal sera enabled the development of a Western blot protocol which could detect two highly virulent and infectious viral pathogens (YHV & CBV). The protocol was able to demonstrate the presence of specific viral proteins in the hemolymph of experimentally-infected animals as early as 43 hours post-injection and at least one day before clinical symptoms of the disease appeared.

The use of the hemolymph as material for detection of the virus allows for a method of sample collection that does not require the sacrifice of the animal or any part of the animal. This is particularly important when monitoring highly valuable broodstock populations for these pathogens. If necessary, both the gill tissues and lymphoid (Oka) organ may serve as highly satisfactory alternate sources of viral antigen.

The Western blot protocol also enabled us to detect specific viral proteins in primary lymphoid cell cultures inoculated with either YHV- or CBV-containing filtrates. This verifies previous data showing that both viruses are able to grow and cause CPE in primary lymphoid cell cultures (Tapay et al., 1996).

Immunological as well as nucleic acid hybridization assays often suffer from poor specificity especially when using tissue extracts or biological fluids as samples. This is mainly due to either cross reactions with non-target proteins (EIA) or nucleic acids (hybridizations) or simply non-specific molecular interactions. The Western blot protocol has a distinct advantage over these other detection protocols in that in addition to the specific antigen-antibody interaction it also separates the antigens on the basis of size. Therefore, a stricter criterion is applied in the identification of positive samples. This makes the Western blot protocol a good confirmatory assay for the less specific but more convenient enzyme immunoassay protocols.

The early detection of CBV and YHV, especially in asymptomatic latent carriers among shrimp populations, will forewarn shrimp farmers not only of the presence of the infectious agent but also of imminent potential disease outbreaks and allow for implementation of contingency management strategies that would prevent or at least minimize losses. Furthermore, a rapid, sensitive and specific detection protocol will enable quarantine officials to restrict importation of infected animals especially into geographic areas free of these viruses.

Both the NC-EIA and WB detection protocols are currently being evaluated in field studies of shrimp samples obtained from various hatcheries and farms in S.E. Asia and the U.S.A. The initial results indicated that the protocols when used in a combined format can yield highly reliable results. The less specific but more convenient NC-EIA served as a presumptive test followed by the WB as the confirmatory assay (Figure 5). The results thus far have been highly accurate and reproducible.

5. SUMMARY

A Western blot protocol capable of detecting YHV and CBV in the hemolymph of infected shrimps was developed. This protocol was highly specific, rapid, and sensitive enough to detect the presence of the viruses before the appearance of overt symptoms. It

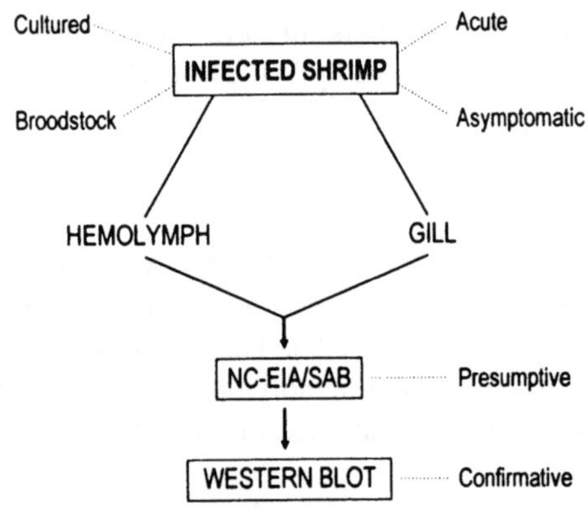

Figure 5. Flow chart for penaeid shrimp virus detection.

was also useful in demonstrating the growth of both viruses in primary lymphoid cell cultures. A flow chart for routine shrimp virus pathogen detection and identification employing the more convenient but less specific NC-EIA as a presumptive test followed by the highly reliable WB protocol as a confirmatory assay was advanced. The combined protocols together with the primary shrimp cell culture system have been successfully used in the field surveillance of shrimp hatcheries and farms.

ACKNOWLEDGMENTS

This study was supported in part by grants from the University of Hawaii Sea Grant and College Program, Institutional Grant No. NA36RG0507, UNIHI-SEAGRANT-CP-98-06, and the Aquaculture Development Program, Department of Land and Natural Resources, State of Hawaii, Contract No. 38066.

REFERENCES

Chang, P.S., Y.C. Wang, C.F. Lo and G.H. Kou. (1996). Infection of white spot syndrome associated non-occluded baculovirus in cultured and wild shrimps. Second International Conference on the Culture of Penaeid Prawns and Shrimps, SEAFDEC, Iloilo City, Philippines, May 14–17, 1996.

Durand, S., D.V. Lightner, L.M. Nunan, R.M. Redman, J. Mari and J.R. Bonami,. (1996). Dis. Aquatic Org. In press.

Laemmli, U.K. (1970). Nature 227:680–685.

Lo, C.F., J-H. Leu, C-H. Ho, C-H. Chen, S-E. Peng, Y-T. Chen, C-M. Chou, P-Y. Yeh, C-J. Huang, H-Y. Chou, C-H. Wang and G-H. Kou. (1996). Dis. Aquatic Org. 25:133–141.

Lu, Y., L.M. Tapay, P.C. Loh, R.B. Gose and J.A. Brock. 1997. The pathogenicity of a baculo-like virus isolated from diseased penaeid shrimp obtained from China for cultured penaeid species in Hawaii. Aquaculture Internatl. 5:277–282.

Mari, J., R.J. Bonami, and D.V. Lightner. (1993). Virmet 74:2637–2643.

Mari, J., D.V. Lightner, B.T. Poulos and J.R. Bonami. (1995). Dis. Aquatic Org. 22:129–134.

Nadala, E.C.B., Jr., Y. Lu, P.C. Loh and J.A. Brock. (1992). Fish Pathol. 27:143–147.

Poulos, B.T., J. Mari, J.A. Bonami, R.M. Redman and D.V. Lightner. (1994). J. Virol. Meth. 49:187–194.

Tapay, L.M., Y. Lu, E.C.B. Nadala Jr., J.A. Brock, and P.C. Loh. (1996). Development of an *in vitro* quantal assay in primary cell cultures for a baculo-like virus of penaeid shrimps. Proc. Second International Conference on the Culture of Penaeid Prawns and Shrimps. Iloilo City, Philippines. p.53.

Tapay, L.M., E.C.B. Nadala, Jr., S. Cao, and P.C. Loh. (1997). The specific detection of shrimp viral pathogens by Western blot. Oceanology Internatl. 97-Pacific Rim 2:35–44.

64

EXPRESSION OF CAPSID PROTEINS FROM INFECTIOUS PANCREATIC NECROSIS VIRUS (IPNV) IN THE MARINE BACTERIUM *VIBRIO ANGUILLARUM*

John T. Singer, Jacqueline H. Edgar, and Bruce L. Nicholson

Department of Biochemistry, Microbiology, and Molecular Biology
University of Maine
Orono, Maine 04469

1. INTRODUCTION

V. anguillarum is a marine fish pathogen with serotype O1 strains comprising the major disease-causing serotype (Crosa et al., 1980). Two components of a high-affinity iron uptake system encoded by the virulence plasmid pJM1 are the secreted siderophore, anguibactin, and the outer membrane protein receptor for iron-siderophore complexes, OM2. Because formalin-inactivated serotype O1 strains are used commercially as vaccines for the prevention of vibriosis in salmonids, we explored the use of *V. anguillarum* as an expression host for protective epitopes of the major capsid protein VP2 from infectious pancreatic necrosis virus (IPNV). IPNV causes primarily a disease of salmonid fry, but survivors become life-long carriers and serve as reservoirs of future infection (Dobos & Roberts, 1983). Segment A of the bisegmented dsRNA IPNV genome is 2.9 kb and is translated from a single open reading frame into a polyprotein which is subsequently cleaved into three proteins, VP2, NS and VP3 (Huang et al., 1986). The major capsid protein of the virus, VP2 (59 kDa), contains most of the neutralization epitopes (Caswell-Reno et al., 1986). NS is the non-structural polyprotein protease and VP3 is a minor structural protein. We show here that IPNV cDNA in a translational fusion with *V. anguillarum* DNA results in production of IPNV and OM2 hybrid proteins expressed in *V. anguillarum*.

2. RESULTS

2.1. Insertion of IPNV cDNA into *V. anguillarum* OM2 DNA

A 2.9-kb *Hin*dIII, *Bam*HI IPNV cDNA fragment containing VP2, NS, and VP3 was cloned into pBR322. A chloramphenicol (Cm) resistance gene from pBR328 was isolated on

New Developments in Marine Biotechnology, edited by Le Gal and Halvorson.
Plenum Press, New York, 1998.

a 1.4-kb *Sca*I fragment and inserted into the *Sma*I site present in the VP3 coding region, resulting in plasmid pBRAC. Based on DNA sequences of om2 and vp2, there were no restriction endonuclease sites suitable for production of an in-frame OM2-VP2 fusion. Therefore, IPNV DNA from pBRAC was isolated and treated with exonuclease Bal-31 to progressively degrade DNA towards the amino-terminal end of VP2. Fragments were made flush ended with mung bean nuclease, and inserted into the *Stu*I recognition site present within OM2 carried by plasmid pSE94, a pBR322 derivative. Restriction endonuclease mapping of plasmids identified those clones containing the vp2 insertion in the correct orientation with respect to om2, and DNA sequencing was used to identify plasmid constructs with in-frame OM2′–VP2 fusions. One of these plasmids, pC9, was chosen for further study.

2.2. Subcloning the OM2–VP2 Fusion into Expression Vector pMMB66EH

To increase and control expression of the OM2–VP2 gene fusion, it was placed under the control of the *tac* promoter in pMMB66EH. Plasmid pMMB66EH is a broad-host-range RSF1010-based expression vector containing the IPTG-inducible *tac* promoter. Plasmid pC9 was digested with *Sal*I and *Pst*I, and the 7.5-kb fragment containing OM2′–VP2–NS–VP3′–Cm–′VP3–′OM2 was ligated directionally into *SalI*, *Pst*I-digested pMMB66EH. The resulting construct, p66C9, was conjugally mobilized from *E. coli* to *V. anguillarum* H775-3 by triparental matings.

2.3. Immunoblotting with Anti-OM2 and Anti-IPNV Antibodies

Total protein from *V. anguillarum* H775-3(p66C9) obtained under IPTG-induced and uninduced conditions was separated by SDS-polyacrylamide gel electrophoresis and transferred to nitrocellulose. Results in Fig. 1 show several high-molecular-weight sets of

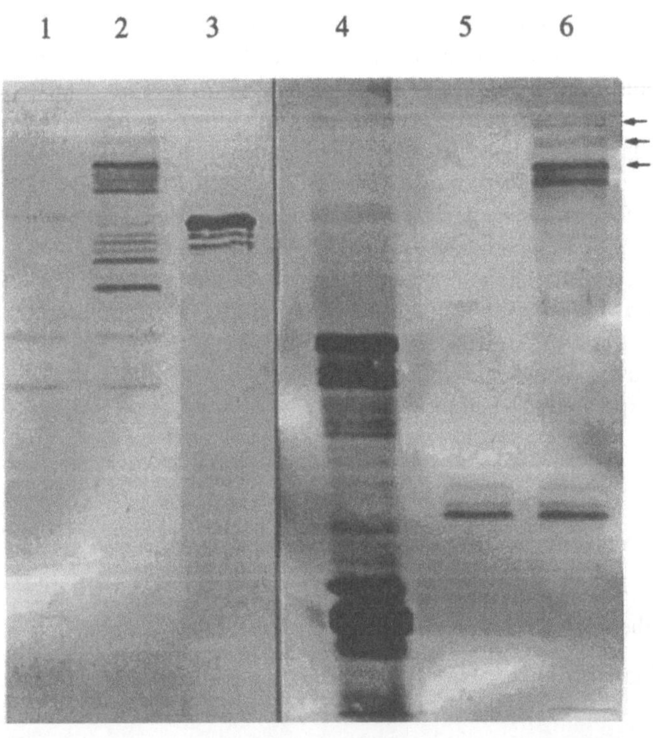

Figure 1. Western blots of OM2′–VP2 fusion proteins. Twenty-five micrograms of protein from lysates of *V. anguillarum* H775-3(p66C9) grown with (lanes 2 & 6) or without (lanes 1 & 5) IPTG induction were separated by SDS-PAGE and blotted to nitrocellulose. Purified OM2, lane 3; IPNV lane 4. Blots were probed with anti-OM2 (lanes 1–3) or anti-Sp IPNV antiserum (lanes 4–6). Matching doublets of proteins are indicated by arrows.

proteins that reacted specifically with anti-IPNV and with anti-OM2 antiserum. Three doublets that reacted with both anti-IPNV and anti-OM2 antiserum (indicated by arrows) were sized by comparison with molecular weight standards. The largest doublet was 148–150 kDa. We compared the estimated size determined from gel mobilities to the sizes of protein products predicted from the DNA sequence determined by PC GENE. The predicted size of a fusion protein beginning at OM2 and terminating at the first translational stop codon prior to the chloramphenicol resistance coding region was 152 kDa. Thus, this doublet likely contains fusion proteins beginning at each of two possible start sites in OM2′ and extending through VP2, NS and VP3′ to a stop codon present just prior to the chloramphenicol resistance coding region. The doublet estimated at 135–139 kDa is consistent with a computer predicted 137-kDa fusion protein starting at OM2′ and continuing through VP2 to the end of NS. Since NS is the polyprotein protease of IPNV, it is possible that this fusion protein is the result of NS-mediated cleavage of VP3. The third doublet measured 113–117 kDa. Based on a computer predicted 110-kDa size, this fusion protein likely contains OM2′ and all of VP2, and also could be the result of NS-mediated cleavage.

Since OM2 is an outer membrane protein in *V. anguillarum*, we thought it possible that the OM2′–VP2 fusion protein was exported and associated with the outer membrane. Western blots of outer membrane protein preparations, however, showed only truncated anti-OM2-reactive bands, and no anti-IPNV-reactive peptides.

2.4. Northern Blotting of Total RNA from *V. anguillarum* H775-3(p66C9)

Total RNA from IPTG-induced and uninduced *V. anguillarum* H775-3(p66C9) and from *V. anguillarum* H775-3 was separated on formaldehyde-containing agarose gels and transferred to supported nitrocellulose. Northern blots were then probed with PCR-produced DNA fragments corresponding to several regions of the expected transcript (Fig. 2). A positive hybridization signal was seen with probes to the OM2–VP2 junction, NS, the beginning and end of VP3, and the intervening chloramphenicol resistance gene. These results indicate that full length transcripts are being produced in *V. anguillarum*. Using a T7 promoter-driven in vitro transcript of the same region as an external standard, it was estimated that the OM2′–VP2–NS–VP3′–Cm–′VP3 transcript comprises 3–4% of the total RNA in IPTG-induced *V. anguillarum* H775–3(p66C9).

3. DISCUSSION

We have expressed IPNV proteins in *Vibrio anguillarum*. VP2 cDNA from IPNV was fused in frame with *V. anguillarum* OM2 DNA and placed under the control of the *tac* promoter. This IPTG-inducible construct, p66C9, produced several proteins which reacted with both anti-IPNV and anti-OM2 antiserum in western blots. The measured sizes of these proteins from acrylamide gels were in close agreement with computer-predicted sizes based on the nucleic acid sequence of the coding DNA in p66C9. We believe the smallest proteins (113–117 kDa) consist of the OM2′–VP2 fusion protein only, while the next largest fusion protein (135–139 kDa) also contains NS, and the largest fusion proteins extend through VP3′ to a translational stop present prior to the chloramphenicol resistance gene inserted in VP3. We found that even though OM2 is an outer membrane protein, the OM2′–VP2 fusion is found only in the cytoplasm and is not exported to the outer membrane.

Expression of IPNV proteins in *V. anguillarum* is the first step required for the purpose of bivalent vaccine production. Recombinant subunit vaccines are capable of provid-

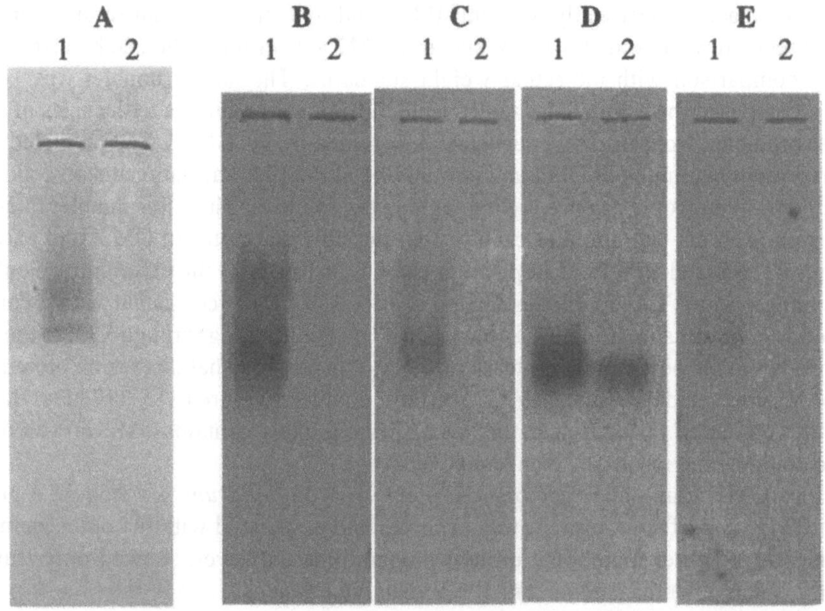

Figure 2. Northern blots of RNA from IPTG-induced and uninduced *V. anguillarum* H775-3(p66C9). Ten micrograms of total RNA from *V. anguillarum* H775-3(p66C9) were transfered to supported nitrocellulose. Lane 1, Induced. Lane 2, Uninduced. Blots were probed with PCR fragments from: **A**, the OM2′-VP2 junction region; **B**, the NS coding region; **C**, the amino-terminal half of the VP3 coding region; **D**, the chloramphenicol acetyltransferase region; and **E**, the carboxy-terminal end of the VP3 coding region.

ing, particularly in the case of viruses, a way of stimulating the immune response without the presence of the pathogen (Amann, 1992). We are performing tests in vivo to determine if the bivalent vaccine strain will induce serum antibodies against VP2 in young adult fish. Also, fry passively immunized with serum from vaccinated adults are being challenged with IPNV to assess protection from disease.

REFERENCES

Amann, E. 1992, *In*: Recombinant DNA Vaccines (Isaacson, R. E., ed.), Marcel Dekker, Inc., N. Y. p. 199
Caswell-Reno, P, Reno, P. W. and Nicholson, B. L. 1986, *J. Gen. Virol.* **67**:2193.
Crosa, J. H., Hodges, L. L. and Schiewe, M. H. 1980, *Infect. Immun.* **27**:897.
Dobos, P. and Roberts, T. E. 1983, *Can. J. Microbiol.* **29**:377.
Huang, M. T, Manning, D. S., Warner, M., Stephens, W. B. and Leong, J. C. 1986, *J. Virol.* **60**:1002.

65

DETECTION OF CULTURABLE AND NON-CULTURABLE *VIBRIO CHOLERAE* 01 IN MEXICO

Marcial Leonardo Lizárraga-Partida,[1] Irma Wong-Chang,[2]
Guadalupe Barrera-Escorcia,[3] Alfonso, and V. Botello[2]

[1]Centro de Investigación Científica y de Educación Superior de Ensenada (CICESE)
Km. 107 Carr. Tijuana-Ensenada, 22860
Ensenada, Baja California, México
[2]Instituto de Ciencias del Mar y Limnologia
UNAM
[3]Universidad Autónoma Metropolitana
Unidad Ixtapalapa
UAM-1

1. INTRODUCTION

The seven pandemic of cholera initiated in Indonesia during 1961 and characterized by a new toxigenic biotype, *V. cholerae* El Tor, has arrived to Peru a South American country, in 1991. Their expansion from Peru to the North and South of the Pacific Ocean was rapid, and in 1992, crossing by the Amazons River, it has arrived to the Atlantic Ocean, and expanded to the North until the Gulf of Mexico (Tauxe et al., 1994). No clear explanation exist how this *V. cholerae* El Tor arrive to the Port of Chancay in Peru, neither for their dispersion thought the American continent, but in Peru, *Vibrio cholerae* 01 has been isolated from plankton, and seafood samples (Tamplin and Castillo Perodi, 1991), suggesting that the marine environment was the source of the disease. Cargo ships ballast and other non-potable waters has been signalled as an important factor in the dissemination of epidemic *V. cholerae* (McCarthy and Khambaty, 1994), but also, according to Colwell (1996), "the association of *V. cholerae* with zooplankton has proven to be a key factor in the deciphering the global nature of cholera epidemics."

In Mexico, the first cholera case was registered in June 1991 since then the Health Department of Mexico (Secretaría de Salud), reported 2831 cases in 1991, 8162 in 1992, 11091 in 1993, 4059 in 1994, 16430 in 1995, 1088 in 1996 and 758 until August, 1997.

New Developments in Marine Biotechnology, edited by Le Gal and Halvorson.
Plenum Press, New York, 1998.

(Secretaria de Salud, 1991, 1993, 1994a, 1994b, 1995, 1996 and 1997). Cholera cases also present a seasonal cycle with an increase from May to August and decrease from August to December (Secretaria de Salud, 1995). This seasonally seems to be correlated with the hot weather of summer months as well as heavy rainfall in the southern regions of Mexico, where we found also, the higher number of cholera cases as well as the lower socio-economic levels (Secretaria de Salud, 1997).

With such information as a background, a plankton sampling was initiated in 1995 in different ports as well as in offshore waters of the Pacific Ocean and at coastal lagoons located in the Gulf of Mexico.

2. METHODS

Plankton sampling (202 nm mesh size) was performing in the Ports and coastal areas of Puerto Peñasco (North Gulf of California), Mazatlán and La Paz, (South Gulf of California) and the offshore waters of the Gulf of California and the Gulf of Tehuantepec, all of them located in the Pacific Ocean. For the Gulf of Mexico, two coastal lagoons were sampled, Pueblo Viejo and Tamiahua, wich are located in the mexican state of Veracruz. In Figure 1, we presented the employed flowchart of operations for the detection of culturable and non-culturable *V. cholerae*. Identification of the serogroup 01 was accomplished with fluorescent monoclonal antibodies (New Horizons), using a ZEISS epifluorescent microscope.

3. RESULTS AND DISCUSION

In Table 1A and B are presented the positive detection of *V. cholerae* non-01 and 01 in their culturable and non-culturable state.

From these tables we observe great differences between the Pacific and the Gulf locaties. No positive results were obteinend in the Pacific in contrast to positive detection of culturable and non culturable *V. cholerae* 01 in the Gulf. The non-01 serogroup, was detected in the Pacific only in the first samplings performed at Puerto Peñasco (02/96) and at Mazatlán (03/96), just after the high incidence of cholera cases of 1995. In the Gulf the non 01 are present when we detect 01 (culturable and non-culturable) but also when 01 were not detected in any of their states, as in the last sampling of October 1996.

Table 1. Detection of *Vibrio cholerae* in plankton samples in the Pacific Ocean (A) and in the Gulf of Mexico (B)[a]

A	Puerto Peñasco		Mazatlán			La Paz	Gulf of California		Tehuan
Vibrio cholerae	02/96	06/96	03/96	09/96	12/96	04/96	11/95	04/96	02/96
01									
Non 01	X		X						

B	Tamiahua				Pueblo Viejo			
Vibrio cholerae	09/95	01/96	05/96	10/96	09/95	01/96	05/96	10/96
01	DFA, X	DFA	DFA		DFA	DFA		
Non 01	X		X	X	X			X

[a]In their states culturable (X) and non-culturable (DFA).

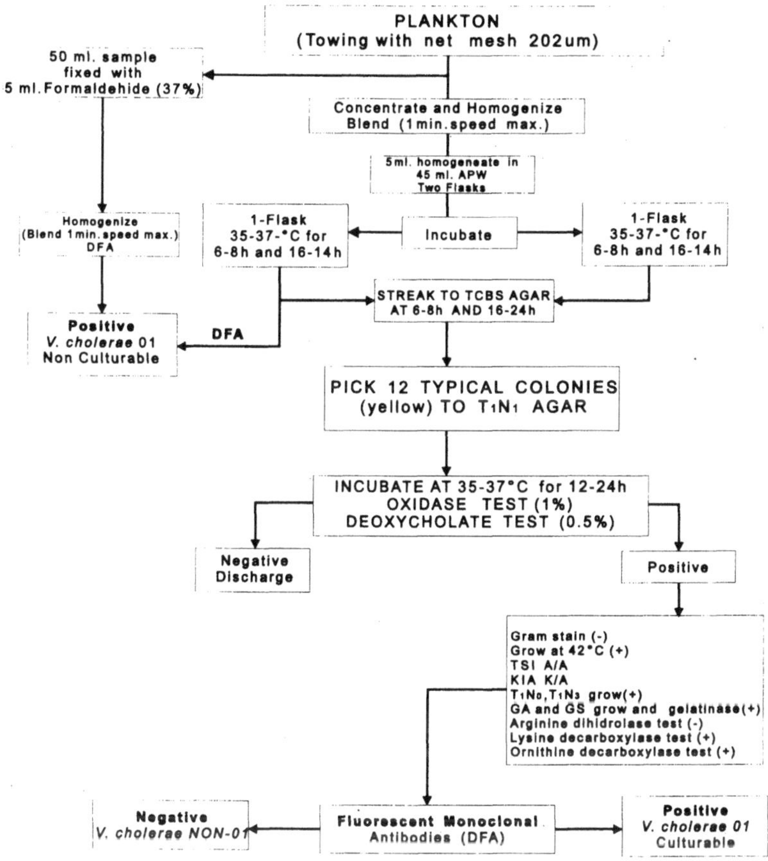

Figure 1.

Direct detection with monoclonal antibodies has been proved to be more sensitive than the culturable method. *Vibrio cholerae* 01 was detected by DFA in plankton samples of Tamiahua and Pueblo Viejo, in september 1995 and January 1996, but the culturable state was only detected in Tamiahua in 1995. In the last sampling (10/96) *V. choleae* 01 was not detected in their culturable or non-culturable state, only the non-01 was present in all of the samples collected in these coastal lagoons.

In September 1995 was registered one of the highest incidence of cholera cases (387 cases) of that year in Veracruz, while for January, May and October 1996, only 5, 4 and 4 cases were registered respectively. This epidemiological pattern seems to correlate with the detection of *V. cholerae* 01 in plancton samples, where *V. cholerae* 01 disappear first in their culturable state and then completely in the last samplig, where the non-01 was the only serogroup detected in plancton samples.

4. SUMMARY

Since 1991 a new epidemic of cholera has been developed in Latin America. Marine organisms such as fish, bivalves and plankton has been indicated as vectors of toxigenic *Vibrio cholerae* 01. In Mexico the cholera has been an acute problem in different states of the country, specially at those located in the coasts of the Gulf of Mexico and the Pacific Ocean. With such information as a background, a joint study has been performed in different localities at the Gulf of Mexico and the Pacific Ocean coasts. The fluorescent monoclonal antibody staining method was employed to detect the presence of culturable and non-culturable *V. cholerae* 01 in plankton (202 μm) samples.

V. cholerae 01 was only detected in coastal lagoons from the Gulf of Mexico in their stages culturable and non culturable, while in the areas studied at the Pacific Ocean, any positive determination was registered for *V. cholerae* 01, only the presence of *V. cholerae* non-01 was registered. We also observed a decrease of positive determinations from 1995 to 1996, which follow the same tendency of the incidence of cholera cases in Mexico, which show a drastic decrease from 13,298 cases in 1995 in regard of 1088 cases registered in 1996.

ACKNOWLEDGMENTS

We thank Amelia Portillo, Sila Najera, Guadalupe Vargas and Cesar Almeda for their technical assistance. The funds for this research were obteined from CONACYT under the project CONACYT/CICESE/UNAM/UAM-I 5167-T9407.

REFERENCES

Colwell, R.R., 1996, *Science*, **274**:2025–2031.
McCarthy, S.A., and F.M. Khambaty, 1994, . *App. Environ. Microbiol.*, **60**: 2597–2601.
Secretaría de Salud, 1991, Boletín cólera/diarreas infecciosas. **1**(16):1–7.
Secretaría de Salud, 1993, Boletín cólera/diarreas infecciosas. **3**(1):218–224.
Secretaría de Salud, 1994a, Boletín cólera/diarreas infecciosas. **4**(1):434–456.
Secretaría de Salud, 1994b, Boletín cólera/diarreas infecciosas. **4**(10):598–601.
Secretaria de Salud, 1995, Epidemiología. semana 52.
Secretaría de Salud, 1996, Epidemiología. semana 52.
Secretaría de Salud, 1997, Epidemiología. semana 23.
Tamplin, M.L., and C. Castillo Perodi, 1991, *The Lancet*, **338**:1216–1217.
Tauxe, R., L. Seminario, R. Tapia, and M. Libel, 1994, *American Society for Microbiology Press, Washington, D.C.* 465 pp.

MOLECULAR CHARACTERIZATION OF METALLOTHIONEIN- AND CYTOCHROME P4501A-CDNAS OF *SPARUS AURATA* AND THEIR USE FOR DETECTING POLLUTION ALONG THE MEDITERRANEAN COAST OF ISRAEL

Moshe Tom, Ophira Moran, Edward Jakubov, Benzion Cavari, and Baruch Rinkevich

National Institute of Oceanography
Israel Oceanographic and Limnological Research
P.O.Box 8030, Haifa, 31080, Israel

1. INTRODUCTION

Members of the fish family Sparidae are abundant in various continental shelf habitats along the Mediterranean, including relatively clean and polluted regions. Fish Metallothionein (MT) and cytochrome P4501A genes are induced by heavy metals and several groups of halogenated and nonhalogenated aromatic hydrocarbones, respectively (George and Olsson, 1994; Bucheli and Fent, 1995). The aim of the presented study is to examine the suitability of the sparid *Lithognathus mormyrus* as bioindicator of pollution in Mediterranean coastal habitats, using its hepatic MT- or P4501A-mRNAs as biomarkers.

2. RESULTS AND DISCUSSION

A complete MT-cDNA (GenBank acc. U58774; Tom et al., 1997), and a partial P4501A-cDNA (GenBank acc. AF005719), were cloned from hepatic cDNA library of the sparid fish *Sparus aurata* and used to evaluate the corresponding hepatic mRNA levels in the abundant coastal sparid *Lithognathus mormyrus* after induction by cadmium or β-naphthoflavone (β-NF), and at time intervals following pulse induction, as well as in fish sampled in the polluted Haifa Bay, the Mediterranean coast of Israel (Fig. 1).

The two specific liver mRNA levels were evaluated by image analysis of dot blotted hepatic RNA samples, visualized on an X-ray film after hybridization to labeled *S. aurata*

New Developments in Marine Biotechnology, edited by Le Gal and Halvorson.
Plenum Press, New York, 1998.

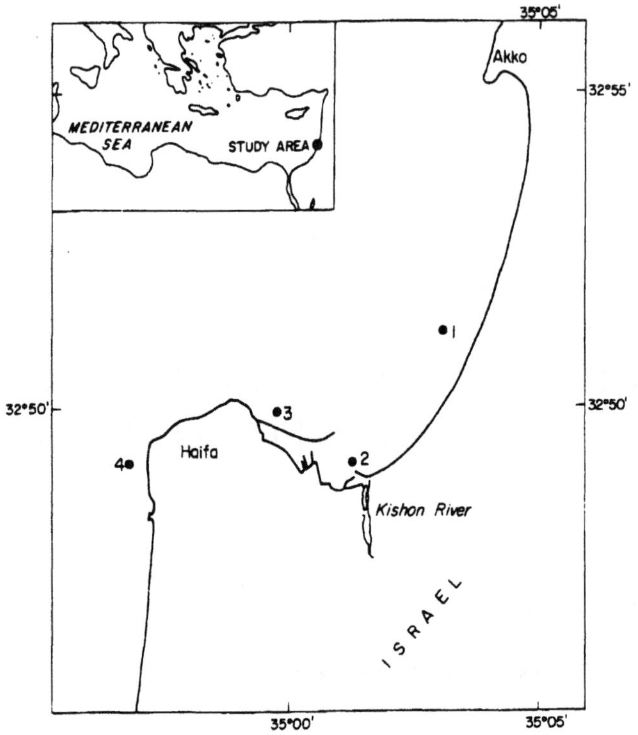

Figure 1. The sampling site. 1–4: sampling stations.

MT- or P4501A-cDNAs. The arbitrary optical density units were normalized per mg total RNA in the sample, and film-to-film differences were adjusted using a standard hepatic RNA sample.

It is concluded that both genes are inducible in the liver of the target species (Fig. 2), although revealing different characteristics. Ambient temperature has no effect on the inducibility of the P4501A. In contrary, the induction of MT-mRNA at high injected cadmium dose was reduced at the higher examined temperature. Nearly maximum induced hepatic MT-mRNA levels were assumedly reached, as the fish did not survived higher cadmium doses (Tom et al., 1997), and in addition, cadmium is known to be a potent MT inducer. Maximal P4501A-mRNA levels have to be further examined using more potent inducer, such as 2,3,7,8-tetrachlorodibenzo-p-dioxin (TCDD). Long depuration period, of several months, is recommended for fish used for induction experiments, to reach low levels of specific mRNAs, hence, increasing resolution of the effect of low inducer levels. The difference in pre-experimental depuration period is reflected in the control mRNA levels of the two 21°C induction experiments, maintained in the laboratory for 3–6 weeks prior to the induction, in comparison to the similar 27°C experiments, maintained for 12 weeks in clean water prior to induction.

The rates of hepatic induction and decay of these two specific mRNAs, determines the efficiency of the bioassay and its ability to distinguish repeating pollution events within small spatial ranges and time intervals. Induction time course was not followed in detail during this study, but even induction in 7 or 10 days observed during the experiments, can be considered fast induction for an environmental bioassay. The depuration patterns of the two mRNAs were more complicated. Hepatic MT-mRNA levels decreased

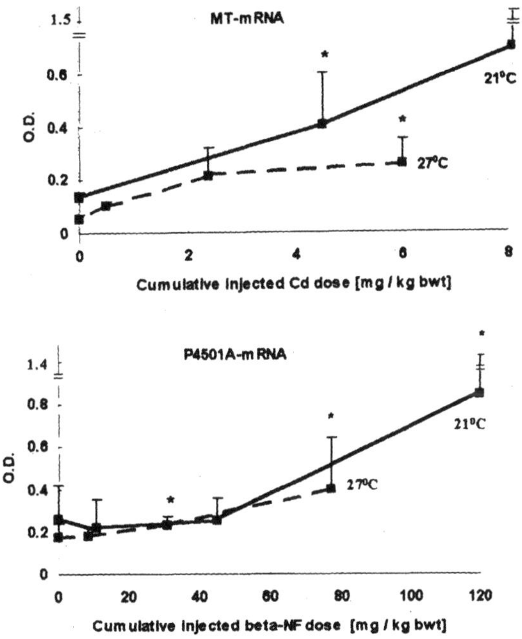

Figure 2. Induction experiments: Dose response of hepatic MT and P4501A-mRNA levels to injected cadmium or β-NF. Relative mRNAs in induced *L. mormyrus* livers is expressed as arbitrary density units (OD), normalized per mg total RNA/ml sample. Each dose was injected to 5 to 7 and 6 to 7 specimens in the MT or P4501A experiments, respectively. * denotes statistically significant optical density difference from control values (t test, P<0.05).

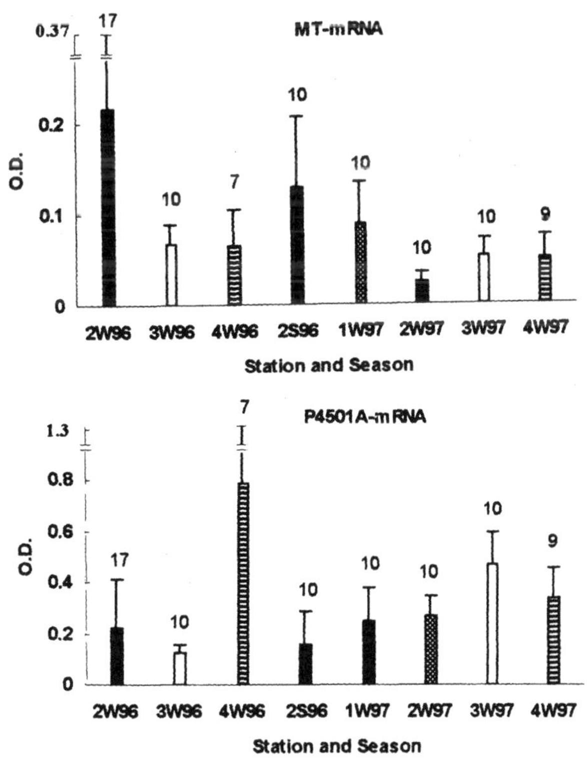

Figure 3. Relative hepatic MT- and P4501A-mRNA levels in sampled *L. mormyrus*. For OD units see Fig. 2. The numbers above the bars determine the sample size. The sampling station designation in x-axis includes station number (see Fig. 1), sampling season, W: winter, S: summer, and sampling year.

only by 36%, 77 days following pulse induction by high dose of cadmium, with no change in the high total hepatic cadmium content during this period (Herut, unpublished results). Sampled fish which were maintained in the laboratory 30 days, revealed more prominent MT-mRNA decrease of 83%, in comparison to the other half of the sample, which was immediately sacrificed. P4501A-mRNA revealed 60% decrease in 77 days following pulse induction by β-NF, and sampled fish showed a decay of 26% during 30 maintenance days in the laboratory. The decrease in MT or P4501A-mRNA levels with time, during maintenance in unpolluted water, is assumed to depend mainly on the inducer depots, and its rate of excretion from the target tissue. This may explain the variability in the rate of MT-mRNA during the decay experiment. The fish liver was not able to eliminate the pharmacologically high doses of injected cadmium during the experimental period, whereas, the sampled fish were able to metabolize their much lower hepatic heavy metal load.

L. mormyrus specimens sampled at Haifa Bay (Fig. 2) revealed relatively high MT-mRNA levels in station 2, the Kishon river mouth, an established point source of heavy metal pollution (Cohen et al., 1992; Herut et al., 1995). Relatively high hepatic P4501A-mRNA levels were related to a more south-western location in the Bay, and at the moment we can not recognize the source of this perturbation.

REFERENCES

Bucheli, T.D., & Fent, K. 1995, Induction of cytochrome P450 as a biomarker for environmental contamination in aquatic ecosystems. *Crit. Rev. Environ. Sci. Technol.* **25**: 201–268.

Cohen Y., Kress, N., & Hornung, H. 1992, Organic and trace metal pollution in the sediments of the Kishon river (Israel) and possible influence on the marine environment. In: *Environmental Quality and Ecosystem Stability* (Gasith, A., Adin, Y., Steinberger, Y., & Garty, J., eds.), Vol. V/B, pp. 503–511.

George, S.G., & Olsson, P.A. 1994, Metallothioneins as indicators of trace metal pollution. In: *Biomonitoring of coastal waters and estuaries* (Kees, J., & Kramer, M., eds). CRC Press, Boca Raton, FL.

Herut, B., Hornung, H., Kress, N., Krom, M.D., & Shirav, M. 1995, Trace metals in sediments at the lower reaches of Mediterranean coastal rivers, Israel. *Wat. Sci. Tech.* **32**:239–246.

Tom, M., Moran, O., Jakubov, E., Cavari, B., & Rinkevich, B. 1997, Molecular characterization of metallothionein-cDNA of *Sparus aurata* used for detecting heavy metal pollution along the Mediterranean coast of Israel. *Mar. Pollut. Bull.* (in press).

WORKSHOP ON FATTY ACID PRODUCTION AND METABOLISM

Synthetic Report

S. A. Poulet[1] and K. Yazawa[2]

[1]Station Biologique
Place Georges Teissier
29682 Roscoff, France
[2]Kanagawa, Japan

The workshop was based on the oral presentations or recent results obtained by eleven speakers, including by alphabetic order: Drs. Bemis D.L., Berdyshev E.V., Carballeira N.M., Garcia D.M., Grizeau D., Nichols P., Samain J.F., Tamaoka J., Tredici M., Yang N. and Yokochi T. The majority of communications covered topic number I, related to natural products, while three other topics were addressing specific questions to aquaculture, cell and molecular biology and developmental biology. Variations, production and characterisation of PUFAs found in marine bacteria, fungi, algae and animals were described. The influence and biological activities of several relevant molecules, belonging to fatty acids, ethanolamides and cannabinoids, conjugated, conjugated tetraenes (AA, DHA, EPA), as well as several key PUFAs, such as DF, LA in addition to sterols, phospholipids and phopholipase were reported. Identification and properties of these molecules were presented for different categories of marine bacteria and microalgae. Their specific interest and relevance as bioproducts were discussed. Their physiological role was presented in relation to several basic biological functions, metabolism or requirements identified in three categories of marine metazoan, including sea urchin, shellfish and fish. Their relevance and potential utilisation as natural products in biological or medical research and for aquaculture were discussed.

The workshop was subdivided into three sub-topics related to bacteria, marine algae and fungi, and marine animals. In the first group, identification of PUFAs found in specific bacteria strains collected in the pelagic and coastal areas, deep-sea mud, or sea-ice was reported. These strains were isolated from the environment and cultured in adequate medium. Fatty acid profiles could be divided into several categories, including DPA, EPA, DHA, LA and AA. The cellular fatty acid composition was in general analysed, using GC or GC-MS methods. Results indicated that PUFA profiles can be used as a kind of taxo-

New Developments in Marine Biotechnology, edited by Le Gal and Halvorson.
Plenum Press, New York, 1998.

nomic tool among *Thrustochytrids*, and helps to understand the phylogenic relation between the strains of these bacteria. It was also shown that temperature ranging from –2 to +16°C, as well as pressure in the 0.1 to 70 MPA range influenced cellular fatty acid composition of deep-sea bacteria stain DSS 12. At both low temperature and high pressure conditions, hexadecenoic and icosapentaenoic acids (EPA) proportions were larger than for the opposite conditions, suggesting some striking adaptation mechanisms. It was further shown that several PUFAs used in the human diet, which are ordinary found in fish oil or algal species, are also present in sea-ice bacteria. Similarly, new strain of *Vibrio alginolyticus*, associated with dinoflagellates or zoanthids, are producing novel octadecenoic and pentadecadienoic acids, which have antibiotic activity against pathogenic streptococci and staphylococci.

In the second group, presentations were oriented towards the metabolism and production of PUFAs by specific marine algae, including *Anadyomene stellata*, *Porphyridium cruentum*, *Rhodosorus marinus* and several microalgae widely used in aquaculture (c.a. *Tetraselmis suecica, Chlorella* sp., *Pavlova lutheri* and *Nannochloropsis* sp.). These algae are likely today among the most important sources of AA and EPA. Methods to produce high biomass of these algae as well as specific protocols to enhance selective PUFA production have been described. For example, it was shown that cells in *Rhodosorus marinus* grown in batch cultures in tubular vertical reactors under continuous illumination, and pH regulation with CO_2 enriched air, induces a large increase in the EPA/AA ratio in the total fatty acid contents.

In the third group, topics were related to the influence of fatty acid ethanolamides on male sea urchin fertilisation, scallop reproduction and larval development, and digestive enzymes PLA2 purified from marine fish. In sea urchin, cannabinoid receptors have been identified on the membrane of spermatozoa. The endogenous ligands (AA ethanolamide) for these receptors, named anandamide, were shown to block the egg jelly-induced acrosome reaction and fertilisation. Additional data in the scallop *Pecten maximus* demonstrate the complex nature of PUFAs in diets and their interference with reproduction and embryogenesis. It was suggested that scallops which face changes in the lipid composition of their food, incorporate PUFAs in their storage lipids without major selection. But strong compensatory mechanisms exist leading to a selective retention of specific essential fatty acids, belonging to the phospholipid classes, suggesting how important they are in the reproduction process. Considering that survival of marine animals is associated with reproduction and growth success, a better understanding of the lipid composition and metabolism deserves further attention in aquaculture research.

This workshop has underlines the importance of several new, exotic or well known marine prokaryotes, including bacteria, fungi and microalgae as new sources of PUFAs. The production of these compounds, known to have several important applications in medical research, pharmacology and aquaculture relies on the development of bioreactor technology, production of biomass under controlled conditions and on the full understanding of their properties, characteristics and activities in plants and animals.

68

WORKSHOP ON BIODIVERSITY

Synthetic Report

J. Frederick Grassle[1] and Jack B. L. Matthews[2]

[1]Rutgers University
New Brunswick, New Jersey
[2]The Scottish Association for Marine Science
Oban, Scotland, United Kingdom

The workshop addressed three main issues in the context of nine scientific contributions which had been assigned to it.

1. METHODOLOGICAL ASPECTS

The following report formed the basis for discussion on the methods and techniques available for the study of biodiversity, in the laboratory and in the field.

Kawauchi introduced work carried out by himself, Sagabe and Amemiya on the "Molecular Cloning of Somatolactins in the Sturgeon and Dogfish". He pointed out that somatolactin may be involved in environmental adaptation of teleost fish but no corresponding molecule had been characterized in other fish or tetrapods, possibly due to the small amounts present relative to growth hormone and prolactin. Cloning was therefore carried out to amplifiy the cDNA by means of the polymerase chain reaction (PCR) method. The results indicate the presence of somatolactin in Actinopterygii and Chrondrichthyes and suggest that it does play a role in environmental adaptation in "primitive" fish as well as in teleosts.

Penna introduced work carried out by herself and Magnani on the "Identification of *Alexandrium* Genus in Culture and in Marine Phytoplankton Samples using Genus-specific rDNA-targeted Probe". A fragment of rDNA was amplified by means of PCR. A ^{32}P-labelled probe was prepared and tested for cross-reactivity with cultured Alexandrium and with phytoplankton samples. Reaction with the cultures and with those samples that contained *Alexandrium*, but not with those that did not, showed that the method made it possible to quantify the presence of the genus, which is known for its toxicity, in samples taken in the field.

New Developments in Marine Biotechnology, edited by Le Gal and Halvorson.
Plenum Press, New York, 1998.

Kato presented work on the "Characteristics of the Obligate Barophilic Bacteria Isolated from the Mariana Trench at a Depth of 11000m". Two kinds of bacteria, related to the genera Shewanella and *Moritella*, were isolated and analysed on the basis of ^{16}S rRNA sequence homology; they showed no sign of growth at pressures less than 50 MPa and grew best at 70–80 MPa. This is the first evidence of an obligate barophilic species of *Moritella*, and has been made possible by the recent development of the "Kaiko" unmanned submersible which is capable of working at maximum ocean depths.

2. APPLICATIONS

The following report formed the basis for discussion on why it is necessary or desirable to study biodiversity.

Hill reported on the "Isolation of Actinomycetes from Diverse Marine Environments for Natural Products Screening". Conventional and molecular techniques showed the isolates, from the Chesapeake Bay, from near-shore waters off Hawaii, and from the vicinity of hydrothermal vents and cold-water seeps, to be highly diverse and that there were distinct differences between them. Optimal growth was obtained at salinities greater than 1%, providing circumstantial evidence for indigenous marine actinomycetes. Isolates are presently being screened for natural pharamaceutical products.

Blackburn reported on "Microalgal Biodiversity in Australian Waters. Importance of Culture Collections and Implications for Biotechnology and Aquaculture", work carried out in collaboration with Bolch, Brown, Jackson, LeRoi and Volkman. This demonstrated that, since the cosmopolitan distribution of many species often masks great diversity at the populational level, it is essential to maintain and study cultured strains of known origin. The uniqueness of Australian strains, as demonstrated through the use of RAPD, RFLP and sequence analysis, has implications for the use of microalgae in aquaculture and in the search for valuable new bioactive compounds.

3. DESCRIPTION

The following report formed the basis for discussion on the need to describe biodiversity, and in what way, for the purposes identified under 2 above.

Thippeswamy described "The Biological Diversity in a Tropical Green Mussel (*Perna viridis* (L)) Bed on the South West Coast of India" as it had changed over a two-year period (1986–1988). For most of the time *P. viridis* was the dominant species in the community but was eliminated annually at the height of the south-west monsoon. It rapidly recolonized the area at the same time as the biodiversity increased with the improved conditions that the mussel provided for infauna and epifauna. Physical factors (air and water temperature, pH, salinity, and tidal range) accounted for 25% of the statistical variance in the community, and biological factors (community structure) for a further 23%.

Buchholz prefaced his report on "Biodiversity at the Rocky Sea-bottom at the Island of Helgoland: Population Development in the European Lobster, Homarus gammarus" with an assessment of German prioriites for studies on biodiversity, which is coupled with a political acceptance of the need for well-founded long-term studies. The character of, and the longstanding interest in, Helgoland is the justification for its designation as a Nature Reserve, and as a site for case studies and for courses in taxonomy and ecophysiology. These advantages are exemplified by the lobster study; this includes analysis of long-term data on catches which show a steady decline since the 1930s, experiments on behaviour and ecophysiology in a large maintenance facility and field studies of population genetics and community structure. Causal analysis is expected to indicate whether restocking of the area is likely to be worthwhile.

Kideys presented the results of "A Study on the Biodiversity off Manavgat River, Northeastern Mediterranean" conducted in collaboration with Bingel, Gucu, Eker and Unsal. The study had been prompted by an impending plan to reduce the fresh-water outflow in the area and is intended to provide baseline information on community composition beforehand. 294 species and 74 other taxa, comprising phytoplankton, zooplankton, macroalgae, benthos and fish, were identified. Some of these were recorded for the first time from Turkish waters and several were Indo-Pacific in origin, probably immigrants through the Suez Canal. These results provide a baseline for future investigations to analyse any community changes that may occur in the region.

Ducrotoy reported on "Marine Bio-diversity Studies in North-east England in Relation to Coastal Management", co-authored by Elliott. Various intertidal and subtidal ecosystem studies, in succession to projects under the European "COST 647" programme on coastal benthic ecology and prompted by the North Sea Quality Status Report (1994) have elucidated mechanisms that generate and maintain biodiversity, have further clarified the species concept using DNA sequencing and allozyme techniques, and have contributed to ideas regarding the apparently global rarity of many species. Studies on the fragmentation of coastal ecosystems (landscape ecology) and the classification of habitats provide a scientific basis for the designation and management of protected areas, and comparison of long-term data with those from other regions is contributing to an understanding of climatic change. Together, they provide a basis for multi-purpose management of coastal areas.

4. CONCLUSIONS

Biodiversity has been the subject of a great deal of discussion in recent years and has been extensively associated in people's minds with the richness of species or habitats, often with the charismatic species whose loss or depletion has caught public and political imagination. This Workshop provided a valuable counterbalance by the attention it paid to biodiversity manifest in molecular terms but operating at supraspecific, specific or infraspecific level.

Molecular methods are complementary to those of traditional taxonomy, but they should not replace it. They have a particular advantage in the light that they can throw on functional biodiversity and its genetic basis. They provide the descriptions necessary to understand the mechanisms of physiological adaptability and help to explain the development of behavioural patterns that determine feeding strategies, migration and many other aspects. They mark a biochemical rather than phenotypic path of evolution and they are particularly important in the case of bacteria where other methods often provide scant information. In practical terms, one of their most valuable applications, to judge by the contributions to the Workshop, is in guiding the search for new natural pharmaceutical and nutritional products.

On their own, however, such methods also pose new questions. What is the nature of the multitude of new species that molecular distinctions seem to indicate? Are they really distinct species? Is DNA sequencing equivalent to the identification of species by conventional taxonomic and ecological means? In other words, what is the "acid test" applicable to molecular material that is equivalent to the reproductive isolation of species identified by traditional means? These scientific questions require to be addressed if the fundamental purpose of extensive cataloguing of genome sequencing is to be widely accepted. The work on actinomycetes, microalgae and barophilic bacteria reported at the Workshop point a way to achieving the complementary benefits of taxonomic and molecular methods.

Furthermore, it was noted that there are other characters or properties relevant to the identification of functional groups and ecotypes on the one hand, and the assessment of the biotechnological potential of marine organisms on the other. It was suggested, for example, that chemotaxonomy and ecophysiological adaptations need to be taken into account in the search for new drugs. This theme deserves full discussion.

It is no coincidence that the four papers dealing with biodiversity for management, conservation, and related purposes described the communities in traditional taxonomic terms. Whatever their shortcomings, faunal and floral data do provide a consistent basis for assessment available to ecologists and environmental managers alike.

Given the enormous hunger for information on biodiversity, and the present-day vigorous assessment of research proposals, the long-term value of biological data, once obtained, is priceless. Not only must the data be stored, it must be known about and made available. The mechanism for this is Networking. Just such marine biological networks exist, in North America, the Caribbean, Europe, and the Pacific Rim, and projects are underway to develop effective storage and collaborative use of faunal and floral data. These require support and resources nationally and regionally.

Internationally, the DIVERSITAS Programme, sponsored by IUBS, SCOPE, UNESCO, ICSU, IGBP-GCTE and IUMS, can provide a focus for this collaboration, in accordance with its aim to foster innovations and partnerships between research institutions and universities of developed and developing countries. A Special Target Area of Research is Marine Biodiversity, where it has been recognised that major advances in understanding ecological processes in intertidal environments and coral reefs have come from a combination of descriptive observations and theory-based hypotheses. These advances need to be extended to other oceanic environments and integrated into the larger framework of oceanic models and patterns of natural physical and chemical variation. In view of the global importance of microbes and considering that less than five thousand of an estimated three million bacterial species have been described, it is essential that microbial diversity features strongly in the Programme.

The ultimate goal is to understand and predict the functioning of marine ecosystems.

5. ACRONYMS

ICSU	International Council of Scientific Unions
IGBP-GCTE	International Geosphere–Biosphere Programme—Global Change and Terrestrial Ecosystems
IUBS	International Union of Biological Societies
IUMS	International Union of Microbiological Societies
SCOPE	Scientific Committee on Problems of the Environment

WORKSHOP ON POLICY*

Bernardino Fantini[1] and Fernando Quezada[2]

[1]Geneva, Switzerland
[2]Boston, Massachusetts

OVERVIEW

Presentations at this workshop provided an overview of the current frontiers in marine biotechnology policy research and described the multiple opportunities for active participation on the part of the scientific and research community. Discussions highlighted how public policy decisions affect public funding for marine biotechnology research, and how and when scientists and researchers can most effectively participate in the public policy process affecting research, development and commercialization. The range of issues covered from economic and legal aspects to questions of social and ethical concerns. The key policy issue areas include access to marine resources/organisms, intellectual property rights, biosafety and regulatory approaches, coordinated efforts/national and international plans and academic/industry/government cooperation.

Opening statements addressed how the ethical and social aspects of biotechnologies and genetically modified organisms and their public image can be taken into account in policy planning. It was pointed out that the respect for future generations should be a predominant concern in any debate on the ethical permissibility of a revolutionary biotechnology procedure which can affect human genetic heritage and biodiversity.

The principle of distributive justice, should be applied to the diffusion of marine biotechnology innovations and products. It will be essential to get a balanced agreement which offers reciprocal benefits to all participants with equitable utilization of the results of marine biotechnology advancements between rich or poor countries.

ACCESS TO MARINE RESOURCES/ORGANISMS

This discussion included reference to the Law of the Sea Convention in relation to the deep seabed portion and other relevant aspects relating to the rights and duties of

* Presenters: Salvatore Arico (Italy), Lori Denno Bouman (USA), Biliana Cicin-Sain (USA), Dosoo Jang (USA), Rollin Johnson (USA), Robert Knecht (USA), William Parish (U.K.), Hank Parker (USA), Cynthia Soumoff (USA), Jack Wiggins (USA).

New Developments in Marine Biotechnology, edited by Le Gal and Halvorson.
Plenum Press, New York, 1998.

nations with regard to their offshore ocean zones. There was also discussion of the various bioprospecting arrangements between the marine biotechnology industry and United States marine protective areas (MPA's) which can serve to promote marine genetic resources, conservation and development. Novel opportunities for marine bioprospecting in MAP's are based on emerging international and national emphases on inventory efforts and partnership approaches in MPA management.

INTELLECTUAL PROPERTY RIGHTS

One presenter pointed out that the issue of patent and trade secret protection for genetic information and modified life forms raises unique and complicated questions about ownership and control of genetic resources. Discord over intellectual property rights in marine biotechnology could potentially have worldwide consequences on the development of the field. The point was made that Western firms generally view the protection of their investments through intellectual property rights as vital to promoting the development, implementation, and transfer of novel technologies. In contrast, traditional societies regard biodiversity and its components, including genetic information, as common property.

Another presenter on this subject highlighted the fact that marine biotechnology patents cover such diverse subjects as compounds from new marine strains, derivatives from marine coral for use as pharmaceuticals, and more.

BIOSAFETY AND REGULATORY APPROACHES

Discussions of biosafety and regulatory control primarily addressed issues relating to proposed releases of genetically modified (GM) fish. It was pointed out that the regulatory process is underpinned by risk assessment, the identification of hazards, the assessment of the likelihood of the hazards being realized and the estimation of the magnitude fo the consequences. The U.K. experience with the regulatory aspects of proposed releases of GM fish was described. It was pointed out that to date, non consent applications for release of GM fish have been made in the U.K.

COORDINATED EFFORTS/NATIONAL AND INTERNATIONAL PLANS

Discussions on this topic described the case of the United States experience with developing a strategic plan for aquaculture. This effort took into consideration several guiding principles in pursuit of clearly established goals and in full recognition of the need to incorporate critical success factors such as regulatory framework, effective coordination among agencies, investment levels, marketing and partnerships. In the ongoing U.S. national aquaculture plan effort, particular attention has been given to stakeholder groups in both the public and private sector, regional and state programs, academic community, various interest groups and the general public. Reaching consensus on various issues has proven to be a challenge in this national aquaculture planning effort.

ACADEMIC/INDUSTRY/GOVERNMENT COOPERATION

Discussions on this topic explored how successful academic, industry and government cooperation efforts can be structured at the national and international level. Collaborative strategies which have spawned a number of innovative approaches have also produced decidedly mixed results in achieving stated missions. A workshop presenter described a survey of U.S. coastal states and territories and a number of selected countries reflecting a range of institutional arrangements that have been employed in promoting marine biotechnology.

Another study of collaborative efforts was described which sought to determine the critical factors affecting the business of aquaculture in the United States: technology, management, product risks, financing, government regulation, permitting, siting, and markets. The most frequently suggested recommendations were to improve the regulatory environment, streamline permitting, increase access to capital for business development, and fund more research.

Studies on perceived impediments to marine biotechnology commercialization presented findings and interpretation of the results which showed lack of scientific understanding to be an important deterrent for the take-off of the marine biotechnology field in the U.S. as is the lack of a comprehensive regulatory framework.

CONTRIBUTORS

L. Abelli
Dipartimento di Scienze Ambientali
Università della Tuscia
I-01100 Viterbo, Italy

Eliane Abou-Mansour
Australian Institute of Marine Science
PMB 3, Townsville, Queensland 4810, Australia

Jagat Adhiya
School of Natural Resources
Ohio St. Univ.
Columbus, OH 43210, USA

Peter Aleström
Department of Biochemistry
Norwegian College of Veterinary Medicine
PO Box 8146 Dep., N-0033 Oslo, Norway

Standish K. Allen, Jr.
College of William and Mary
School of Marine Science
Virginia Institute of Marine Science
Gloucester Point, VA 23062-1346

Dmitry L. Aminin
Pacific Institute of Bioorganic Chemistry of the Russian Academy of Sciences
Vladivostok-22, Russia

Neil K. Armstrong
Ottawa Civic Hospital Loeb Research Institute
725 Parkdale Avenue
Ottawa, Canada K1Y 4E9

S. Auzoux-Bordenave
Laboratoire de Biologie des Invertébrés Marins et Malacologie
URA CNRS 699
Muséum National d'Histoire Naturelle
57, rue Cuvier, 75005, Paris, France

Elena Barbieri
Marine Biological Laboratory
Woods Hole, USA

Guadalupe Barrera-Escorcia
Universidad Autónoma Metropolitana
Unida Ixtapalapa, UAM-1

Valadmir Belenky
Departments of Chemistry and Biology
William Paterson University
Wayne, New Jersey 07470, USA

Marlies Betka
Department of Biology
Boston University
5 Cummington St.
Boston, MA 02215, USA

David J. Bourne
Australian Institute of Marine Science
PMB 3, Townsville
Queensland 4810, Australia

Alfonso V. Botello
Instituto de Ciencias del Mar y Limnologia, UNAM

C. Brennan
Dept. of Biological Sciences
University of Southern California
Los Angeles, CA 90089-0371

Xiao-Hua Cai
Depts. of Biochemistry and Plant Biology and Graduate Program in Biophysics
Ohio St. Univ.
Columbus, OH, 43210 USA

Ian P. Callard
Department of Biology
Boston University
Boston, MA 02215 USA

Gloria V. Callard
Department of Biology
Boston University
5 Cummington St.
Boston MA 02215, USA

I. Cancre
Station de Biologie Marine
Muséum National d'Histoire Naturelle et Collège de France BP 225
29900 Concarneau, France

Karen Casciotti
Marine Biology Research Division and Center for Marine Biotechnology and Biomedicine
Scripps Institution of Oceanography
University of California
San Diego, La Jolla CA 92093-0202

Benzion Cavari
National Institute of Oceanography
Israel Oceanographic and Limnological Research
P.O.B. 8030, Haifa, 31080, Israel

E. Cesar
Department of Microbiology
University of Hawaii
Honolulu, Hawaii 96822, USA

Devi Chakravorty
Department of Microbiology
University of Maryland (UMPC)

Songlin Chen
Physiologische Chemie I
Biozentrum der Universität Würzburg
97074 Würzburg, FRG

D. Cheney
Seaweed Biotechnology Laboratory
Marine Science Center
Northeastern University
Nahant, MA 01908, USA

Seinen Chow
National Research Institute of Far Seas Fisheries
Shimizu 424, Japan

Ka-Hou Chu
Department of Biology
The Chinese University of Hong Kong
Shatin, Hong Kong, China

Maria Ciaramella
Institute of Protein Biochemistry and Enzymology
Naples, Italy

Philippe Collas
Department of Biochemistry
Norwegian College of Veterinary Medicine
PO Box 8146 Dep., N-0033 Oslo, Norway

Maria Rosaria Coscia
Institute of Protein Biochemistry and Enzymology
CNR, Naples, Italy

Regina T. Counihan
Department of Zoology
University of Queensland
Brisbane, Qld 4072 Australia

Vernon E Coyne
Department of Microbiology
University of Cape Town
Private Bag, Rondebosch 7701, South Africa

B. Dale
Stazione Zoologica 'Anton Dohrn'
Villa Comunale 1, 80121 Napoli, Italy

Seana K. Davidson
Marine Biology Research Division and Center for Marine Biotechnology and Biomedicine
Scripps Institution of Oceanography
University of California, San Diego
La Jolla, CA 92093-0202

Heather L. Davis
Ottawa Civic Hospital Loeb Research Institute
725 Parkdale Avenue
Ottawa, Canada K1Y 4E9

Bernard M. Degnan
Department of Zoology
University of Queensland
Brisbane, Qld 4072 Australia

Luigi De Martino
Zoological Station
Naples, Italy

M. L. De Simone
Stazione Zoologica 'Anton Dohrn'
Villa Comunale 1
80121 Napoli, Italy

Anna Di Cosmo
Dept. Zoology
University of Naples Federico II
Naples, Italy

Jocelyne Di Ruggiero
Center of Marine Biotechnology
University of Maryland Biotechnology Institute
701 E. Pratt St., Baltimore, MD, USA

J. L Ding
School of Biological Sciences
National University of Singapore
Kent Ridge, Singapore 119260

D. Doumenc
Laboratoire de Biologie des Invertébrés Marins et Malacologie
URA CNRS 699
Muséum National d'Histoire Naturelle
57, rue Cuvier, 75005, Paris, France

Shaojun Du
Department of Clinical Biochemistry and Biochemistry
University of Toronto
Toronto, Canada

W. C. Dunlap
Australian Institute of Marine Science

H. Edgar
Department of Biochemistry, Microbiology and Molecular Biology
University of Maine
Orono, ME 04469, USA

Deeanne B. Edwards
Marine Biology Research Division and Center for Marine Biotechnology and Biomedicine
Scripps Institution of Oceanography
University of California, San Diego
La Jolla, CA 92093-0202

Susan M. Efler
Ottawa Civic Hospital Loeb Research Institute
725 Parkdale Avenue
Ottawa, Canada K1Y 4E9

Marc Ekker
Loeb Inst. for Medical Research
Ottawa, Canada

Anthony E. Ellis
SOAFD Marine Laboratory
PO Box 101, Victoria Road
Aberdeen AB1 9DB, Scotland

Rebecca Ericson
Forsyth Dental Center
Boston, MA, USA

Mario P. Estrada
Mammalian Cell Genetics Division
Centro de Ingenierìa Genètica y Biotecnologìa
P.O.Box 6162, Havana, Cuba

Milton Fingerman
Tulane University

Chris A. Francis
Marine Biology Research Division and Center for Marine Biotechnology and Biomedicine
Scripps Institution of Oceanography
University of California, San Diego
La Jolla, CA 92093-0202

José de la Fuente
Mammalian Cell Genetics Division
Centro de Ingenierìa Genètica y Biotecnologìa
P.O.Box 6162, Havana, Cuba

Masahiro Furutani
Marine Biotechnology Institute
Kamaishi, Iwate, Japan

Erwin A. Galinski
Institute for Biochemistry
Westfälische Wilhelms-Universität
Wilhelm-Klemm-Straße 2
48149 Münster, Germany

Adriana A. Gambarotta
University of Ottawa Heart Institute

Jonathan B. Geller
Department of Biological Sciences
University of North Carolina at Wilmington
Wilmington, NC 28403, USA

William H. Gerwick
College of Pharmacy
Oregon State University
Corvallis, OR 97331, USA

Ida Gesualdo
Zoological Station
Naples, Italy

Genny Giroux
Loeb Inst. for Medical Research
Ottawa, Canada

J. Frederick Grassle
Rutgers University
New Brunswick, NJ, USA

Patrik Gröhn
Department of Biotechnology
University of Würzburg
Biocenter-Am Hubland
D-97074 Würzburg, Germany
kloeck@biozentrum.uni-wuerzburg.de

L. Grumetto
Stazione Zoologica 'Anton Dohrn'
Villa Comunale 1
80121 Napoli, Italy

Isabel Guillén
Mammalian Cell Genetics Division
Centro de Ingenierìa Genètica y Biotecnologìa
P.O.Box 6162, Havana, Cuba

Ximing Guo
Rutgers University
Haskin Shellfish Research Laboratory
Port Norris, NJ 08349, USA

Christan Hasse
Department of Surgery
University of Marburg
Baldinger Strasse
D-35043 Marburg, Germany

Margo G. Haygood
Marine Biology Research Division and Center for Marine Biotechnology and Biomedicine
Scripps Institution of Oceanography
University of California, San Diego
La Jolla, CA 92093-0202

Liming M. He
Marine Biology Research Division and Center for Marine Biotechnology and Biomedicine
Scripps Institution of Oceanography
University of California, San Diego
La Jolla, CA 92093-0202

Joël Heppell
Ottawa Civic Hospital Loeb Research Institute
725 Parkdale Avenue
Ottawa, Canada K1Y 4E9

Choy L. Hew
Department of Clinical Biochemistry and Biochemistry
University of Toronto
Toronto, Canada

Russell T. Hill
Australian Institute of Marine Science
PMB 3, Townsville, Queensland 4810, Australia

B. Ho
Department of Microbiology
National University of Singapore
Kent Ridge, Singapore 119260

Yunhan Hong
Physiologische Chemie I
Biozentrum der Universität Würzburg
97074 Würzburg, FRG

Deborah Hughes
Scripps Institution of Oceanography
La Jolla, CA, USA

Toshii Iida
Marine Biotechnology Institute
Kamaishi, Iwate, Japan

Kaori Inoue
Department of Biotechnology
Tokyo University of Agriculture and Technology
Koganei, Tokyo 184, Japan

Ann E Jaffray
Department of Microbiology
University of Cape Town
Private Bag, Rondebosch 7701, South Africa

Edward Jakubov
National Institute of Oceanography
Israel Oceanographic and Limnological Research
P.O.B. 8030, Haifa, 31080, Israel

Kei Kamino
Marine Biotechnology Institute
1900 Sodeshi, Shimizu, Shizuoka 424, Japan

I. Karube
Research Center for Science and Technology
University of Tokyo

J. G. I. Khoo
Department of Zoology
University of Canterbury
Private Bag 4800, Christchurch, New Zealand

Tae Hyun Kim
Korea Ocean Research & Development Institute
Ansan P.O. Box. 29, 425-600, Korea

Sang-Jin Kim
Korea Ocean Research & Development Institute
Ansan P.O. Box. 29, 425-600, Korea

B. Klein
Station de Biologie Marine
Muséum National d'Histoire Naturelle et Collège de France, BP 225
29900 Concarneau, France

Gerd Klöck
Department of Biotechnology
University of Würzburg
Biocenter-Am Hubland
D-97074 Würzburg, Germany
kloeck@biozentrum.uni-wuerzburg.de

Horst H. Klump
Department of Biochemistry
University of Cape Town
Rondebosch, South Africa 7700.K

David P. Knight
Department of Biological Sciences
King Alfred's College, Winchester
Hants, United Kingdom

Thomas J. Koob
Skeletal Biology Section
Shriners Hospital for Children
Tampa, FL 33612, USA

Arthur Kornberg
Department of Biochemistry
Stanford University Medical Center
Beckman B400
Stanford, CA 94305-5307, USA

U. Kuhnlein
Department of Animal Science
McGill University
21 111 Lakeshore Road
Sainte Anne de Bellevue, Québec, H9X 3V9, Canada

Kae Kyoung Kwon
Korea Ocean Research & Development Institute
Ansan P.O. Box. 29, 425-600, Korea

Keiichiro Kyozuka
Asamushi Marine Biological Station
Asamushi, Aomori 039-34, Japan

L. Lamarcq
PBRC, University of Hawaii
41 Ahui St.
Honolulu, HI 96813

Hans Laufer
Department of Molecular and Cell Biology
University of Connecticut
Storrs, CT 06268, USA

R. M. Larsen
Sir George Fisher Centre
James Cook University

C. Le Boulay
Station de Biologie Marine
Muséum National d'Histoire Naturelle et Collège de France BP 225
29900 Concarneau, France

Patrick SC Leung
Division of Rheumatology/Allergy and Clinical Immunology
School of Medicine
University of California
Davis, CA 95616, USA

I. Levine
Coastal Plantations International, Inc.
Poland, Maine 04274, USA

Alison Lewis
Loeb Inst. for Medical Research
Ottawa, Canada

Lei Liu
Department of Molecular and Cell Biology
University of Connecticut
Storrs, CT 06268, USA

Zongzhu Liu
Institute of Oceanology
Academia Sinica
Qingdao 266071, China

Marcial Leonardo Lizárraga-Partida
Centro de Investigación Científica y de Educación Superior de Ensenada (CICESE)
Km. 107 Carr. Tijuana-Ensenada
22860, Ensenada B.C., México

Lynda Llewellyn
School of Molecular and Medical Biosciences
University of Wales
P.O. Box 911
Cardiff, CF1 3US, United Kingdom

Philip C. Loh
Department of Microbiology
University of Hawaii
Honolulu, Hawaii 96822, USA

Jose Lopez
Harbor Branch Oceanographic Institution, Inc.
Fort Pierce, Florida 34946, USA

Niels Lorenzen
Danish Veterinary Laboratory
Hangøvej 2, DK8200 Århus, Denmark

Yuanan Lu
Department of Microbiology
University of Hawaii
Honolulu, Hawaii 96822, USA

Leon Mc Clusky
Department of Biology
Boston University
5 Cummington St.
Boston, MA 02215, USA

Norman Maclean
Division of Cell Science
School of Biological Sciences
Bassett Crescent East
University of Southampton
Southampton SO16 7PX, United Kingdom

Tatiana N. Makarieva
Pacific Institute of Bioorganic Chemistry of the Russian Academy of Sciences
Vladivostok-22, Russia

M. McFall-Ngai
PBRC, University of Hawaii
41 Ahui St.
Honolulu, HI 96813

Dennis L. Maeder
Center of Marine Biotechnology
Columbus Center
University of Maryland Biotechnology Institute
Baltimore, MD, USA

M. Marino
Stazione Zoologica 'Anton Dohrn'
Villa Comunale 1
80121 Napoli, Italy

M. Martin
Laboratoire de Biologie des Invertébrés Marins et Malacologie
URA CNRS 699
Muséum National d'Histoire Naturelle
57, rue Cuvier, 75005, Paris, France

Tadashi Maruyama
Marine Biotechnology Institute
Kamaishi, Iwate, Japan

K. Masaki
Toyo Suisan Kaisha Ltd

L. Mastrolia
Dipartimento di Scienze Ambientali
Università della Tuscia
I-01100 Viterbo, Italy

Jack B. L. Matthews
The Scottish Association for Marine Science
Oban, Scotland, UK

Tadashi Matsunaga
Department of Biotechnology
Tokyo University of Agriculture and Technology
2-24-16 Naka-cho, Koganei, Tokyo 184, Japan

M. Mazzini
Dipartimento di Scienze Ambientali
Università della Tuscia
I-01100 Viterbo, Italy

Josefina Méndez
Department of Cell and Molecular Biology
University of La Coruña
Campus de Zapateira
La Coruña 15071, Spain

J. E. Merril
Department of Microbiology
Michigan State University
East Lansing, Michigan 48824, USA

B. Metz
Seaweed Biotechnology Laboratory
Marine Science Center
Northeastern University
Nahant, MA 01908, USA

Alexander T. Mikhailov
Institute of Health Sciences
University of La Coruña
Campus de Zapateira
La Coruña 15071, Spain

Bradley S. Moore
Department of Chemistry
Box 351700
University of Washington
Seattle, WA 98195

Marco Moracci
Institute of Protein Biochemistry and Enzymology
Naples, Italy

Ophira Moran
National Institute of Oceanography
Israel Oceanographic and Limnological Research
P.O.Box 8030, Haifa, 31080, Israel

U. K. Mukherjee
Department of Zoology
University of Canterbury
Private Bag 4800
Christchurch, New Zealand

Peter T. Murphy
Australian Institute of Marine Science
PMB 3, Townsville, Queensland 4810, Australia

E. Cesar B. Nadala, Jr.
Department of Microbiology
University of Hawaii
Honolulu, Hawaii 96822, USA

Rachakonda Nagabhushanam
Tulane University

Hideki Nakayama
Department of Biotechnology
Tokyo University of Agriculture and Technology
2-24-16 Naka-cho, Koganei, Tokyo 184, Japan

Bruce L. Nicholson
Department of Biochemistry, Microbiology and Molecular Biology
University of Maine
Orono, ME 04469, USA

Bradley Noren
Department of Biology
Boston University
Boston, MA 02215, USA

Umberto Oreste
Institute of Protein Biochemistry and Enzymology
CNR, Naples, Italy

A. Orta-Ramirez
Department of Food Science and Human Nutrition

Anna Palumbo
Zoological Station
Naples, Italy

Marina Paolucci
Department of Biology
University of Naples Federico II
Naples, Italy

Laurence H. Pearl
Structural Biochemistry Section
Dept. of Biochemistry and Molecular Biology
University College London
United Kingdom

Claudia Pecorella
Harbor Branch Oceanographic Institution, Inc.
Fort Pierce, Florida 34946, USA

Harald Petermann
Fachbereich Geowissenschaften
University Bremen
Postfach 330440
2800 Bremen 33, FRG

Shirley A. Pomponi
Harbor Branch Oceanographic Institution, Inc.
Fort Pierce, Florida 34946, USA

Ludmila P. Ponomarenko
Pacific Institute of Bioorganic Chemistry of the Russian Academy of Sciences
Vladivostok-22, Russia

S.A. Poulet
Station Biologique
Place G. Teissier, 29682 Roscoff, France

Deborah M. Power
Unidade de Ciencias e Tecnologias Agrarias
Universidade do Algarve
Campus de Gambelas, Faro 8000, Portugal

Nigel P. Preston
CSIRO Division of Marine Research
Cleveland, Qld 4116, Australia

A. W. M. Pui
School of Biological Sciences
National University of Singapore
Kent Ridge, Singapore 119260

Azizur Rahman
Division of Cell Science
School of Biological Sciences
Bassett Crescent East
University of Southampton
Southampton SO16 7PX, United Kingdom

Vimi P. Ramsurn
School of Molecular and Medical Biosciences
University of Wales
P.O. Box 911, Cardiff, CF1 3US, UK

David W. Rice
Krebs Institute
University of Sheffield
Sheffield, UK

Baruch Rinkevich
National Institute of Oceanography
Israel Oceanographic and Limnological Research
P.O.B. 8030, Haifa 31080, Israel

S. D. Roopashree
School of Biological Sciences
National University of Singapore
Kent Ridge, Singapore 119260

Frank T. Robb
Center of Marine Biotechnology
Columbus Center
University of Maryland Biotechnology Institute
Baltimore, MD, USA

K. Roberts
Seaweed Biotechnology Laboratory
Marine Science Center
Northeastern University
Nahant, MA 01908, USA

N. Romano
Dipartimento di Scienze Ambientali
Università della Tuscia
I-01100 Viterbo, Italy

Gregory L. Rorrer
Department of Chemical Engineering
Oregon State University
Corvallis, OR 97331, USA

Mosé Rossi
Institute of Protein Biochemistry and Enzymology
Naples, Italy

Edward G. Ruby
Pacific Biomedical Research Center
University of Hawaii
Ahui Street, Honolulu, HI 96813

B. Rudolph
Copenhagen Pectin
DK-4623 Lille Skensved, Denmark

G. L. Russo
Institute of Food Science and Technology
National Research Council
83100 Avellino, Italy

B. Sainte-Marie
Division des invertebrés et de la biologie expérimentale
Institut Maurice-Lamontagne
Ministère des Pêches et des Océans
850 route de la Mer, C.P. 1000
Mont-Joli, Québec, G5H 3Z4, Canada

Mukesh K. Sahni
Departments of Chemistry and Biology
William Paterson University
Wayne, New Jersey 07470, USA

Gail Samples
Harbor Branch Oceanographic Institution, Inc.
Fort Pierce, Florida 34946, USA

Rachakonda Sarojini
Tulane University

Thomas Sauer
Chiron Behring GmbH & Co.
Preclinical Development Vaccine
P.O.Box 1630, 35006 Marburg, Germany

Richard Sayre
Departments of Biochemistry and Plant Biology and Graduate Program in Biophysics
Ohio State University
Columbus, OH, 43210 USA

G. Scapigliati
Dipartimento di Scienze Ambientali
Università della Tuscia
I-01100 Viterbo, Italy

Manfred Schartl
Physiologische Chemie I
Biozentrum der Universität Würzburg
97074 Würzburg, FRG

Joachim Schorr
Qiagen GmbH
Max-Volmer-Strasse 4
4010, Hilden, Germany

Joshua R. Schultz
University of Ottawa Heart Institute

D. Sellos
Station de Biologie Marine
Muséum National d'Histoire Naturelle et Collège de France BP 225
29900 Concarneau, France

Susan H. Sennett
Harbor Branch Oceanographic Institution, Inc.
Fort Pierce, Florida 34946, USA

J.-M. Sévigny
Division des invertebrés et de la biologie expérimentale
Institut Maurice-Lamontagne
Ministère des Pêches et des Océans
850 route de la Mer, C.P. 1000
Mont-Joli, Québec, G5H 3Z4, Canada

S. Sharma
Departments of Chemistry and Biology
William Paterson University
Wayne, New Jersey 07470, USA

Yoshikazu Shizuri
Marine Biotechnology Institute
1900 Sodeshi, Shimizu, Shizuoka 424, Japan

Doo Suep Sim
Korea Ocean Research & Development Institute
Ansan P.O. Box. 29, 425-600, Korea

F. Y. T. Sin
Department of Zoology
University of Canterbury
Private Bag 4800, Christchurch, New Zealand

I. Sin
Department of Zoology
University of Canterbury
Private Bag 4800, Christchurch, New Zealand

John T. Singer
Department of Biochemistry, Microbiology and Molecular Biology
University of Maine
Orono, ME 04469, USA

D. M. Smith
Department of Food Science and Human Nutrition

Jae Hak Sohn
Korea Ocean Research & Development Institute
Ansan P.O. Box. 29, 425-600, Korea

Valentine A. Stonik
Pacific Institute of Bioorganic Chemistry of the Russian Academy of Sciences
Vladivostok-22, Russia

D. Sud
Laboratoire de Biologie des Invertébrés Marins et Malacologie
URA CNRS 699
Muséum National d'Histoire Naturelle
57, rue Cuvier, 75005, Paris, France

Satoru Suzuki
Department of Aquaculture
Kochi University
Nankoku, Kochi 783, Japan

Glen E. Sweeney
School of Molecular and Medical Biosciences
University of Wales
P.O. Box 911
Cardiff, CF1 3US, United Kingdom

Haruko Takeyama
Department of Biotechnology
Tokyo University of Agriculture and Technology
2-24-16 Naka-cho, Koganei, Tokyo 184, Japan

Lourdes M. Tapay
Department of Microbiology
University of Hawaii
Honolulu, Hawaii 96822, USA

Bradley M. Tebo
Marine Biology Research Division and Center for Marine Biotechnology and Biomedicine
Scripps Institution of Oceanography
University of California, San Diego
La Jolla, CA 92093-0202

Andreas Teske
Woods Hole Oceanographic Institution
Woods Hole, USA

Nicola J. Tolliday
Center of Marine Biotechnology
Columbus Center
University of Maryland Biotechnology Institute
Baltimore, MD, USA

Moshe Tom
National Institute of Oceanography
Israel Oceanographic and Limnological Research
P.O.B. 8030, Haifa, 31080, Israel

Mario Torrado
Institute of Health Sciences
University of La Coruña
Campus de Zapateira
La Coruña 15071, Spain

E. Tosti
Stazione Zoologica 'Anton Dohrn'
Villa Comunale 1
80121 Napoli, Italy

Samuel Traina
School of Natural Resources
Ohio St. University
Columbus, OH 43210, USA

Nicola Urbani
Department of Animal Science
McGill University
21 111 Lakeshore Road
Sainte Anne de Bellevue, Québec, H9X 3V9, Canada

Kensal E. Van Holde
Department of Biochemistry and Biophysics
Oregon State University
Corvallis, OR 97331-7305, USA

Lorraine Van Waasbergen
Marine Biology Research Division and Center for Marine Biotechnology and Biomedicine
Scripps Institution of Oceanography
University of California, San Diego
La Jolla, CA 92093-0202

A. Van Wormhoudt
Station de Biologie Marine
Muséum National d'Histoire Naturelle et Collège de France BP 225
29900 Concarneau, France

Costantino Vetriani
Center of Marine Biotechnology
Columbus Center
University of Maryland Biotechnology Institute
Baltimore, MD, USA

Karen L. Visick
Pacific Biomedical Research Center
University of Hawaii
Ahui Street, Honolulu, HI 96813

Miryam Z. Wahrman
Departments of Chemistry and Biology
William Paterson University
Wayne, New Jersey 07470, USA

L. Z. Wang
Seaweed Biotechnology Laboratory
Marine Science Center
Northeastern University
Nahant, MA 01908, USA

K. Watson
Seaweed Biotechnology Laboratory
Marine Science Center
Northeastern University
Nahant, MA 01908, USA

Ronald Weiner
Department of Microbiology
University of Maryland (UMPC)

V. Weis
Deptartment of Zoology
Oregon State University
Corvallis, OR 97331-2914

Trevor Wigham
School of Pure and Applied Biology
University of Wales
P.O. Box 915
Cardiff, CF1 3TL, UK

Lynne Whitehead
Department of Microbiology
University of Maryland (UMPC)

M. Wilding
Stazione Zoologica 'Anton Dohrn'
Villa Comunale 1, 80121 Napoli, Italy

Robin Willoughby
Harbor Branch Oceanographic Institution, Inc.
Fort Pierce, Florida 34946, USA

Christoph Winkler
Physiologische Chemie I
Biozentrum der Universität Würzburg
97074 Würzburg, FRG

Irma Wong-Chang
Instituto de Ciencias del Mar y Limnologia
UNAM

Amy E. Wright
Harbor Branch Oceanographic Institution, Inc.
Fort Pierce, Florida 34946, USA

Tong Wu
Ottawa Civic Hospital Loeb Research Institute
725 Parkdale Avenue, Ottawa, Canada K1Y 4E9

Yuan Xiang
Institute of Oceanology
Academia Sinica
Qingdao 266071, China

Yongli Xu
Institute of Oceanology
Academia Sinica
Qingdao 266071, China

Y. Yamamoto
Research Center for Science and Technology
University of Tokyo

Sigeyuki Yamano
Marine Biotechnology Institute
Shimizu, Shizuoka, Japan

Akira Yamazawa
Department of Biotechnology
Tokyo University of Agriculture and Technology
2-24-16 Naka-cho, Koganei, Tokyo 184, Japan

K. Yazawa
Kanagawa, Japan

Kitty S. P. Yip
Krebs Institute
University of Sheffield
Sheffield, United Kingdom

D. Zadworny
Department of Animal Science
McGill University
21 111 Lakeshore Road
Sainte Anne de Bellevue, Québec, H9X 3V9, Canada

Peijun Zhang
Institute of Oceanology
Academia Sinica
Qingdao 266071, China

Ted Zerucha
Loeb Inst. for Medical Research
Ottawa, Canada

Ulrich Zimmermann
Department of Biotechnology
University of Würzburg
Biocenter—Am Hubland
D-97074 Würzburg, Germany
kloeck@biozentrum.uni-wuerzburg.de

INDEX

MIX
Papier aus verantwortungsvollen Quellen
Paper from responsible sources
FSC® C105338

If you have any concerns about our products,
you can contact us on
ProductSafety@springernature.com

In case Publisher is established outside the EU,
the EU authorized representative is:
Springer Nature Customer Service Center GmbH
Europaplatz 3, 69115 Heidelberg, Germany

Printed by Libri Plureos GmbH
in Hamburg, Germany